Protein Structure – Function Relationship

Protein Structure – Function Relationship

Edited by

Zafar H. Zaidi

H. E. J. Research Institute of Chemistry
University of Karachi
Karachi, Pakistan

and

David L. Smith

Department of Chemistry
University of Nebraska
Lincoln, Nebraska

Plenum Press • New York and London

Library of Congress Cataloging-in-Publication Data

International Symposium on Protein Structure-Function Relationship
 (4th : 1995 : Karachi, Pakistan)
  Protein structure--function relationship / edited by Zafar H.
Zaidi and David L. Smith.
      p.   cm.
  Includes bibliographical references and index.
  ISBN 0-306-45285-5
  1. Proteins--Structure-activity relationships--Congresses.
I. Zaidi, Zafar H. II. Smith, David L. III. Title.
QP551.I5535 1995
574.19'245--dc20   96-12998
                                                        CIP

Proceedings of the Fourth International Symposium on Protein Structure – Function Relationship, held January 20 – 25, 1995, in Karachi, Pakistan

ISBN 0-306-45285-5

© 1996 Plenum Press, New York
A Division of Plenum Publishing Corporation
233 Spring Street, New York, N. Y. 10013

DEDICATED TO

Professor Muhammad Akhtar, S.I., FRS

PREFACE

Although many pursue understanding of the relationship between protein structure and function for the thrill of pure science, the pay-off in a much broader sense is the ability to manipulate the Earth's chemistry and biology to improve the quality of life for mankind. Immediately goals of this area of research include identification of the life-supporting functions of proteins, and the fundamental forces that facilitate these functions. Upon reaching these goals, we shall have the understanding to direct and the tools required to implement changes that will dramatically improve the quality of life. For example, understanding the chemical mechanism of diseases will facilitate development of new therapeutic drugs. Likewise, understanding of chemical mechanisms of plant growth will be used with biotechnology to improve food production under adverse climatic conditions.

The challenge to understand details of protein structure/function relationships is enormous and requires an international effort for success. To direct the chemistry and biology of our environment in a positive sense will require efforts from bright, imaginative scientists located throughout the world. Although the emergence of FAX, e-mail, and the World Wide Web has revolutionized international communication, there remains a need for scientists located in distant parts of the world to occasionally meet face to face. This need is the basis for this series of symposia, which have brought together leading scientists from Europe, North America, and Southeast Asia to exchange in a very personal sense results of their current research and plans for future research. These proceedings will provide to those who were unable to attend the essence of the meeting.

This symposium was dedicated to Professor Muhammad Akhtar, S.I., FRS, for his services to promote science and technology in Pakistan. The organizers are pleased to acknowledge the following organizations for financial support: Chemical Society of Pakistan, Federation of Asian and Oceanian Biochemists and Molecular Biologists (FAOMB), Hamdard Foundation, Husein Ebrahim Jamal Foundation, National Science Foundation (USA), Pakistan Society of Biochemists, Third World Academy of Science, and the University of Karachi. We also wish to acknowledge our indebtedness to the students and staff of the institute who worked tirelessly and selflessly in organizing the symposium and workshop.

David L. Smith
Zafar Zaidi

CONTENTS

HOW IS THE DIOXYGEN-IRON BOND MANIPULATED BY TRANSPORT AND CATALYTIC HAEM PROTEINS?

M. Akhtar

Department of Biochemistry
University of Southampton
Bassett Crescent East, Southampton SO16 7PX, United Kingdom

Mankind! If ye have a doubt about the Resurrection (consider) that we created you out of dust, then out of sperm, then out of a leech-like clot, then out of a morsel of flesh, partly formed and partly unformed, in order that We may manifest (Our power) to you; and We cause whom We will to rest in the wombs for an appointed term, then do We bring you out as babes, then (Foster you) that ye may reach your age of full strength; and some of you are called to die, and some are sent back to the feeblest old age, so that they know nothing after having known (much). And (further), thou seest the earth barren and lifeless, but when We pour down rain on it, it is stirred (to life), it swells and it puts forth every kind of beautiful growth (in pairs).

This is so, because God is the Reality: it is He who give life to the dead, and it is He who has Power over all things.

—(Al-Quvair Sura Al-Haj 5 & 6)

ABSTRACT

The Chemistry which underpins corrosion has been borrowed by evolutionary biology and used to perform a myriad of cunning feats. The lecture will exemplify this by showing that the first stage of the reaction involving corrosion forms the basis of the mechanism of transport of oxygen by haemoglobin. In turn, the chemistry underpinning the latter process represents a fundamental reaction from which more complex biological oxidative reactions have evolved. These reactions are catalysed by drug-metabolising hydroxylases of the cytochrome P-450 family but more importantly, from the viewpoint of the lecture, by multifunctional cytochrome P-450s involved in the biosynthesis of sex hormones.

In all these biological processes, the initial reaction requires the participation of the Fe^{II} form of haem-iron which transfers an electron to O_2 forming a hypothetical superoxide equivalent ($\overline{\text{O}} - \dot{\text{O}}$) that is then converted into an Fe^{III}-OOH species (Scheme 3, 10). The $\overline{\text{O}} - \dot{\text{O}}$ bond in the latter is labile and easily cleaved to produce an important intermediate, commonly called the oxo-derivative for which a number of canonical structures are possible. It is the oxo-derivative (Scheme 3, 11) that is involved in the otherwise difficult hydroxylation reaction. At the active site of certain multifunctional P-450s the Fe^{III}-OOH species can also be trapped by electrophilic intermediates to produce an adduct (Scheme 5, 23) that undergoes an acyl-carbon cleavage. A cytochrome P-450 enzyme, 17α-hydroxylase-17,20-lyase (P-450$_{17\alpha}$) lies at the crossroad of androgen and corticoid biosynthesis. The manner in which the Fe^{III}-OOH species is directed towards the two alternative pathways, hydroxylation and cleavage, is of crucial importance to the endocrinological well-being of all animals but particularly the human female. Our recent results sheeding light on this problem will be described.

It is indeed a great honour to have been invited to give this inaugural lecture and for that I thank you most warmly. By their very nature, the inaugural lectures rarely live up to the high expectations of the audience - they are trying for both parties and maturity requires that this inevitable fact is accepted graciously.

If there is an underlying theme in this lecture, it is to link the chemistry of the physical world to that of the living cell. Let us begin at the beginning. When the earth was red-hot and the temperatures were high, all types of energy-requiring chemical reactions might well have been possible. However, as the earth cooled to produce the temperate conditions of today, only those reactions occurred which conform to the requirements of the 2nd law of thermodynamics; which simply means the reactions which are attended by a negative Gibbs free energy change - put even more plainly, the reactions which are exothermic in nature.

My task tonight is to take you on a journey and show how the chemical principles borrowed from the physical world by the replicating living cell have been preserved and practised with subtlety as well as elegance. To illustrate my viewpoint, I have selected two common elements: oxygen and iron.

Iron has a strong affinity for oxygen, the consequence of which in the physical world is two-fold: first, that iron is found on this planet not in a metallic form but as an ore combined, principally, with oxygen; second, that the ultimate fate of manufactured metallic iron is to suffer corrosion. The inevitable reaction of iron and oxygen leading to corrosion has attracted fanciful descriptions. For example, it is said: *"in the aerobic atmosphere of this planet, that iron should rust is the obligatory requirement of the second law of thermodynamics"*. The bases of such statements are the strongly opposite redox potentials of the two reactions:

$$\tfrac{1}{2}FE^{++} + 1e \; \rightleftharpoons \; \tfrac{1}{2}Fe \quad E'_0 = (\text{circa}) - 0.44 \text{ volt} \tag{1}$$

$$\tfrac{1}{2}O_2 + 2e + 2H^+ \; \rightleftharpoons \; H_2O \quad E'_0 = 0.80 \text{ volt} \tag{2}$$

(E'_0 is standard redox potential at pH 7.0)

These half-reactions can therefore be coupled to drive a thermodynamically favourable oxidation of iron. If, for the sake of simplicity, we represent the corrosion process by eq. 3, it would have $\Delta G^{0'}$ of about -50 kcal^{-1}. This is only an approximate value since the redox

$$Fe + \tfrac{1}{2}O_2 + 2H^+ \; \rightleftharpoons \; Fe^{++} + H_2O \tag{3}$$

potentials of the constituent half reactions, and particularly of the reaction of eq. 2 are dependent on pH and also, as we shall see below, the exact stoichiometry of the corrosion reaction may significantly depart from the one implied in eq. 3.

Nonetheless, thermodynamics having endorsed our everyday experience, of the inevitability of corrosion, let us explore the molecular details of the process in terms of structural chemistry. In its ground state, oxygen has a diradical structure and is prone to undergo two one-electron reductions. From the negative redox potential of the half-reaction in eq. 1, iron is a good electron donor and one may envisage an initial reaction, between Fe and O_2, in which an electron transfer from iron leads to the formation of an Fe^I-superoxide ion-pair as shown in Structure 1. The donation of the second electron by another iron atom will then produce a highly unstable di-iron peroxide species (2). The fission of the O-O bond in the latter, followed by an intramolecular electron transfer produces $Fe^{II}=O$ (4). It is possible to envisage several other closely-related routes for the decomposition of the species (2) but the merit of the reaction sequence selected here is that it produces $Fe^{II}=O$ (4) which has been found to be the earliest product of corrosion[1]; it should though be remembered that the composition of a mature sample of rust[1] is Fe_2O_3 which must be the result of several competing processes. The above analysis presents a completely speculative but, I believe, not an irresponsible view. The individual steps of the model above involve plausible chemistries. The main merit of the exercise, however, is that it allows some basic chemical principles to be enunciated using a rather simple system involving merely four atoms (O_2 + 2Fe) before these are extended to the daunting structures of proteins comprising several thousand atoms. My task tonight is to show how without losing their simplicity, these chemical reactions are harnessed at the active site of proteins and used for sophisticated functions.

$$Fe + \overset{\cdot}{O}\text{-}\overset{\cdot}{O} \longrightarrow \left[Fe^I + \overset{-}{O}\text{-}\overset{\cdot}{O} \right] \rightleftarrows Fe^I\text{-}O\text{-}\overset{\cdot}{O} \downarrow Fe$$
$$\underset{4}{2Fe^{II}=O} \longleftarrow \underset{3}{Fe^I\text{-}\overset{\cdot}{O} \quad \overset{\cdot}{O}\text{-}Fe^I} \longleftarrow \underset{2}{Fe^I\text{-}O\text{-}O\text{-}Fe^I}$$

The Oxygen Transporting Haemoglobin System

Haemoglobin gathers O_2 in the lung and then distributes it to remote parts of the body. Our current view of the mechanism through which this process occurs has progressively evolved from a range of physico-chemical discoveries made during this century. Often, the impact of these discoveries has extended beyond the specialist field of oxygen transport. Haemoglobin and myoglobin were the first proteins to be successfully subjected to structure determination by X-ray diffraction[2], thus marking a trend, in the elucidation of the 3-dimensional structures of proteins, which has now reached an explosive proportion. From the viewpoint of bioinorganic chemistry, however, the most outstanding contribution, overshadowing any other by a long way, was made by Pauling and Coryll in 1936[3]. Using magnetic susceptibility measurements, it was found that the iron in haemoglobin had four unpaired electrons, an arrangement found in $FeCl_2$. However, its oxygen adduct, oxyhaemoglobin, was diamagnetic like ferrocyanide, $Fe(CN)_6^{4-}$, and other co-ordination complexes of haem formed when the latter reacts with 2 molecules of a nitrogen base. This observation suggested that O_2 binding to haemoglobin is attended by the conversion of a high spin iron into a low spin iron[3]. There was, however, a long period of indecision before the precise chemistry through which O_2 becomes a co-ordination ligand for Fe^{II} could be deduced. The evidence which had been accumulating from a range of techniques ever since 1964[4] and ignored for

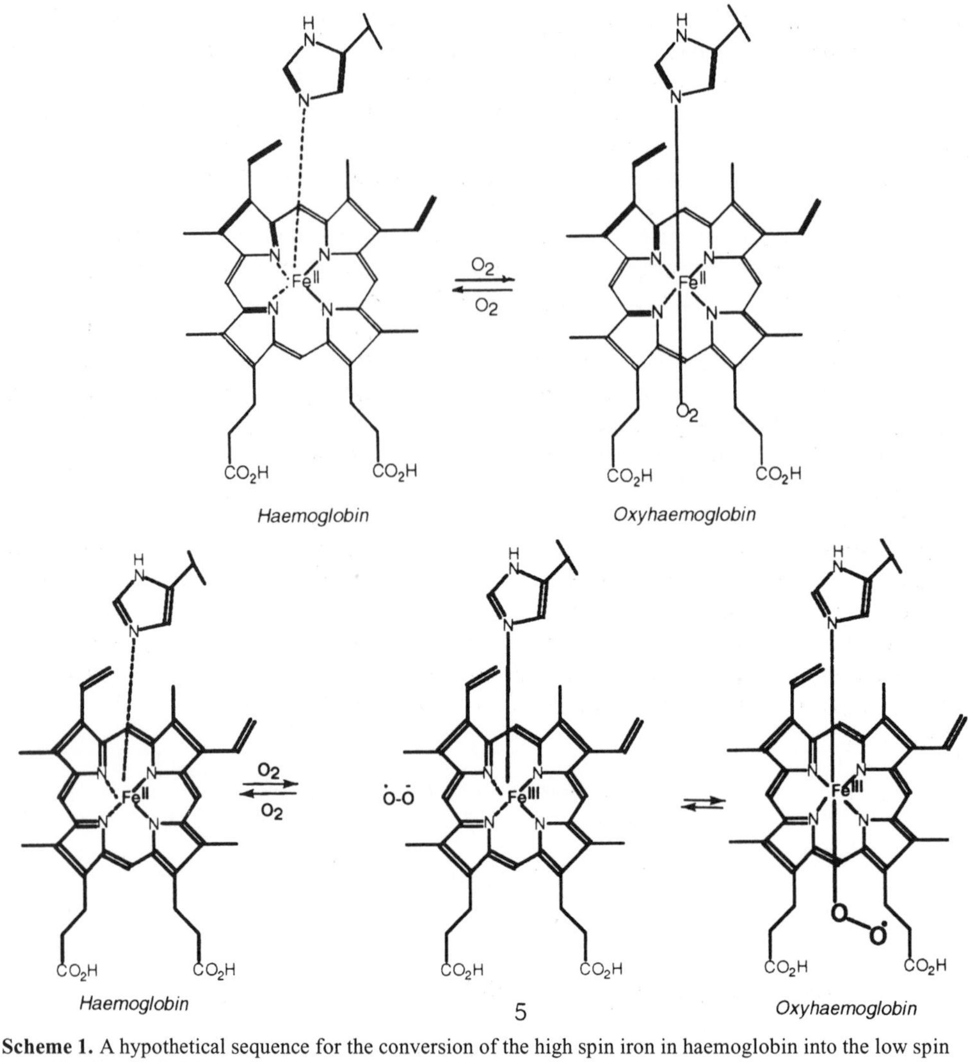

Scheme 1. A hypothetical sequence for the conversion of the high spin iron in haemoglobin into the low spin iron of oxyhaemoglobin.

nearly two decades (but *not* by the writer) is now gaining general acceptance that oxyhaemoglobin is formulated as shown in Scheme 1. [This aspect is discussed in detail in Ref. 5; for comparison see Ref. 6]. The structure may be viewed to arise from the initial interaction of Fe^{II} with O_2 which causes a redox reaction producing Fe^{III} and the superoxide ion, $\overline{O} - \dot{O}$, the latter then acts on the sixth ligand to give the final co-ordination complex.

The high spin to low spin transition accompanying the conversion of haemoglobin into oxyhaemoglobin has a profound effect on the protein structure as revealed by X-ray diffraction data. In haemoglobin iron is found to occupy a 'domed' configuration and is forced into planar orientation in oxyhaemoglobin. During this process, the axial histidine ligand is dragged towards the porphyrin causing protein conformational changes, which may contribute to allosteric behaviour of this protein[2] (Scheme 2).

If the cause of physiology is to be served, then oxyhaemoglobin must unload its O_2 faithfully without engaging in side reactions. The potentially reactive superoxide ligand of

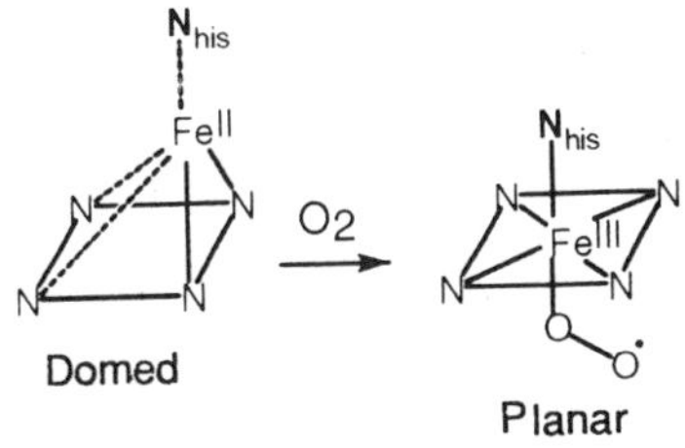

Scheme 2. A stylised diagram showing the conversion of a 'domed' structure of haemoglobin into a planar orientation in oxyhaemoglobin.

oxyhaemoglobin therefore must be prevented from random redox reactions and this presumably is achieved by the architectural design of the protein which makes provision for a protective layer of amino acid side chains shielding the oxygen ligand. Another chemical reaction having high propensity is the direct dissociation of the oxygen ligand producing the superoxide ion and methaemoglobin. This side reaction must form a low background activity and to deal with the consequences of which have evolved enzymes such as superoxide dismatase and methaemoglobin reductase.

Before I leave the haemoglobin story, I must make two points. First, the mode of binding of O_2 to the Fe^{II} of haemoglobin represents a general phenomenon and is applicable to a wide variety of systems involved in the manipulation of O_2, Second, it has given me much pleasure that this lecture has required me to include a brief description of haemoglobin, a subject which also features in the main programme of this meeting, and to which our organiser, Professor Zafar Zaidi has contributed richly.

The P-450 Hydroxylases

Let us now move to the next level of complexity and deal with reactions in which oxygen is trapped not merely for transport but is utilised in oxidative reactions. A large number of enzymes found in all life forms, from bacteria to man, catalyse the hydroxylation reaction of eq. 4[5-9] in which a strong aliphatic C-H bond is cleaved and replaced by an hydroxyl group.

$$R–H + O_2 + NADPH + H^+ \longrightarrow R–OH + H_2O + NADP^+ \qquad (4)$$

The largest group of enzymes catalysing the hydroxylation reaction are cytochromes P-450[7]. These contain haem as the prosthetic group and derive their name from the difference spectrum obtained by the addition of CO to the reduced form of the enzyme when a peak at 450nm is produced. The P-450 group of hydroxylases are most abundantly present in the mammalian liver and participate in the metabolism of drugs and other xenobiotics[7]. The protein moiety of P-450s contains an essential cysteine whose thiolate group serves as one of the axial ligands for the haem iron. Two features of the reaction of eq. 4 are worthy of note. First, the hydroxyl oxygen atom of the product is derived from molecular oxygen and second, the overall reaction involves the participation of the reduced form of pyridine nucleotide, usually NADPH. Our current view[5-9] of the mechanism of hydroxylases may be enunciated as follows[5]. In the resting state, cytochromes P-450 have the haem iron as Fe^{III}. The first step in catalysis (Scheme 3) is the compulsory binding of the substrate to the enzyme, giving a binary complex[5]. Since this seemingly prosaic feature is retained in a wide

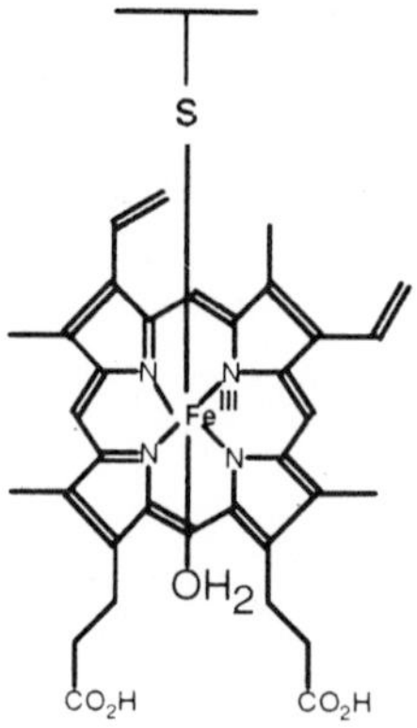

Structure 1. P-450 (Prosthetic group).

variety of P-450 enzymes, it is likely to confer some sort of advantage which is considered elsewhere in the paper.

We have already seen with the haemoglobin system that it is the ferrous form of the iron that is able to directly coordinate with O_2. This feature is retained in P-450s and for the conversion of Fe^{III} into Fe^{II} using NADPH as a reductant, another enzyme, NADPH-cytochrome P-450 reductase, is required. The reductase has two roles: to separate the electron pair of NADPH into two single electrons and then to interact with specific intermediary forms of P-450 to transfer each electron in a stepwise fashion. The transfer of the first electron to the P-450-substrate complex produces the Fe^{II} form of the enzyme which immediately reacts with O_2 giving a species which by analogy with oxyhaemoglobin may be formulated as a co-ordination complex of Fe^{III} containing superoxide as one of the axial ligands (Scheme 3). The addition of the second electron plus a proton then gives the Fe^{III}-OOH species (10). Thereafter, the cleavage of the O-O bond in the peroxide furnishes an iron mono-oxygen species designated as the oxo derivative. Several structures

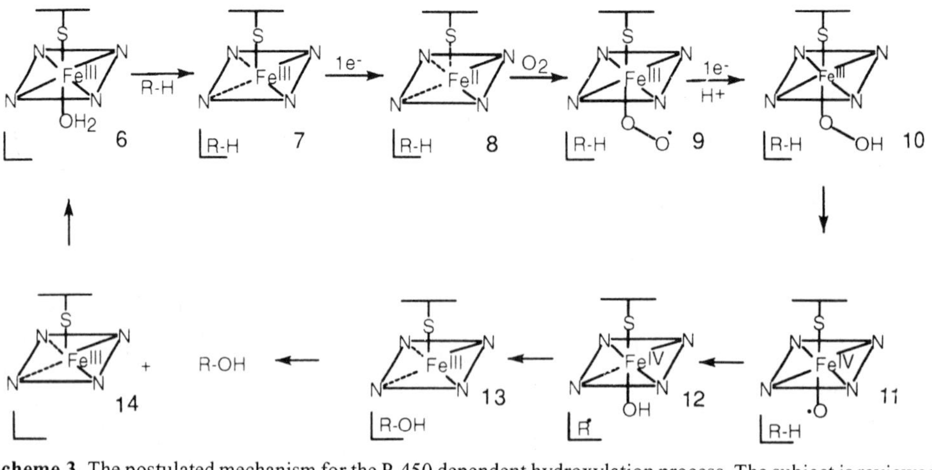

Scheme 3. The postulated mechanism for the P-450 dependent hydroxylation process. The subject is reviewed in Ref. 5 by the author and in 6-9 by other scientists.

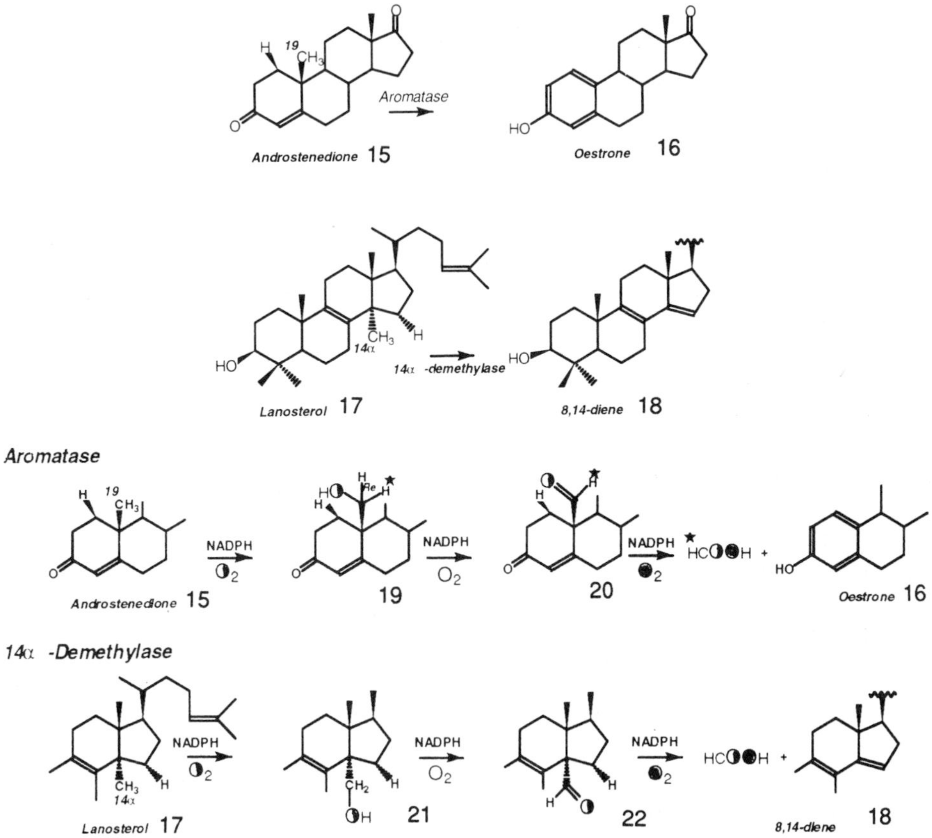

Androstenedione **15**   *Aromatase* →   Oestrone **16**

Lanosterol **17**   *14α -demethylase* →   8,14-diene **18**

Aromatase

Androstenedione **15** $\xrightarrow[O_2]{NADPH}$ **19** $\xrightarrow[O_2]{NADPH}$ **20** $\xrightarrow[O_2]{NADPH}$ HCOOH +   Oestrone **16**

14α -Demethylase

Lanosterol **17** $\xrightarrow[O_2]{NADPH}$ **21** $\xrightarrow[O_2]{NADPH}$ **22** $\xrightarrow[O_2]{NADPH}$ HCOOH +   8,14-diene **18**

Scheme 4. The conversions catalysed by aromatase (15)→(16) and 14α-demethylase (17)→(18). These may be viewed to represent the same generic process in which a double bond is produced by the cleavage of two syn-oriented substituents. The status of the 3 oxygen atoms in the aromatase catalysed process was determined by the methods in Ref. 11 and a similar approach has recently been used to obtain the results implied in the diagram for the reaction catalysed by 14α-demethylase (Zare-Shyadehi, Akhtar, Kelley). Half-filled, unfilled and filled oxygen atoms identify the oxygen atoms at each stage of their subsequent fate.

for the oxo derivative are possible (for example Scheme 6 of Ref. 5) and one of its canonical forms, $Fe^{IV}\ddot{A}O\cdot$ (11) behaves like an alkoxy radical and participates in the hydroxylation reaction via hydrogen abstraction, (11) → (12), followed by an oxygen rebound reaction (12) → (13)[5-9].

Carbon-Carbon Bond Cleaving Hormonal P-450s

Yet another property of certain P-450s was discovered during the course of our extensive studies primarily directed to the elucidation of the chemical details through which 14α-methyl group is removed in sterol biosynthesis[10] (15 → 16) and 19-methyl group in the formation of the female hormone (17 → 18), oestrogen[11,] (also see)[5]. We showed that each of the conversion of Scheme 4 occurs via three discrete steps and that in both cases, a single protein catalyst is involved in promoting reactions which belong to entirely different generic types. The two enzymes are aromatase and 14α-demethylase, each of which contains a single polypeptide complexed to 1mol of haem. The three reactions catalysed by these enzymes are hydroxylation, oxidation of an alcohol into a carbonyl compound and, most notably, an acyl-carbon cleavage reaction, represented by eq. 5.

We have suggested that these reactions are catalysed at the same active-site by using two distinct forms of activated oxygen species[5,12]. The hydroxylation reaction must occur by the same process which has been described for conventional P-450s and involves the use of the oxo derivative (11). The key step in the conversion of an alcohol into a carbonyl compound is the same as for the hydroxylation reaction and involves the cleavage of a C-H bond. Therefore, this process can also be rationalised by a mechanism using the oxo derivative[5,12].

The third reaction in the two sequences belongs to an entirely different generic type (eq. 5, p. 9) in which a C-C bond is cleaved. The mechanism of this process was rationalised by assuming (Scheme 5) that in this case, it is the Fe^{III}-OOH that is the key performer and instead of being converted to the oxo derivative, it directly reacts with the aldehyde group of the substrate, producing an adduct (Scheme 5, 23) that then undergoes a fragmentation giving rise to the final product[5,12]. That two distinct oxygen species be selected for catalysing reaction at the same active site was originally viewed with scepticism. We have, however, drawn attention to the fact that in P-450 reaction the formation of binary complex between substrate and enzyme is the obligatory first step and only then does the binding of O_2 to the haem iron occur. In such a situation, when a substrate containing a strategically located carbonyl group is correctly positioned at the active site, it is not unreasonable to expect that the strongly nucleophilic Fe^{III}-OOH should be intercepted by an equally strongly electrophilic carbonyl group to produce the adduct of the type 23 by the operation of Path B in Scheme 5.

When such a scenario is not feasible, because of the chemically inert nature of the target substituent, then, of course, the Fe^{III}-OOH decomposes to the oxo derivative (by Path A, Scheme 5) that participates in an initial hydrogen abstraction followed by oxygen-rebound or an equivalent reaction. The catalytic pluralism proposed by us for multicatalytic P-450s and illustrated in Scheme 5 is now gaining acceptance and has been supported by a wide range of experimental observations[9].

The Control of Androgen Biosynthesis in Human

An even more interesting improvisation on the theme of multicatalysis is provided by another enzyme, 17α-hydroxylase-17,20-lyase (P-450$_{17\alpha}$, the new systematic name is CYP17). This enzyme lies at the crossroad of androgen and corticoid biosynthesis. It

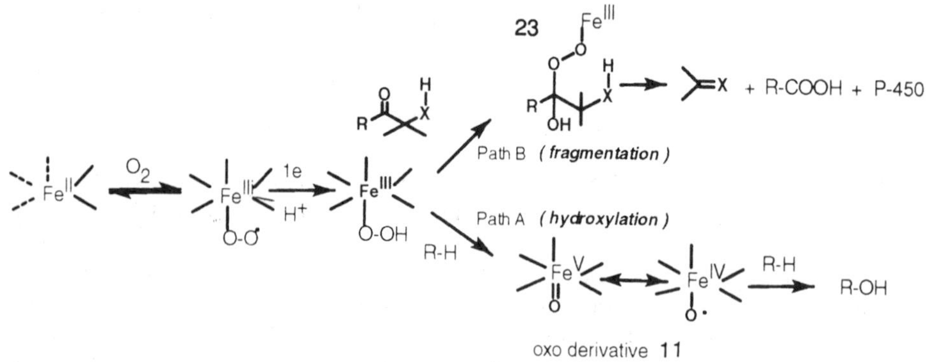

Scheme 5. The Fe^{III}OOH species at the cross road of hydroxylation (Path A) and acyl-carbon fission (Path B).

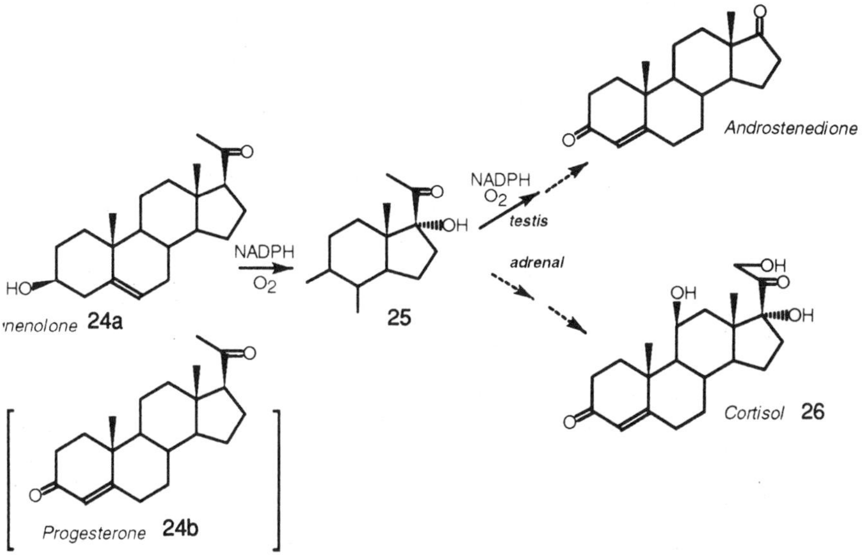

Scheme 6. The differing roles of P-450$_{17\alpha}$ in testis and adrenal. In both the tissues the hydoxylation reaction must occur but acyl-carbon cleavage leading the formation of androgenic steroids is only required in the testis.

catalyses not only an hydroxylation reaction, which is common to both pathways, but also the cleavage of a carbon-carbon bond[13] required only for the formation of androgens.

A careful examination of the 'landscape' around the scissile C-C bond (see structure 25) reveals that the side chain cleavage process, (25) → androstenedione (Scheme 6), can be modelled on the acyl-carbon cleavage reaction, already considered for aromatase and 14α-demethylase, and represented in eq. 5, if the X denotes an oxygen rather than a carbon atom.

$$\mathrm{R{-}\overset{O}{\overset{\|}{C}}{-}\overset{H}{\overset{|}{C}}R_1{-}X{-}H + O_2 + NADPH \xrightarrow[\text{X, O or } \rangle C \langle]{} RCOOH + H_2O + \begin{matrix} R_1 \\ \\ H \end{matrix}\!\!\rangle C{=}X + NADP^+} \tag{5}$$

The three enzymes (aromatase, 14α-demethylase and P-450$_{17\alpha}$) therefore belong to the same general class with one important difference. In the cases of aromatase and 14α-demethylase, only the terminal compounds, produced at the end of the three-step sequence, are physiologically important: the hydroxy and aldehyde derivatives have no biological significance and so long as these are, eventually, converted to the C-C cleaved products, the physiological purpose is served. In other words, the hydroxy and aldehyde intermediates may dissociate from and reassociate with the enzyme several times until the formation of the final product. P-450$_{17\alpha}$ is, however, emburdened with a more complex role. In the adrenal, it must catalyse only the hydroxylation reaction producing 17α-hydroxypro-gestogen (25) which is required for corticoid biosynthesis *but* prevented from promoting the acyl-carbon cleavage. This restriction is particularly important in the human female when excessive formation of androgens will have undesirable consequences. On the other hand, in the testis, both the reactions, hydroxylation and C-C bond cleavage, must occur for the production of androgens required by the male of the animal species. How may this diametrically opposite objective be achieved? The clue to this dilemma has been provided by our studies which were originally directed to the evaluation of the mechanistic duality hypothesis considered in Scheme 5, which uses the oxo derivative and Fe^{III}-OOH in two different roles.

The two steroidal P-450 enzymes considered above (aromatase and 14α-demethylase) are available only in microgram amounts. However, bovine and porcine $P-450_{17\alpha}$ can be prepared in an homogenous state in somewhat larger amounts (2-5mg). This has enabled us to perform a number of critical experiments. Not all of these can be described in a general lecture of this type but suffice it to mention that these mechanistic experiments have provided compelling evidence in support of the proposition that the acyl-carbon cleavage indeed occurs using an iron-peroxy species, according to the stoichiometry of eq. 5^{14}.

In the meantime, human $P-450_{17\alpha}$ was cloned and expressed by Professor Michael R. Waterman in the United States[15]. We have used a plasmid constructed by his group to obtain an *E. coli* clone which expresses the human $P-450_{17\alpha}$ in sufficient amounts to be purified to homogeneity. In a comparative study using homogenous $P-450_{17\alpha}$ from human, bovine and porcine, we have extended the mechanistic studies to examine the interaction of $P-450_{17\alpha}$ with two redox proteins. Attention has already been drawn to the fact that the electron pair from NADPH is transferred to P-450 in a step-wise fashion through the participation of NADPH-cytochrome-P-450 reductase. Another redox protein cytochrome b_5 (a haem protein) has been known to deputise for the reductase in delivering the second electron: in this case the reductase transfers an electron to cytochrome b_5, converting its haem Fe^{III} into Fe^{II} that then converts $Fe^{III}OO^.$ into Fe^{III}-OOH (Scheme 7). In general, these two alternative modes of electron transfer have no significant bearing on the catalytic behaviour of P-450s. However, studies with porcine $P-450_{17\alpha}$ have shown that in this case the product profile is influenced[13] when the second electron is transferred via cytochrome b_5.

In the light of this background, we have studied[16] aspects of the behaviour of porcine and human $P-450_{17\alpha}$ using either reductase only for the transfer of both the electrons or a mixed system involving the reductase and cytochrome b_5. With the porcine enzyme, as is also the case with the bovine $P-450_{17\alpha}$, the hydroxylation of progesterone and pregnenolone occurs smoothly when the reductase is used as the sole electron mediator and this process is stimulated with the mixed system in which cytochrome b_5 is also present. Likewise, the cleavage of the hydroxylated product to androgens occurs without requiring cytochrome b_5 though the latter stimulates the reaction significantly. The human enzyme, however, was found to display dramatically different behaviour compared to its animal counterparts[16]. In this case, the hydroxylations of pregnenolone and progesterone occur effectively with the

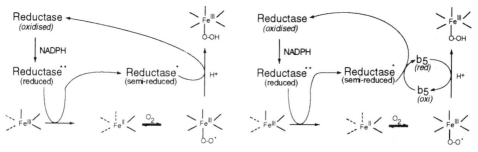

Scheme 7. The two-electron reduction of oxygen during the catalytic cycle of P-450. The diagram on the left shows that both the electrons are delivered by NADPH-cytochrome P-450 reductase (Reductase) while on the right, the transfer of the second electron occurs via cytochrome b_5.

reductase only but the 17α-hydroxyprogestogens are not cleaved at all under these conditions. The cleavage of 17α-hydroxyprogestogens is entirely dependent on the involvement of cytochrome b_5[16].

In the light of these observations, it is possible that a low level of cytochrome b_5 in the human adrenal is one of the important factors in ensuring that this gland does not contribute to unwanted androgen production under normal physiological conditions. A corollary from this deducation is that the production of androgens by the male ought to depend on the presence of an adequate supply of cytochrome b_5 by the testis. The levels of b_5 in these tissues have not yet been accurately determined but indirect evidence suggests that such is likely to be the case.

The conclusion that the side-chain cleavage activity of the lyase necessary for androgen biosynthesis may be suppressed by the low levels of cytochrome b_5 in human adrenal has an important aesthetic consequence. It is this safety mechanism which enables the female of the human species to escape from the unwanted effects of androgen and display the feminine attributes involving the absence of facial hair, deep voice and other physiological ramifications produced by the male hormone. The dogma of the classical biology is that most evolutionary changes are directed to improving the survival chances of the species. Perhaps, on a light-hearted note, I may suggest that in the evolution of human lyase, an aesthetic element might well have been at play.

The salient features of the message I wished to convey in this lecture may be summarised as follows:

1. Evolutionary biology was a wonderful pupil; whatever chemical lessons it learnt from the physical world have been put to excellent use.

2. One active site, one chemical reaction is a norm for enzymes but we have now seen that Nature has departed from this practice in the case of hormonal enzymes.

3. In general, the regulation of the activities of enzymes is carried out by sophisticated foolproof mechanisms and that androgen versus corticoid biosynthesis should have been left to the levels of a peripheral protein in tissues is surprising.

Why should these unusual options in (2) and (3) have been selected? I do not know the answer, nor do I lose sleep on such questions. I belong to the school of scientists who believe that the proper function of science is to find out what is the nature of the physical and biological world and not to ponder about why things are as they are.

ACKNOWLEDGMENTS

During the last two decades many colleagues have contributed to the research work that forms the basis of this lecture. I acknowledge their contributions with gratitude. I am particularly thankful to J. Neville Wright, who has collaborated with me as an equal now for over 15 years, and has made numerous contributions. The recent work described in the lecture was carried out by Peter Lee-Robichaud, Dr Akbar Zare-Shyadehi and Monika E Akhtar, whom I thank warmly.

REFERENCES

1. Mellor, J.W. (1942) In: *A Comprehensive Treatise on Inorganic and Theoretical Chemistry*, Vol. XIII (part 2), pp.432, Longman, Green & Co., London.

2. Perutz, M.F. (1989) In: *Mechanisms of Cooperativity and Allosteric Regulations in Proteins*. University Press, Cambridge; M.A. Gilles-Gonzalez, G. Gonzalez and M.F. Perutz, *Biochemistry,* 1995, 34: 232 and references to Perutz's previous work cited therein.

3. Pauling, L. and Coryell, C.D. (1936) *Proc. Natl. Acad. Sci.*, USA, 22: 210.

4. Weiss, J.J. (1964) *Nature*, 203: 183 and references cited therein; J. Peisach, W.E. Blumberg, W.A. Wittenberg and J.B. Wittenberg, (1968) *J. Biol. Chem.*, 243: 1871.

5. Akhtar, M. and Wright, J.N. (1991) *Nat. prod. Rep.* 527.

6. Korth, H.G., Sustmann, R., Thater, C., Butler, A.R. and Ingold, K.U. (1994) *J. Biol. Chem.,* 269: 17776.

7. Ortiz de Montellano, P.R. (ed.), (1986) *Cytochrome P-450: Structure, Mechanism & Biochemistry*, Plenum Press, New York & London.

8. McMurray, T.J. and Groves, J.T. (1986) In: *Cytochrome P-450: Structure Mechanism & Biochemistry*, (Ortiz de Montellano, P.R. ed.), pp.1. Plenum Press, New York & London.

9. Coon, M.J., Ding, X., Pernecky, S.J. and Vaz, A.D.N. (1992) *FASEB J.* 6: 669.

10. Akhtar, M., Alexander, K., Boar, R.B., McGhie, J.F. and Barton, D.H.R. (1978) *Biochem. J.* 169: 449.

11. Akhtar, M., Calder, M.R., Corina, D.L. and Wright, J.N. (1982) *Biochem. J.* 201: 569.

12. Stevenson, D.E., Wright, J.N. and Akhtar, M. (1988) *J. Chem. Soc., Perkin Trans. I.* 2043.

13. Nakajin, S., Takahashi, M., Shinoda, M. and Hall, P.F. (1985) *Biochem. Biophys. Res. Commun.* 132: 708.

14. Akhtar, M., Corina, D.L., Miller, S.L., Shyadehi, A.Z. and Wright, J.N. (1994) *Biochemistry* 33: 4410.

15. Imai, T., Globerman, H., Gertner, J.M., Kagawa, N. and Waterman, M.R. (1993) *J. Biol. Chem.* 268: 19681.

16. Lee-Robichaud, P., Wright, J.N., Akhtar, M.E. and Akhtar, M. (1995) *Biochem. J*, (in press).

A NEW FLOW MODULATED CONTINUOUS BIOREACTOR BASED ON INVERSED CAPILLARY MEMBRANES FOR LOW RISK GENE TECHNOLOGY

Hermann Bauer,[*1] Udo Promm,[1] Wilhelm Schumann,[2] and Peter Bendzko[3]

[1] Fachbereich Technische Chemie
   Fachhochschule Nürnberg, Germay
[2] Institüt für Organische Chemie
   University of Tübingen, Germany
[3] Invitek, Berlin, Germany

ABSTRACT

Bioreactors are widely used in biotechnology for numerous applications. Typically such bioreactors are used with isolated enzymes, enzyme-complexes, whole cells or even only cell extracts without defined composition. The immobilisation of these producing systems may be done by different techniques. Beneath the physical or chemical bonding to non soluble matrices the compartimentation of the reactors with semipermeable membranes becomes more and more important. The membranes are similar to those, which are used in ultra filtration and dialysis techniques.

Conventional membrane based bioreactors for continuous operation consist commonly of a flow through system of which the outflow is equiped with a semipermeable membrane. Flat membranes are more often used than hollow fiber membranes.

The reactor chamber filled with the producing bio system is normally feeded while adding the substrate, the coenzymes and other essential substances, continuously. The equivalent amount is pressed through the terminating semipermeable membrane into the outflow. The producing system is retained, whereas the low molecular weight reaction products are eluted. The retained high molecular weight resp. large size components are enriched on the surface of the membrane. Thus an osmotic gradient is formed. If it is not possible to transfere these macromolecules via diffusion or convection resp. erosion back into the centre of the reactor chamber, the solubility product may be exceeded easily. Thus the socalled secondary layer is formed which results in several problems:

---

[*] Represents person presenting Paper

- the required pressure is enhanced; resp. the flow rate is reduced, the whole system may come to a standstill
- commonly the rate of flow is continuously decreased according to the duration of operation.
- the macro molecules which form the secondary layer are separated from the reaction solution and thus commonly are no more able to act as active biocatalysts. The productivity of the system is diminished.
- the precipitated macro molecules may be denaturated and lose their physiological activity.

To overcome these problems several techniques are available.

- increasing of the transfer rate of the macromolecules from the secondary layer into the centre of the producing system with rigorous stirring. This may cause severe problems because of share forces, which denaturate the producing system.
- drastically increasing of the surface area of the membranes to reduce the local solvent velocity
- reduction of the flow rate (reduction of the reactor efficiency).

Conventional membrane based bioreactors lack more or less of clogging of the membranes and of the by this caused problems. Furthermore the ratio in between the volume of the producing system and the reasonable rate of flow is disadvantageous.

All above discussed problems are eliminated by the developed new flow trough bioreactor with the application of flow modulation.

The new type of a bioreactor consists of a reaction space which is separated from a by pass flow system, consisting of the feeding system as well as of an outflow system, with semipermeable membranes. The exchange in between the two phases may be done by an alternating gentle pressure which results in a modulated flow pattern. This modulation causes a restricted flow from the feeding side into the reaction space and under reverse mode to the

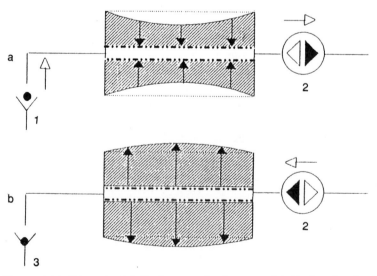

Figure 1. Flow modulated bioreactor. 1: ball valve, open; 2: micro pump; 3: ball valve, closed; a: flow direction from reaction space to outflow; b: flow direction into reaction space.

outflow side. Pressure compensation is done by an elastic jacket. All flow and pressure modulations are formed by a computer controlled micro pump.

Figure 1 shows a schematic drawing of the developed new biorector. For better understanding only one hollow fiber membrane capillary is drawn in. Usually it consists of up to 30 hollow fiber membranes of inversed structure with dimensions of L= 50 mm. OD=1.5mm, ID=0,8 mm. The exclusion limit is almost in the range of MW 10 000 - 25 000. The volume which is available for the producing system in the reaction space varies from 1 to 5 ml.

The volume of modulation should be at least in the same range as the deadvolume of the membranes which are used. Thus the diffusion barrier in the membranes is overcome and at least exchange characteristics similar to liquid-liquid interfaces are found. During this procedure a gentle flow occurs, which generates flow characteristics in the reaction space which support the active transport of molecules and particles from the surface of the membrane into the centre of the space and reverse. In addition the whole system is gently mixed. Larger modulation volumes are applied, if more efficient mixing is required or if very fast reaction kinetics are assumed. Thus the efficiency of this new system is controlled by the frequency and the amplitude of the modulation.

Bioreactors with flat membranes have been investigated as well. However they show almost unfavourable phase ratios between the producing and the bypass system. The application of capillary membranes increases the efficiency of the new bioreactor drastically.

Modulated flow exchange bioreactors of different sizes and phase ratios have been produced and tested. The basic exchange characteristics are similar to that of conventional reactors. However even under conditions with living cells (in contrary to conventional ones) no clogging of the membranes resp. decreasing of flow is observed.

Figure 2 shows the characteristic pattern of the modulated flow under starting conditions using an inert dye for visualisation.

The new bioreactor was applied to the continuous cell free translation to produce polypeptides.

Under classic conditions the cell free translation lacks of low yield due to the restricted translation ratio of only one to some moles polypeptide per mole mRNA.

With the new type of bioreactor represented here it gets firstly possible to work under continuous conditions. As the formed products are removed continuously from the reaction space, the ratio of translation depends solely on the stability of the producing system. Flow, modulation amplitude and -frequency are used as parameters for the optimization of the whole system.

In figure 3 the synthesis of human interleukine-2 via translation from its mRNA is shown. The translation is carried out with a wheat germ extract. After 24 hours reaction time already a ratio of translation of more than 1 000 is found.

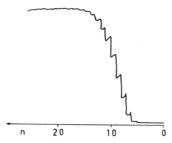

Figure 2. Characteristic pattern of the modulated flow. Application of an inert dye for visualisation. n: number of modulated cycles.

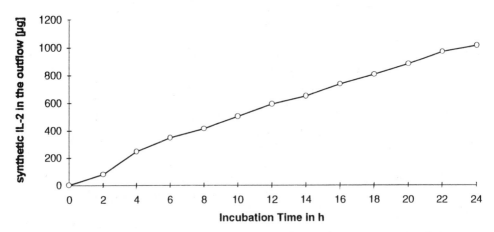

Figure 3. Preparative in vitro synthesis of human IL-2 in a continuous wheat germ extract translation system.

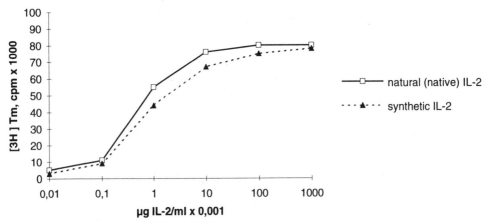

Figure 4. Bioassay of native and synthetic human IL-2 on the IL-2 dependant CTTL-2 line.

Figure 4 shows the bioassay of the synthesized IL-2 in comparison with natural IL-2.

Our results demonstrate clearly the applicability and the advantages of the new bioreactor presented here.

In addition it gets firstly possible to work under sterile conditions without the well known restrictions, as the producing system is separated via the semipermeable membranes from environmental influences.

REFERENCES

1. Bauer, H. *et al.*, (1988) 6th international symposium on synthetic membranes in science and industry, Tübingen, FRG, proceedings 141 - 144
2. Spirin, A.S. *et al.*, (1988) *Science* 242: 1162 - 1164
3. Kigawa, T. *et al.*, (1989) *J.Biochem* 110: 166 - 168
4. Sarkar, G. *et al.*, (1989) *Science* 244: 331 - 334
5. Mackov, E.R. *et al*, (1990) *Poc.Natl.Sci.USA*, 87: 518 - 522
6. Bauer, H. *et al*, (1995) Regionalkonferenz Nürnberg, abstracts.

CHARACTERIZATION AND SUBSTRATE SPECIFICITIES OF VARIOUS MOUSE KALLIKREINS

Obaid U. Beg and Mukarram Uddin

Department of Anatomy and Cell Biology and Molecular Biology Core
  Facility
Division of Biomedical Sciences
Meharry Medical College
1005 D.B. Todd. Blvd., Nashville, Tennessee 37208

ABSTRACT

Kallikrein gene family in mouse consists of 25 gene members. Only few members of this family have been characterized. These include epidermal growth factor binding proteins A, -B, -C, γ-renin, α- and γ- nerve growth factors (α- and γ-NGF's) and prorenin converting enzyme. These proteins show specificity towards various growth factors and enzymes. We have carried out the characterization of mK11, a product of clone mKlk-11. This protein was present at higher levels in the mice genetically selected for high blood pressure. However, the protein was not detected in the mice selected for low blood pressure. The substrate specificity of this protein was tested on tetradecapeptide (TDP, a renin substrate) and β-lactoglobulin (β-lg). The protein shows specificity toward Tyr-Ile (4-5) bond in TDP and Tyr-Ser (20-21) and Arg-Val (40-41) bonds in β-lg . Another member of the kallikrein family, γ-NGF was found to cleave the Phe-His (8-9) bond in TDP. The cleavage specificity of these proteins raises the possibility of the involvement of these proteins in the processing of angiotensin-II (AT-II). Furthermore, these proteins may play a role in the regulation of local blood pressure/flow. The characterization of these kallikrein family members and determination of their substrate specificities is presented.

INTRODUCTION

The mouse glandular kallikreins are a group of serine proteases which exhibit substrate specificity. Enzymes of this subgroup show homology of their nucleotide sequences and are encoded by tightly clustered and closely related genes[1]. So far , the gene products of only few members of this family have been characterized[2]. Activity profile of the characterized members indicate that this group of proteins is involved in the processing and conversion of several inactive proteins and peptides to their biologically active form. The characterized members include the family of epidermal growth factor binding proteins; EGF-BP -A, -B and -C[3], prorenin converting enzyme; (PRECE[4]), -α and -γ NGF's[5] and γ-renin[6].

Participation of these structurally related members in the processing of a variety of substrates demonstrate their diverse biological function(s). The structural relatedness and functional diversity of this family of proteases makes them very valuable for analyzing their specificities for natural and synthetic substrates.Elucidation of the substrate specificity of each member of this gene family would help us understand their physiological role.

In vitro, kallikreins have been shown to selectively process vasointestinal peptide, procollagenase, proinsulin, atrial natriuritic factor, low density lipoprotein and prorenin[7].These studies also tend to suggest involvement of kallikreins in the processing of inactive peptides to their biologically active forms.

Some members of the kallikrein family of enzymes participate in the activation of renin angiotensin system. Examples include γ-renin, which is a kallikrein and is capable of cleaving synthetic renin substrate[6] to form angiotensin-I (AT-I).Rat tonin is another kallikrein which can cleave renin substrate to form the active angiotensin-II (AT-II)[8]. Some investigations suggest the presence of renin-angiotensin system in the brain, however, the enzyme, which generates angiotensin from its substrate angiotensinogen has not been detected[9]. It has been suggested that γ-renin or perhaps other kallikreins may be involved in the processing of angiotensinogen[6].

The objective of this study is to determine the specificity of additional different mouse submandibular gland kallikreins for angiotensin. With this in mind, in the present investigation we have characterized and analyzed the effect of: a) mK11 on TDP which corresponds to the N-terminal residues 1-14 of the renin substrate angiotensinogen[10] and β-lg and b) γ-NGF on TDP.

MATERIALS AND METHODS

Isolation of Proteins

Crude aqueous extracts from mouse submandibular gland of normal mice were separated by reverse phase high performance liquid chromatography (RP-HPLC) on a C-18 (A300,Waters) column using a linear gradient of aqueous 0.1% trifluoroacetic acid (TFA) and acetonitrile containing 0.1% TFA at a flow rate of 1ml/min. The isolated peaks were lyophilized and stored at -20°C for further analysis.

Peptide Synthesis

The peptide corresponding to the first fourteen residues of the human angiotensin (peptide sequence; DRVYIHPFHLVIHN) was synthesized on an Applied Biosystems peptide synthesizer (431A) using standard fmoc protocol.The peptide was cleaved from the resin using a mixture of TFA:Thioanisole (95:5 V/V) for 4h at room temperature.The peptide was precipated using five volumes of ethyl ether and allowed to dry. The peptide was dissolved in 0.1% TFA for further purification by reverse phase HPLC.

Modification of Proteins

The isolated peaks were subjected to carboxy- methylation in 100 mM Tris-HCl, pH 8.0 containing 6M Guanidine-HCl. The proteins were reduced with 100mM dithiothreitol for 2h at room temperature and carboxymethylated with 1M iodoacetic acid at 37°C for 2h in the dark. The mixture was adjusted to a concentration of 0.1% TFA prior to HPLC. The HPLC was carried out under conditions described for the proteins.

Sequence Analysis of Proteins

The HPLC peaks were subjected to amino acid sequence analysis on a solid phase protein sequencer (MIlligen-BioSearch) using procedures described by the manufacturer.

Peptide Cleavage

The synthetic TDP (1nmol) was digested with 25pmol of the mouse submandibular gland HPLC purified peaks. These peaks were previously identified by their sequence analysis.The digests were carried out in 0.1M ammonium bicarbonate, pH 8.0 or 10mM Tris-HCl, pH 8.0 for 4-24h. The digests were separated by HPLC under conditions similar to that of the intact and carboxymethylated proteins.For controls TDP was digested with chymotrypsin under similar conditions.

RESULTS

Isolation of mK-11 and γ-NGF and Identification by Amino Acid Sequence Analysis

HPLC separation of aqueous extracts of the mouse submandibular gland on a reverse phase column is shown Fig.1. Following amino acid sequence analysis of all major peaks in Fig.1 and comparison with known amino acid sequences of the members of mouse kallikrein family we have identified mK11 to be present in peak-2 (Fig.1) and γ-NGF as a major

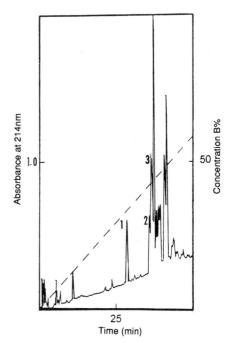

Figure 1. The separation of soluble aqueous extract of submandibular proteins from genetically selected nor-motensive mouse on a reverse phase HPLC column. Peak 2 represents mK11 and peak 3 contains both γ-NGF and mK11. The other peaks represent various members of the mouse kallikrein family. The conditions for the separation are described in the text.The broken line represents gradient.

Table 1. a) The sequence obtained for the intact and that of the carboxymethylated mK11 protein.The sequence shows the presence of three chains due to the internal cleavage of the protein.The position of the three chains according to the predicted protein sequence are shown. Thr in chain B (position 3) is only present in mK11 sequence among all known mouse kallikreins. (b) The sequence obtained for the γ-NGF from the analysis of the intact and carboxymethylated chains.The presence of three chains is due to internal cleavages.The residue positions according to the predicted sequence are shown. Tyr in chain B (position 8) is unique among all known mouse kallikreins

<u>a</u>

1

Chain A   I V G G F N C E K N S Q P W H V A V Y R - - - -

141

Chain B   F Q T P D D L Q C V S I K L L P N E V C - - - -

98

Chain C   M L L R L S E P A D - - - -

<u>b</u>

1

Chain A   I V G G F K C E K N S Q P W H V A V Y R Y T Q -

141

Chain B   F Q F T D D L Y C V N K L L P N E D C - - - -

88

Chain C   F L E Y D Y S N D L M L L - - - -

component in peak-3 (Fig.1). The results from the sequence analysis of these proteins and of their carboxymethylated and isolated chains are shown in Table Ia and Ib. Most of the mK11 elutes in peak-2 (Fig.1) and small portion of this protein is present in the adjacent peak-3 (Fig.1). Furthermore, we analyzed the mK11 and γ-NGF by sequence analysis of the carboxymethylated chains (Table Ia and Ib). Since γ-NGF and mK11 differ in certain residue positions from each other and from the established sequences of the other members of this family, we have positively identified γ-NGF in peak-3 (Fig.1) .No other proteins apart from mK11 were detected in peak-3 (Fig.1). Additionally, EGFBP-A,-B and -C and mK11 were also tested for their activity on the synthetic renin substrate (TDP).

Specificity of mK11 and γ-NGF For TDP

The HPLC separation of TDP peptides obtained after cleavage by mK11 (peak-2, Fig. 1) and γ-NGF fraction (peak-3, Fig.1) is shown in Fig's. 2a and 2b respectively. The cleavage of β-lg by mK11 is shown in Fig.2c.

The sequence of each peak was confirmed by amino acid sequence analysis as shown in Fig's.3a and 3b. In the digest of TDP by mK11 (Fig. 2a) peak -1 constitutes residues 1-4 of TDP and peak-2 contains residues 5-14 of TDP. The TDP digest by γ-NGF fraction generated three peaks (Fig. 2b) ; peak-1 correponds to a peak of TDP digest by mK11 (peak-1, Fig. 2a) which contains residues 1-4 (DRVY) of the TDP. The presence of peak-2 (HLVIHN;

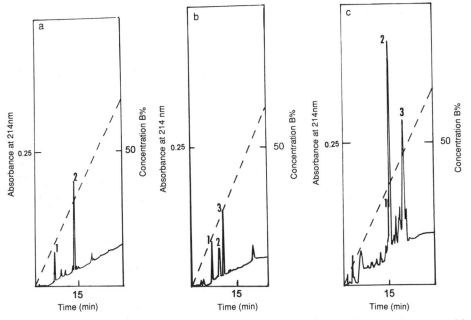

Figure 2. The separation of peptides from the TDP digest by; (a)mK11, (b) γ-NGF fraction and that of β-lg by mK11(c). The conditions for the separation are described in the text. The peaks were identified by sequence analysis to determine the cleavage points.

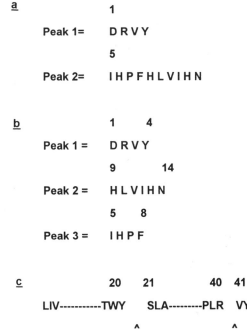

Figure 3. The amino acid sequence of the TDP peptides (Fig2a-2c) obtained after cleavage with the (a) mK11, (b) γ-NGF fraction and (c) that of β-lg by mK11. The positions of the peptides are mentioned to show their position in the TDP and β-lg.

residues 9-14 in TDP) and peak-3 (IHPF;residues 5-8 in TDP) following digestion of TDP by γ-NGF fraction could only be due to the γ-NGF's specific cleavage between Phe-His bond.

The sequence analysis of the peaks from β-lg digest by mK11 is shown in Fig. 3c. The cleavage between Tyr-Ser (residues 20-21) and Arg-Val (residues 40-41) in β-lg is noted.

DISCUSSION

In the present study we have shown that :

a) mK11, a protein product of gene mKlk-11 in mouse submandibular gland specifically cleaves between Tyr-Ile (residues 4-5) in TDP and in β-lg between Tyr-Ser(residues 20-21) and Arg-Val (residues 40-41) respectively. These cleavages are indicative of tryptic and chymotryptic type cleavage. The cleavage of β-lg (as a substrate) was carried out to determine the cleavage preference of mK11 because β-lg offers combination of variety of bonds cleaved by different proteases.Various other kallikreins which exhibit protease like activity have been shown to exhibit similar cleavages[11]. Since other members of the kallikrein family have been involved in cleavage of biologically active peptides, it is likely that mK11 may have a similar function, especially in relation to the cleavage of renin substrate (TDP) between Ile-Tyr (residues 4-5). The observation that mK11 cleaves TDP between residues 4-5 (Fig. 4) might indicate a physiological role of mK11 in the degradation of AT-II. Another enzyme, a neutral endopeptidase which cleaves Tyr-Ile bond in AT-II (Fig. 4) has been characterized[12]. However, this enzyme also cleaves Pro-Phe (residues 6-7) in AT-II. Since mK11 does not cleave any other bond besides Tyr-Ile (positions 4-5) therefore mK11 appears to be more specific in the in vitro processing of AT-II.

b) γ-NGF which belongs to kallikrein family and co-elutes with another kallikrein, mK11 on reverse phase HPLC specifically cleaves synthetic TDP between Phe8-His9 residues. TDP represents a sequence which correspond to residues 1-14 of angiotensinogen and matches with the precursor sequence of AT-I and AT-II. Our in vitro results indicate that γ-NGF may be involved in the formation of biologically active AT-II directly from the precursor molecule (Fig. 4).This action of γ-NGF in the mouse submandibular gland is similar to that of rat tonin[8]. Since tonin has not been detected in mice glandular tissue it is likely that in mice γ-NGF acts like tonin (Fig. 4). This notion is further supported by the fact that there is high sequence homology between γ-NGF and rat tonin.

Additionally, we tested the activity of three other kallikreins namely EGFBP-A, -B and -C on TDP and β-lg. No cleavage of TDP was observed, however, these proteins were able to cleave other bonds in β-lg. These results further suggest specificity of structurally related kallikreins in terms of their action on protein bonds and support the notion of our finding that mK11 and γ-NGF as a member of kallikrein family function as highly specific enzymes. γ-NGF's in vitro cleavage of Phe-His bond of the TDP (which represents the precursor sequence of AT-I and AT-II) is the first demonstration of γ-NGF involvement in

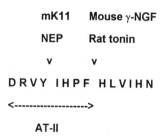

Figure 4. The representation of TDP sequence and the position of bonds which are cleaved by mK11, Angiotensinase B (NEP=neutral endopeptidase 24.11), γ-NGF and rat tonin. The line below the sequnce represents the amino acid sequence of AT-II.

renin angiotensin system. However, whether this *in vitro* cleavage also occurs *in vivo* as well is not clear and remains to be evaluated.

γ-NGF has been shown to be involved in the processing of other precursor proteins to their active forms.These include β-NGF[13], complement C-1, urokinase type plasminogen activator and macrophage stimulating protein[14]. These findings are consistent with our *in vitro* results of the processing of angiotensin precursor molecule to its active form.

In conclusion, our *in vitro* results implicate mK-11 in the degradation and γ-NGF in the formation of angiotensin-II from a synthetic renin substrate and emphasize a more diverse and significant role of these proteases in important biological processes.

ACKNOWLEDGMENT

We would like to thank Prof.G.Schlager, University of Kansas, Lawrence for providing mice.

REFERENCES

1. Evans,B.A., Drinkwater,C.C. and Richards, R.I. (1987) *J.Biol.Chem*. 262: 8027-8034.
2. Carretero, O.A., Carbini, L.A. and Scicili, A.G. (1993) *J.Hypertension* 11: 693-697.
3. Drinkwater, C.C., Evans, B.A. and Richards, R.I. (1987) *Biochemistry* 26: 6750-6756.
4. Kim, W.S., Nakayama, K., Nakagawa, T., Kawamura, Y., Haraguchi, K. and Murakami, K. (1991) *J.Biol.Chem*. 266: 19283-19287.
5. Evans, B.A. and Richards, R.I. (1985) *EMBO.J*. 4: 133-138.
6. Drinkwater, C.C., Evans, B.A. and Richards, R.I. (1988) *J.Biol.Chem*. 263: 8565-8568.
7. Clements, J.A. (1989) *Endocrine Rev*. 10: 393-419.
8. Moreau,T., Brillard-Bourdet, M., Bouhnik, J. and Gauthier, F. (1992) *J.Biol.Chem*. 267:10045-10051.
9. Darby, I.A. (1986) Location of renin and angiotensinogen gene expression using hybridization histochemistry, Ph.D. thesis, University of Melbourne, Australia.
10. Kageyama, R., Ohkubo, H. and Nakanishi, S. (1984) *Biochemistry* 23: 3603-3609.
11. Drinkwater, C.C., Evans, B.A. and Richards, R.I . (1988) *TIBS*, 169-172.
12. Erdos, E.G. and Skidgel, R.A. (1990) *Kidney International* 38: S24-S27.
13. Edwards, R.H., Shelby, M.J., Garcia, P.B. and Rutter, W.J. (1988) *J.Biol.Chem*. 263: 6810-6815.
14. Wang, M.H., Gonias, S.L., Skeel, A., Wolf, B.B., Yoshimura, T. and Leonard, E.J. (1994) *J.Biol.Chem*. 269: 13806-13810.

4

NATURAL OCCURENCE OF ENZYMES LINKED TO INORGANIC SUPPORTS

Venice Lagoon and Internal City Canals

N. Sabil,[1] Y. Aissouni,[1] B. Pavoni,[2] D. Tagliapietra,[2] and
M.-A. Coletti-Previero[*1]

[1] Centre de Recherche INSERM
70 Rue des Navacelles, 34090 Montpellier, France
[2] Dipartimento di Scienze
Ambientali dell' Università, Venezia, Italy

ABSTRACT

Powerful enzyme activities were found in the sediments of the Venice lagoon and internal city canals. No detectable enzyme activity was present in the aqueous phase even after centrifugation. These insolubilized enzymes showed remarkable heat stability and an increased resistance to severe environmental conditions. They were probably of bacterial origin, mostly immobilized on the inorganic component of the sediment, so that they could survive the organisms from which they were generated, since their lifespan is prolonged by insolubilisation. As a consequence they resist to conditions where the same enzyme under soluble form would be rapidly inactivated. They are useful diagnostic factors of the ecosystem, since their presence is related to the waste products. A study on the linkage between enzymes and clays suggested that a spontaneous and specific affinity between them, previously unsuspected, allowed the formation of "activated" sediments, the leading force in the first step of organic matter degradation in ecosystems.

INTRODUCTION

The "*Palude della Rosa*", which is a delimited area of the Venice lagoon, and the town center canals (rii) were intensively studied in the framework UNESCO-MURST Venice Lagoon and Internal Canals Projects. As a model ecosystem, it can be grossly subdivided in three phases, gaseous (air), aqueous (water) and solid (sediments). Each one participates to the biodegradation as such and at the interphases between them. Scheme 1 represents the

---

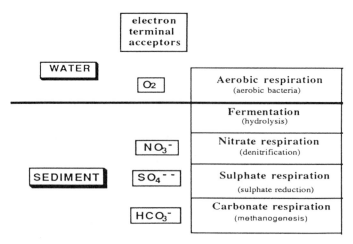

Scheme 1. Schematic representation of a sedimentary ecosystem vertical profile.

vertical profile of a sedimentary ecosystem with the ideal subdivision of the degradative events. This ecological sequence decomposes the organic matter toward the terminal acceptors, successively oxygen, nitrate, sulphate and carbonate, and is achieved by distinct populations of microorganisms. The molecular events underlying this degradation are the actions of catalytic systems (enzymes) produced by the microorganisms, and this is the leading force in the degradative transformations.

Since biological treatment has been around for millions of years, organisms exist that can degrade any compound present in the environment. However, xenobiotic compounds include products which are, at best, slow to degrade, since they arise from the combustion of natural elements. Considerable evidence shows that enzymes, evolved for the degradation of biogenic compounds can be recruited to degrade xenobiotics and that microorganisms are also able to modulate their enzymatic synthesis to transform large quantities of new compounds. While an extended literature is available on the pollution of two of the three components of the *habitat*, namely the air and the water layer, much less has been done on the sediment. This is rather surprising, since of the three component it is the one less subject to changes, such as those induced by winds and tides/currents. Only studies on the bacterial development are available, which, although interesting, overlook an important parameter: the enzymatic potential. We focused our effort on the insoluble phase of the *Palude* by large scale monitoring of the enzymatic activity as well as by evaluating the biodegradative potential of the insoluble part of the system and the relative importance of anthropogenic influences. Enzyme-mediated hydrolytic reactions were selected as investigation matter, as hydrolysis is usually the first step in organic matter degradation.

In previous studies on anaerobic sediments from septic tank and therapeutic sludges[1-4], we were able to show that the enzymatic activities responsible for the degradation of phosphate derivatives, proteins, cellulose, lipids and urea, were mostly linked to the insoluble material. Since this insolubilisation protects the enzymatic activity, it seemed interesting to determine whether the same situation occurred in other biological systems that include both a solid and a liquid phase to ascertain if this enzymatic process could be a general system[5-7]. We also investigated the evolution of the enzyme activities linked to the sediment with temperature, pH, time and seasonal variation in order to evaluate their reaction to environmental changes, even extreme ones, and compared it to the behaviour of the same catalysts, when solubilized.

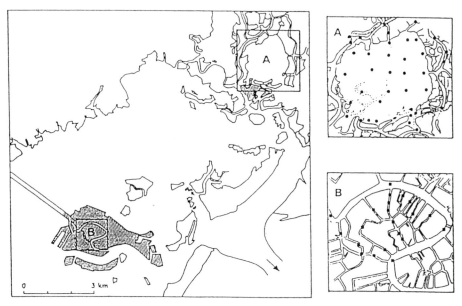

Figure 1. Venice and the Venice lagoon. A): *Palude della Rosa*; B): San Polo district: Inner canals.

EXPERIMENTAL

Samples

Sediment cores were randomly collected in the *Palude della Rosa*, a round shaped pond situated in the northern Venice lagoon (2 km diameter, 80 cm depth in the center, 10 to 50 cm at the boundaries) surrounded mainly by strips of land regularly flooded by the tide, as well as from the internal canals of the Venice district of S. Polo. This last sampling area was selected as it is a well delimited unit, being bordered by the Grand Canal and by the canal Rio Nuovo (Fig. 1B). The sediment in all cores was a relatively uniform mix of 80% silt clay (ø < 62 micron) and sand, with some shell fragments, which were roughly removed. This composition confirmed the known sediment composition of the Venice lagoon sediment, which is reported to be clay and silted clay at 75 to 90 %[8].

Enzyme Activities

The insoluble material was centrifuged 30 min at 40 500 g and washed twice in an isotonic solution. The enzymatic activities were determined in triplicate on suitable quantities of the centrifuged sediment (from 0.1 to 1.0 g in the same volume) and of the combined supernatants, as already described[3,4]. The sediment samples were suspended in the buffer required for each enzymatic activity and the kinetic measured as a function of time. The enzymatic activities present in the sediments of *Palude* della Rosa, were examined in four campaigns/year, in winter (December), spring (March), summer (July) and autumn (October). The sediments were heated at different temperatures (from 37° to 100°C) for 15 min and then rapidly cooled to room temperature. The different residual activities were determined as described above. The enzymatic reactions were determined at pH values ranging from 3 to 9.5.

Table 1. Inorganic supports and active proteins analysed for interaction

S u p p o r t s							
(A) Zeolithe 4 Å	(B) Zeolithe 5 Å	(C) Zeolithe 9 Å	(D) Zeolithe Y	(E) Kaolin G1/91	(F) Montmorillonite K10	(G) Bentonite, (Bentonite Na)	(H) Bentonite (Bentonite Ca)
L i g a n d s				A c t i v i t y			
Aprotinin				Trypsin Inhibitor			
Trypsin				--X-CO-NH-Y-- $\longrightarrow$ --X-COOH + H$_2$N-Y-- protein peptides			
Cellulase				cellulose $\longrightarrow$ (glucose)n			
Phosphatase				PNPP $\longrightarrow$ PNP			
Urease				urea $\longrightarrow$ 2 NH$_4$OH ammonia			

Desorption

The isotonic solution washing and centrifugation, 30 min at 40 500 g, did not release any enzymatic activity. The sediments were then treated with the following solutions: 0.5 M NaCl, 0.5 M NaHCO$_3$, 0.5 M phosphate buffer at pH 8, and 1% Triton X-100. A suitable quantity of each sediment (0.1 to 1.0 g) was suspended in each solution (5 ml) and allowed to react for 15 min, room temperature. After centrifugation, the activity was determined in the supernatant.

Isolation of Phosphatase and Cellulase

The enzymes, released from the sediment by phosphate buffer (15 min), were first chromatographed on a Sephadex G 75 (molecular sieve). The active fractions were than charged in the same buffer on a Mono Q column (basic ion exchange), where they eluted by rising salt concentration.

Heavy Metals

Sediments for metal analyses were placed in glass jars and refrigerated. Water was later removed by lyophilization and analysis was carried out with dry sediment. For the analysis of all metals except mercury, five grams of lyophilized sediment were treated, drop by drop and shaking, with 30 ml of 8 N HNO$_3$ in a 100 ml beaker: the suspension was then heated to boiling point for 4 hours on a controlled-temperature stirring heater. Analysis were carried out with an Atomic Absorption Perkin-Elmer 5000 spectrophotometer with an air-acetylene burner and provided with an electronic integrator and recorder. For the analysis of mercury, sediment samples were digested in an Erlenmeyer flask for 2 hours at 60°C with concentrated H$_2$SO$_4$, HNO$_3$, HCl. Mercury was then oxidized with a 5% solution of KMnO$_4$

and $K_2S_2O_8$. After reduction with $NaBH_4$, samples were analyzed by AAS in flameless mode[9].

Interactions Enzymes/Inorganic Supports

Aprotinin (M.W. 6500) which inhibits trypsin activity, trypsin (M.W. 24000) which hydrolyses the peptide bond near basic residues, cellulase (M.W. 57000) which cleaves bonds in polysaccharides, phosphatase (M.W. 69000) which hydrolyses phosphate derivatives, urease (M.W. 469000) which transforms urea to ammonia were chosen as model active proteins. Zeolithe (4 Å, 5 Å, 9 Å, Y) and clay (Kaolin, Montmorillonite, Bentonite) were chosen as model inorganic supports (Table I). Active proteins were allowed to react *in vitro* and the insolubilisation rate and percentage were determined. Protein-support coupling yield was estimated by differential spectral determination and activities were measured both on the soluble and insoluble phase of the reaction mixture. Optimum protein concentration, desorption behaviour, activity yield, time-resistance and temperature effect were determined for each inorganic support.

THE LAGOON (*Palude della Rosa*)

The Enzymatic Activities

It is usually assumed that enzymatic activity is confined to the soluble phase or to the pore-water phase of the upper layer of the sediment column. However the aqueous phase of the sample cores of the *Palude della Rosa* was found to be devoid of enzymatic activity, at least a measurable one. By contrast, the insoluble phase showed distinct and sometimes unexpectedly strong enzymatic activities, responsible for the degradation of urea, cellulose and phosphate derivatives, mostly linked to the insoluble material. Protease and lipase activities were undetectable: they are probably present in the ecosystem but at a dilution impossible to determine with any degree of certainty. Table II summarizes the situation.

The enzyme activities are, however, distributed unequally. When represented over the geographical map of the *Palude*, as shown in Fig. 2. for phosphatase activity, distinct areas of activity can be seen.

Cellulase and urease enzymatic maps have also been drawn[4] and showed the same unequal distribution. These patterns were obtained by joining the sample positions with nearly identical to identical activity by lines, hereafter called iso-enzymatic. The values of the internal samples were of comparable strength, the shaded areas are the ones with higher activity, while the stripped ones are the zones of lowest activity.

Table 2. Enzymatic activities of the different components from the *Palude della Rosa*. (Spring, means of thirty experiments) Controls were from the south of the Venice lagoon. (From Ref. 4)

		Phosphatase[a]	Cellulase[b]	Urease[a]	Protease	Lipase
Supernatant	Controls			undetectable		
	Palude d. R					
Interstitial liquid	Controls	traces	traces	traces	undetectable	undetectable
	Palude d. R	0.24 ± 0.04	0.11 ± 0.03	0.075 ± 0.005		
Solid Phase	Controls	0.82 ± 0.12	2.26 ± 0.18	2.44 ± 0.09	traces	traces
	Palude d. R	9.14 ± 0.82	9.35 ± 0.55	7.39 ± 0.57		

[a] µmoles / min / g
[b] mg CMC / h / g

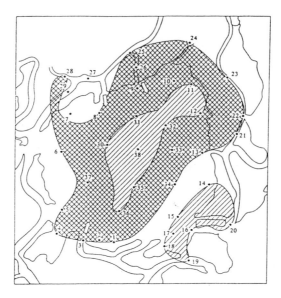

Figure 2. Enzymatic map of *Palude della Rosa* sediment. Lower activity = striped zone. Higher activity = grid zone. (From Ref. 4).

Binding Between Enzymes and Sediment

In order to distinguish between the enzymes from intact microorganisms, those entrapped in microorganism membranaceous fragments and those bonded to the inorganic material, selective washing were used. If only Van der Waals forces were involved in the adsoption of the enzymes on the insoluble matrix, $NaHCO_3$, NaCl or any salt solution would desorb them. Alternatively, if the enzymes were entrapped in membranes or fragments of microbial membranes, the non-ionic detergent Triton X-100 should desorb them, at least in part. To differentiate between bindings, several experimental approaches were chosen.

1) Analysis by scanning electron microscopy did not show intact microorganisms. Since only a small part of the sediment samples could be submitted to this analysis, the presence of whole organisms could not be excluded, but certainly they could not in such density to account for the very high enzymatic activities detected. DNA analysis[10] also failed to show consistent presence of microorganisms. In the samples with the highest DNA content (internal canals sediments), only 5 to10% activity could be attributed to the biomass. The absence of intact microorganisms, in addition to the persistance of enzymatic activities in the collected sediment (several weeks at 4°C and twenty days at 25°C, see Fig. 13) and its resistance to increasing temperatures (see § 4), is of crucial importance. It was a first indication suggesting that most of the activity detected was from enzymes present in insoluble form, since in comparable conditions a soluble enzyme would be completely inactivated.

2) The sediments were submitted to differential extraction tests. Solutions of 0.5 M $NaHCO_3$ and 0.5 M NaCl at neutral pH were unable to release any urease and phosphatase activities, while less than 10% of the cellulase activity was extracted. However, when a phosphate solution at the same molarity was used, most of the phosphatase and a significant part of the cellulase activity were extracted. This indicated that the increase in ionic strength

Table 3. Release of enzymatic activities from *Palude della Rosa* with different extracting solutions (15 min extracting time)

Activity (1 g, d.w.)	Extracting Solutions		
	NaCl 0.5 M NaHCO$_3$ 0.5 M	Phosphate 0.5 M pH 8	Triton 1%
Phosphatase*	0	1.93 ± 0.2 (yield 84,6%)	0.19± 0.01 (yield 8,3%)
Cellulase**	0.13 ± 0.005 (yield 3,7%)	1.4 ± 0.2 (yield 40%)	2.11 ± 0,18 (yield 60,3%)
Urease§	0	0	0

*μmoles PNP. min^{-1}. g^{-1}
**μmoles Reducing sugars.h^{-1}.g^{-1}
§ μmoles Urea min^{-1}.g^{-1}

was not responsible by itself for the release of the enzymes from the sediment and that the type of ion was the key factor. This specific ability of phosphate anions to release enzymes bound to inorganic material has been extensively studied in our laboratory during a research on biotechnological procedures that include immobilized enzymes[11,12].

3) Analysis of enzymatic activities after treatment of the sediment with detergents, showed that a fraction of immobilized cellulase activity could be liberated by the use of a 1% solution of Triton X-100, a non-ionic detergent often used to solubilize membrane enzymes (Table III). Urease activity remained insoluble under all treatments. This enzyme has a tendency to adsorb to inorganic supports and, because of its stability, has been one of the first insolubilized enzymes to be used for therapeutical purposes.

Since the phosphate anions are present in the *Palude della Rosa* at concentration far lower than 0.5 M, we studied the effect of the concentration on the release of the enzymatic activity (Fig. 3A). As can be seen, the desorption was proportional to the concentration in phosphate anions but, given enough time, even the lower (Fig. 3B) concentrations were effective in releasing the enzyme activities. The increase of phosphate concentration in the ecosystem (0.2 to 0.3 mM), due to its common use in the surrounding agricultural lands, probably stimulated phosphatase synthesis in the microorganisms. However, this effect was

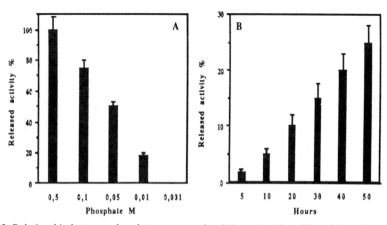

Figure 3. Relationship between phosphate concentration (A), contact time (B) and the release of enzymatic activity. (From Ref. 4).

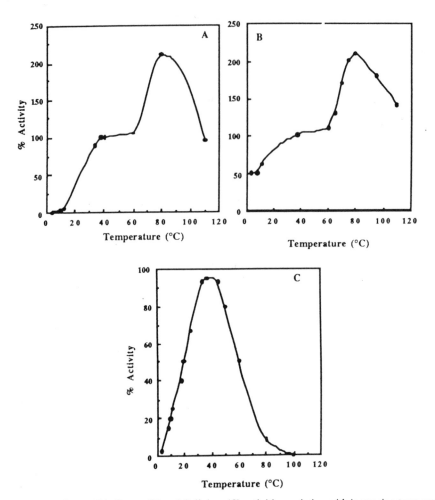

Figure 4. Phosphatase (A), Urease (B) and Cellulase (C) activities variation with increasing temperature.

counterbalanced by the ability of phosphate anions to wash out the enzymatic activity from the sediment.

4) Finally, the sediment samples were heated and the residual activity tested after cooling. The temperature dependence of the sediment enzymatic activities was examined from 4 to 100°C. The results reported in Fig 4 (A, B & C) showed that urease and phosphatase activity were enhanced by temperature exposure up to 80°C and still fully active after 15 min at 100°C. These results are in agreement with the hypothesis that these enzymes are bound to the solid support since resistance to temperature increase is a typical feature of immobilized enzymes. Cellulase was much less resistant and this was thought a confirmation of the insolubilisation in membrane fragments rather than on the inorganic material. When the phosphatase activity of a given sample was released from the insoluble support and heated (Fig. 5), it showed the typical behaviour of a soluble enzyme, thus refuting the hypothesis of thermostable catalysts produced by thermostable bacteria.

The pH dependence for phosphatase and cellulase insoluble enzymatic activities is reported in Fig. 6 A and 6 B respectively and shows two optimum pH values, one at slightly acid and one at bacic pH. This last one is at the *Palude della Rosa* found value

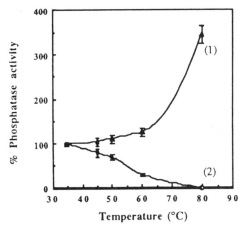

Figure 5. Temperature effect on the Phosphatase activity of immobilized (1) and released (2) enzyme. The samples were heated 15 min and after cooling, the activities were measured at 37°C (From Ref. 4).

and is considered the relevant one. The pH dependence for the urease acitivity (Fig.6 C) could be determined only on the immobilized enzyme since all our attempts to release this activity were unsuccessful.

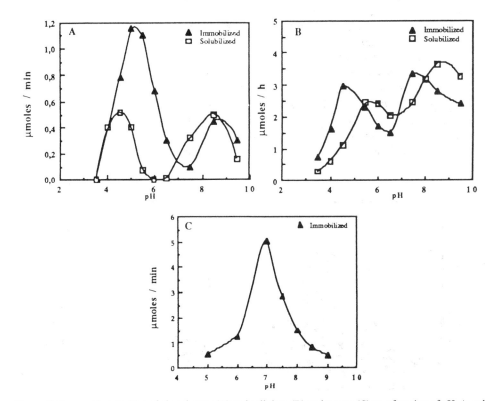

Figure 6. Enzymatic activities of phosphatase (A) and cellulase (B) and urease (C) as a function of pH. A and B present both immobilized and solubilized activities from *Palude della Rosa* sediment. (From Ref. 3)

Table 4. Enzymatic activities of the sediment from the whole surface of *Palude della Rosa*

Activity	Palude della Rosa			
	Winter	Spring	Summer	Autumn
Phosphatase[1]	0.686 ± 0.046	3.132 ± 0.256	7.236 ± 0.426	2.52 ± 0.2
	54,582 Kg /min	246,902 Kg /min	578,862 Kg /min	194,036 Kg /min
Cellulase[2]	2.61 ± 0.02	7.09 ± 0.37	4.08 ± 0.52	1.29 ± 0.19
	2,680 Kg /min	9,453 Kg /min	5,440 Kg /min	1,717 Kg /min
Urease[3]	0.008 ± 0.0005	0.29 ± 0.02	0.47 ± 0.04	0.21 ± 0.02
	624 Kg /min	23,760 Kg /min	37,728 Kg /min	16,464 Kg /min

[1]mg PNPP/ min / g
[2]mg CMC / h / g
[3]mg Urea / min/g

The intensity of the immobilized enzymatic activity, especially when compared with the very low values found in the supernatant and the interstitial fluids, indicated that the main part of the organic material initial degradation in Venice lagoon is brought about by insoluble catalysts. Whereas an extended literature is available on the contamination of sediments from the chemical point of view, and some papers have also dealt with the microorganism development within the sediment, the important aspect of the immobilized enzymatic biodegradative potential has been up to now completely ignored. It should be noted that such an activity will transform efficiently any chemicals susceptible to the enzyme actions and that, as a consequence, the real contaminating agents of the ecosystem will be the products of such reactions rather than the original compounds. In order to allow an evaluation of this biodegradative capacity, the following table (Table IV) translates the kinetic values in more common orders of magnitude.

The formula used for calculating the activity of the whole *Palude* is the following: a x M_s x S x q = A. a= activity; M_s= substrate M.W.; S= surface; q= sediment weight (1m², 1cm deep); A = whole *Palude* activity (4 Km², 1cm depth).

The strength of the enzymatic activity is clearly remarkable, even if it depends of course from the availability of enough dissolved substrate. This situation where the enzyme remains in a latent form until a sufficient quantity of substrate diffuses, is a fairly common

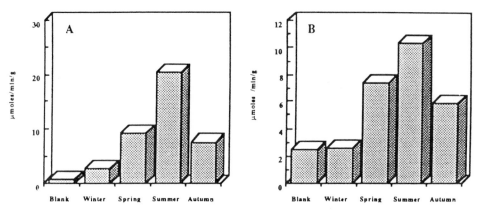

Figure 7. Seasonal variation of phosphatase (A) and urease (B) activities of *Palude della Rosa* sediment. (From Ref. 4)

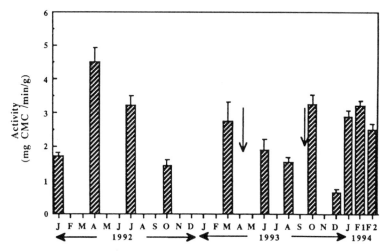

Figure 8. Seasonal variation of Cellulase activity of *Palude della Rosa* sediment. Arrows indicate *Ulva* decrease.

one. Data of Table IV, however, show that enough enzyme activity is present to digest its substrates even when their concentration is low.

The seasonal variation of phosphatase and urease are presented in Fig. 7 and, as awaited, all activities increase from winter to spring.

Cellulase synthesis is induced by the cellulose arising from the degradation of walls from algae and in particular *Ulva*, whose overwheming presence in the Venice lagoon is due to contamination and is a major environmental problem. The seasonal variation of insoluble cellulase from the sediments of *Palude della Rosa* was followed for two years and showed changes related to the algal variations in the lagoon (Fig. 8). It was concluded that the specific bio-degradative capacities could be considered, to some extent, of diagnostic importance as it varied as a function of the availability of digestible products.

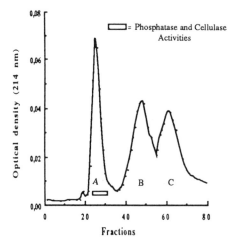

Figure 9. Molecular sieve of extracted enzymatic activity from the sediment of *Palude della Rosa*. A: Phosphatase activity; B: Inactive peak; C : Cellulase activity.

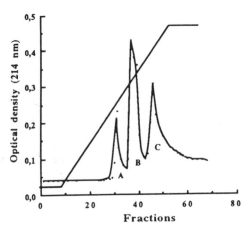

Figure 10. Ion exchange chromatography of the active fractions from G-75 Sephadex.

Isolation of Phosphatase and Cellulase

The enzymes, namely the phosphatase and the cellulase, once extracted from the sediment of *Palude della Rosa,* were examined in order to further demonstrate their presence, to characterize their efficiency as catalysts and to gain perhaps an insight as to which microbial entities are responsible of their synthesis.

The sediment was extracted and the supernatant chromatographed on Sephadex G 75 (Pharmacia). Both activities eluted in an area of the column, corresponding to a M. W. between 55 and 70 kD. The cellulase molecular weight was estimated to be 57 000 and the phosphatase 69 000 (Fig. 9).

The active fractions were then charged on a ion exchange column (Mono Q) where the active material needed a salt gradient to elute. This shows that both enzymes are acidic, cellulase slightly more than phosphatase. As seen in Fig.10, a satisfactory separation was achieved, which confirmed the ability of phosphate anions to release these enzymes from the sediment of *Palude della Rosa.*

VENICE INNER CANALS

Sediments of internal canals (rii) of Venice's historical centre (Fig. 1), were analyzed for the concentrations of heavy metals and for enzymatic activity.

Heavy Metals

The concentrations of heavy metals of the sediments in the inner canals of Venice are normalized on the background values found in the deepest strata of radioisotopically dated sediment cores and related to reference levels reported in the literature for non-polluted sediments[13]. The following values were used (µg/g d.w.): Hg, 0.1; Pb, 25; Cd, 1; Ni, 20; Co, 15; Zn, 70; Cr, 20; Cu, 20; Fe, 20000.

The highest accumulation factor was observed for mercury, with an average value of 41. Fairly high factors were also observed for Zn (17.11), Cd (7.34), Pb (7.26), Cu (11.44). Mercury concentrations found in the sub-surface sediments of this urban area are comparable

Table 5. Immobilized enzymatic activities in 1 g sediment from the *San Polo* district of Venice

Samples (N°)	Phosphatase*	Cellulase**	Urease***	Protease**** (-) EDTA	Protease**** (+) EDTA
1	26.6 ± 2.5	117.8 ± 12.0	6.05 ± 0.55	0.70 ± 0.04	3.60 ± 0.30
2	15.4 ± 0.8	106.0 ± 9.5	1.50 ± 0.12	0.15 ± 0.01	8.40 ± 0.70
3	40.4 ± 3.5	94.1 ± 8.4	2.50 ± 0.18	2.43 ± 0.12	4.53 ± 0.35
4	25.6 ± 1.3	105.6 ± 10.1	3.50 ± 0.41	0.35 ± 0.02	13.5 ± 1.2
5	44.0 ± 4.0	99.8 ± 9.5	2.30 ± 0.15	0.03 ± 0.002	3.23 ± 0.25
6	33.8 ± 2.8	99.8 ± 9.4	1.20 ± 0.11	0.11 ± 0.01	8.30 ± 0.80
7	41.0 ± 3.2	94.6 ± 9.0	2.05 ± 0.15	0.09 ± 0.004	5.58 ± 0.6
8	47.8 ± 5.0	88.2 ± 7.3	3.20 ± 0.42	0.69 ± 0.04	2.41 ± 0.13
9	39.8 ± 3.9	82.6 ± 8.0	1.80 ± 0.20	3.60 ± 0.15	8.32 ± 0.8
10	37.6 ± 3.5	127.8 ± 11.0	2.01 ± 0.15	0.68 ± 0.07	14.5 ± 1.5
11	36.2 ± 3.0	100.8 ± 9.5	5.31 ± 0.25	0.45 ± 0.02	6.1 ± 0.38
12	33.0 ± 2.5	97.0 ± 8.5	2.72 ± 0.16	0.53 ± 0.03	3.62 ± 0.27
13	25.8 ± 2.4	80.7 ± 7.9	2.50 ± 0.25	1.83 ± 0.15	3.90 ± 0.40
14	20.8 ± 1.5	91.6 ± 9.2	2.04 ± 0.11	0.43 ± 0.02	2.64 ± 0.25
15	27.8 ± 2.9	88.8 ± 6.5	1.30 ± 0.12	0.43 ± 0.02	2.53 ± 0.20
16	39.2 ± 4.0	88.6 ± 7.2	3.00 ± 0.15	0.31 ± 0.03	2.4 ± 0.18
17	38.2± 2.8	91.6 ± 8.9	3.11 ± 0.16	0.77 ± 0.08	11.4 ± 0.9

* μmoles PNP. min^{-1}. g^{-1}
** μmoles Reducing sugars.h^{-1}.g^{-1}
***μmoles Urea min^{-1}.g^{-1}
****mg Hb.min^{-1}.g^{-1}

with those found in the lagoon close to the industrial district[14]. Mercury contamination was derived from a large chloralkali plant employing mercury cathodes since the fifties. An additional source of mercury can be an *in situ* leakage from antifouling boat paintings. They have been banned recently, but there is reasonable evidence that they might be still used due to large amounts stocked.

The Enzymatic Activities

Metal ions are involved in a variety of enzymatic reactions. They control catalysis by binding directly to the active site of the enzyme or indirectly by maintaining the enzyme structure in a poised conformation. They can also act as inhibitors. In order to establish the influence of metals on the enzymatic activities of the canal sediments, we analysed these activities in the same samples used for heavy metal determinations. Protease, phosphatase, urease and cellulase activities were present in the insoluble part of the sediment at a significant level, whereas lipase was practically absent (Table V). No activity was detectable in the supernatants, even after prolonged centrifugation, with the exception of cellulase in very small amounts (about 5%).

These sediments were treated with leaching solutions and detergents, as described for the sediments of the lagoon, and showed similar behaviour. When a phosphate solution was used, most of the phosphatase and a significant part of the cellulase activity were extracted while other solutions were ineffective. An important fraction of immobilized cellulase (40%) and some protease activity (15%) could be released by the use of Triton X-100 (Table VI). As for the lagoon sediments, urease activity remained insoluble under all treatments.

The sediment samples from inner canals of Venice town center were heated and the residual activity tested after cooling. The results reported in Fig.11 showed that urease,

Table 6. Release of enzymatic activities from 1 g of sediment samples
(N° 1,6,8,10,11,16 from Table V) with different extracting solutions
and 15 min. extracting time

Activity	Extracting Solutions		
	NaCl, NaHCO$_3$ 0.5 M	Phosphate 0.5 M	Triton 1%
Phosphatase*	0	21.4 ± 1.2 (yield 61%)	11.4 ± 0.5 (yield 33%)
Cellulase**	9.2 ± 0.5 (yield 9%)	38.5 ± 1.8 (yield 37%)	95.8 ± 5.0 (yield 93%)
Urease***	0	0	0
Protease****	0	trace	1.1± 0.05 (yield 15%)

*μmoles PNP. min^{-1}. g^{-1}
**μmoles Reducing sugars.h^{-1}. g^{-1}
***μmoles Urea min^{-1}. g^{-1}
****mgHb min^{-1}. g^{-1}

protease and phosphatase activities were enhanced by temperature exposures up to 60-80°C and still fully active after 15 min. at 100°C. Cellulase was much less resistant, a confirmation of the insolubilisation on membrane fragments rather than on the inorganic material.

Due to the high metal content, the sediment was treated with a chelating agent (0,2 M EDTA). No variation of activities was observed except for protease, which was enhanced by almost one order of magnitude (Table V). It is well known that cysteine proteases are strongly and reversibly inhibited by mercury[15]. Therefore the protease activity increase was interpreted as induced by the leaching of the metal chelated by EDTA and therefore to a reactivation of HS-proteases: this effect was confirmed by restoring the concentration of free metal, that caused an activity drop to almost exactly the initial low value. From a comparison between the insoluble enzymatic activities found in the inner canals of the city and in the open lagoon, significant differences were detected. In the lagoon, only three activities were present in measurable amounts: phosphatase, cellulase and urease. Protease and lipase were below the detection limits. In any case, the detected activities were lower than in the city

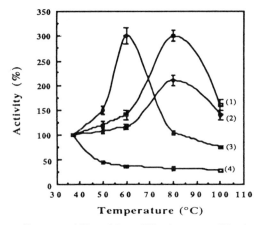

Figure 11. Temperature effect on stability of immobilized protease (1), phosphatase (2), urease (3) and cellulase (4) activities of the sediment. The activity at 37°C is taken as 100%.

Table 7. Comparison between enzymatic activities in the sediment of canals, lagoon and septic tank

Immobilized activity (1g)	Lagoon#	Canal (rii)	Septic tank##	Artificial lagoon
Phosphatase[B]	2.5 ± 0.05	33.8 ± 2.1	65.0 ± 0.5	132 ± 12
Lipase*	no	traces	24.0 ± 0.9	0.42 ± 0.30
Protease***	no	0.79 ± 0.23	1.1 ± 0.2	0.4 ± 0.3
+*EDTA*		*6.17 ± 0.95*		
Cellulase**	4.0 ± 0.01	97.4 ± 2.9	105.4 ± 2.0	86.4 ± 3.5
Urease****	2.4 ± 0.1	2.71± 0.31	2.3 ± 0.2	0.75 ± 0.27

Ref. 3, 4; ## Ref. 1; [B]μmoles p-nitrophenyl phosphate, min^{-1}; *μmoles tributyrin, min^{-1}; ** μmoles glucose, h^{-1}; ***mg hemoglobin, min^{-1}; ****μmoles urea, min^{-1}

canals. A much higher similarity, both quantitatively and qualitatively, was found between the enzymatic activities of the Venice canals and those of the sediment of a septic tank and of an artificial lagoon (Table VII).

Table 8. Yield, activity and specific activity in enzymes spontaneously linked to zeolithes (A to D) and clays (E to H)

Support	Immobilized activity				
	Aprotinin❖	Trypsin*	Cellulase**	Phosphatase*	Urease*
A					
Immobilization %	12,5	26	11	23	50
Activity	19	85	15,3	6	5
Specific activity	152	340	138	26	10
B					
Immobilization %	30	25	14	18	43
Activity	6,5	52	15,5	8,4	7
Specific activity	21,6	215	110	47	16
C					
Immobilization %	21	32	11	25	54
Activity	76	108	20,8	7,2	20
Specific activity	345	310	189	28,5	36
D					
Immobilization %	24	37	12	25	54
Activity	7	93	35	8,4	20
Specific activity	29	245	233	34	36
E					
Immobilization %	18	31	25	26	64
Activity	53	74	49,6	39,4	67
Specific activity	294	245	198	152	103
F					
Immobilization %	49	88	31	48	80
Activity	6,4	500	47	54	108
Specific activity	12,8	568	150	113	135
G					
Immobilization %	55	44	10	34	56
Activity	71	36	50	49	117
Specific activity	129	80	500	144	212
H					
Immobilization %	78	80	23	78	85
Activity	45	70	45	240	600
Specific activity	100	88	193	306	706

❖ % Inhibition (10 mg support-BPTI; Trypsin 0,1nmole)
* μmoles /min / 100 mg support ** μmoles /h / 100 mg support

Enzyme/Inorganic Supports Interactions

Since the Venice lagoon and canals sediment are rich in clay, we studied the behaviour *in vitro* of enzymes in presence of different kinds of inorganic materials (Table I). Clays are silicates with a lamellar structure and the alternance and composition of strata define the type clay. Zeolithes are tectosilicates either natural or synthetic, formed by a network of SiO_4 and Al_2O_3 units with cavities ranging from 4 to 8Å: they were included in this study since they are now largely used (and thus released in the environment) in detergents and animal foods.

Experiments using representative active proteins of increasing M.W., showed that a spontaneous linking to inorganic supports occurs with different affinities. The time of contact (15 min, 1, 2 & 16 h, at 25°C) had little effect on immobilisation yields, while the initial concentration affected the binding. The affinity of proteins is higher for clays (Table VIII, E to H), while zeolithes, whose pore dimensions are too small compared to the size of enzymes, are poor supports (Table VIII, A to D) for binding enzymes. When the reaction mixture enzyme/ support was heated, only cellulase yield increased (4 times at 37°C). The protein concentration selected to study other variables was 1 mM, except for urease whose high molecular weight prevented solubilization and was studied at 0.1 mM. Enzyme activities were measured after each addition of support both in the supernatant and the insoluble material and compared with the one of in initial solution. The resulting "activated" powders showed unexpected differences in their specific activities, suggesting that the mode of binding and intramolecular interactions affected the performances of the linked bio-molecule. For instance, trypsin linked with practically the same yield to Montmorillonite and to Bentonite but in the first case the specific activity was more than 6 times higher. Cellulase linked very poorly on Bentonite Na but this activated support was the one with the best specific activity, phosphatase and urease also showed similar behaviour. Table VIII compares yields, activities and specific activities of the different enzymes.

Attempts to release the enzymatic activities from clays by selected solutions showed that the coupling is quite stable and hardly affected by ionic strength. Table IX shows the release of Bentonite coupled cellulase, phosphatase and urease, compared to the "natural" clay-linked enzymes of the Venice ecosystem.

Phosphate and triton were, for both supports, the only effective in releasing part of the cellulase and phosphatase activity, while urease was stable under all conditions. This behaviour is quite similar to the one of the Venice ecosystem sediments. These results are an additional support to hypothesis that enzymes produced in large quantity in the lagoon, can spontaneously link to the sediment inorganic material and behave as an efficient long-lasting degradative agent.

Time resistance of clay-linked enzymes was measured and activities remained practically unchanged for 7 weeks at 4°C and 4 weeks at 25°C. This behaviour is similar to the one of enzymes linked to Venice lagoon sediment, while the same molecules in soluble form were completely inactivated at 25°C in few days (Fig.12).

Table 9. Release of enzymatic activities with different extracting solutions

Activities (% of release)	Venice ecosystem sediments			Inorganic support (Bentonite)		
	NaCl 0.5 M NaHCO$_3$ 0.5 M	Phosphate 0.5M	Triton 1%	NaCl 0.5 M NaHCO$_3$ 0.5 M	Phosphate 0.5M	Triton 1%
Cellulase	2.8	57.1	71.4	0	72.4	36.3
Phosphatase	0	85.3	2.4	0	2.2	5.1
Urease	0	0	0	0	0	0

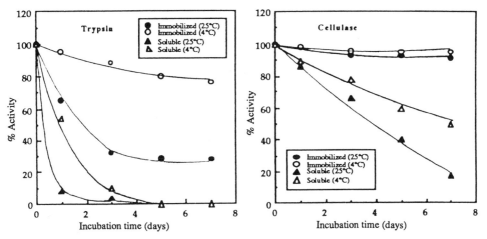

Figure 12. Time resistance of synthetic and natural clay-linked enzymes.

As in the lagoon sediments, activities were also proved to be temperature-resistant without any loss after heating at 100°C for 15 min. Since the surface-located residues are known to be critical for thermostability, it can be assumed that a rigidification in the enzyme structure had occured, that stabilised the molecule *versus* temperature-induced unfolding. Therefore, the interaction with insoluble supports probably induced the protection of the biomolecule toward denaturation.

When more than one molecule is allowed to react concomitantly with an inorganic support a competition, dependent of course on the protein and on the support nature, is clearly evident (Fig.14). Trypsin and phosphatase, which linked to bentonite with similar yield (Å 80%), when alone competed with an advantage for trypsin. This behaviour is not due to the coupling rate alone, since phosphatase did not increase its insolubilisation yield even after 24 h.

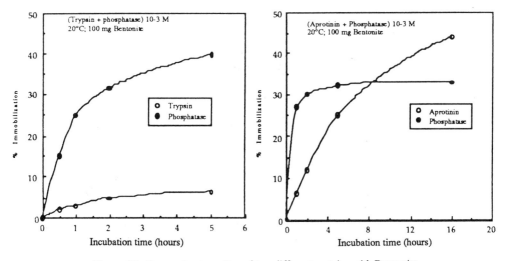

Figure 13. Concomitant reaction of two different proteins with Bentonite.

When aprotinin and phosphatase are allowed to react with the same support, the phosphatase is not hindered as in the first example and showed a better rate of coupling. However, aprotinin was able to link with a better yield in spite of its lower coupling rate.In the present study the complex between the inorganic material and the enzymes present in the Venice ecosystem have been compared with a similar supramolecular system, constructed by self-assemblage *in vitro* between clays and enzymes. These mineralized structures were in both cases more adapted to extreme environmental conditions and are less susceptible to denaturating conditions. Synthetic inorganic supports, such as zeolithes, were also examined as supports of enzymatic activities since they are released in the environment in quantities which, given time, could change the sediment composition with unknown consequences.

CONCLUDING REMARKS

Large amounts of material including dielectric fluids, refrigerants, lubricants, pesticides etc., are released in the environment and enter the ecosystem. An increasing need to assess the safety of waste repositories and the fate of agricultural/industrial chemicals has appeared, due to their accumulation in land, ponds, lakes, rivers, groundwater and soils. The present research shows that one of the parameters, the degradation of chemicals by enzymes accumulated on the inorganic part of the sediment, had completely escaped attention. In summary:

- A biodegradative capacity able to survive the microorganisms, from which it was generated, represent an important and perhaps vital share of the degradative power of an ecosystem. It can be considered, up to a certain point, of diagnostic importance, since enzymes accumulate in reply to increased nutrients, released in the ecosystem.
- The enzymatic activity, if any, is found in the insoluble part of the system. This is worth considering since it is well known that the soluble enzymatic activities are easily destroyed even in a protected environment. The insolubilization has a protective effect giving to the enzymatic activity of a bio-sediment the possibility to resist to otherwise denaturating conditions.
- A new understanding of phosphates action on the degradative potential of a lagoon sediment starts to unravel. In fact, even very low concentrations of phosphate ions can wash out the insoluble enzymatic activity if they are allowed to stay long enough in contact.
- The biodegradative potential of any ecosystem is a function of the environment: we have shown here that it is also a function of the previous history of the sediment, since the immobilisation of enzymes allows their accumulation and prolongs their lifespan. When a natural or anthropogenic event lowers the microorganisms density in the lagoon, the insoluble enzymatic activity can take over while the population develops again to its optimum density. If, however, the negative effect goes beyond the enzyme breakpoint, the two-step relay collapses sometimes dramatically.
- The regenerating power of the sediment can be at danger from contaminants which might not be particularly dangerous per se but could simply poison these activities, thus initiating an irreversible chain of events.
- The relation between the bio-molecules and the insoluble material could be inferred by the results on the release of the enzymes by choosen solutions. An *in vitro* study on the interactions between inorganic support and enzymes corroborated the hypothesis of a direct bond between enzymes and clay. The insolubilisa-

tion of enzymes is a spontaneous event. It takes place *in vitro* by simple interaction, when enzymes and clays are allowed to react together.

- The sediment composition is thus an essential parameter of this immobilization of enzymes and close attention should be payed to any interference that tends to change the inorganic composition of the sediment. For instance, new channels from the sea, especially if they are deep, can facilitate an increase of the sand level as compared to clay and/or silted clay. It is forseenable that the recent replacement of phosphate by zeolite in the commercial washing powders and animal feed materials will, given enough time, affect the sediment composition and therefore its enzymatic capacity both quantitatively and qualitatively.

ACKNOWLEDGMENTS

This work was carried out in the framework of the UNESCO-MURST project "Venice Inner Canals". The authors are profoundly grateful to Prof. A. Marzollo, Coordinator of the project, UNESCO-ROSTE, for his help and support throughout this research.

REFERENCES

1. Maunoir, S., Sabil, N., Rambaud, A., and Coletti-Previero, M-A. (1991) *Env. Technol.* 12: 313-323.
2. Haddane, M., Baudinat, C., Rambaud, A. and Coletti-Previero M-A. (1986) *J. Fra. Hydr.* 17: 253-262.
3. Sabil, N., Tagliapietra, D. and Coletti-Previero, M-A. (1993) *Env. Technol.* 14: 1089-1095.
4. Sabil, N., Cherqui, A., Tagliapietra, D. and Coletti-Previero, M-A. (1994) *Wat. Res.* 28: 1, 77-84.
5. Aissouni, Y., Sabil, N., Pavoni, B. and Coletti-Previero, M-A. In: *Perspective in Protein Engeneering,* (Geisow, M. ed.), in press.
6. Sabil, N., Tagliapietra, D., Aissouni, Y. and Coletti-Previero, M-A. In: *Venice Lagoon Ecosystem,* Chapter 1.3. (UNESCO ed.) , in press.
7. Sabil, N., Aissouni, Y., Coletti-Previero, M-A., Donazzolo R., D'Ippolito R. and Pavoni B. *Env. Technol.* in press.
8. Froelich, P.N. (1980) *Limnol. Oceanogr.* 25: 564-572.
9. Agemian, H. and Chau, A.S.Y. (1976) *Analyst.* 101: 91-95.
10. Brunk, C.F., Jones, K.C. and James, T.W. (1979) *Anal. Biochem.* 92: 497-500.
11. Coletti-Previero, M-A. and Previero, A. (1989) *Anal. Biochem.* 180: 1-10.
12. Coletti-Previero, M-A., Pugnière, M., Favel, A.and Previero, A. (1988) In: *Protein Structure-Function Relationhip,* (Zaidi, Z.H. ed.), pp. 39-58.
13. Pavoni, B., Donazzolo, R., Marcomini, A., Degobbis, D. and Orio, A.A. (1987) *Mar. Poll. Bull.* 18: 18-24.
14. Donazzolo, R., Orio, A.A., Pavoni, B. and Perin, G. (1984) *Oceanol. Acta.* 7: 25-32.
15. Shaw, E. and Green, G.D.J. (1981) In: *Methods in Enzymology,* 80: pp. 820-826, Acad. Press, New York.

5

NEW DEVELOPMENTS ON HUMAN MILK PROTEIN COMPOSITION AND FUNCTION

Human Milk as a Model for the Production of Dietetic Milks

Amedeo Conti,[*][1] Gabriella Giuffrida,[1] Maria Cavaletto,[2] and Carlo Giunta[2]

[1] National Council of Research (CNR)
Centro Studi Alimentazione Animale
Turin, Italy
[2] Dipartimento di Biologia Animale
University of Turin, Italy

There are two main aims to the scientific research underlying the production of powdered milk (formulas) destined for feeding infants who are not breast-fed: to make such formulas as similar as possible to breast milk, and to obviate against the increasingly frequent phenomena of intollerance of cow's milk, generally used as a substitute for breast milk. Biotechnological techniques are thought to have a possible useful application above all in the production of hypoallergenic milks.

But the true limit to the use of biotechnologies in this field lies a step further back: research carried out at our laboratories over the last several years have concerned the protein composition of human milk at different periods of lactation and with different mothers' diets; we have seen that the first problem to be solved in the production of suitable powdered milk is that a better and scientifically more exacting knowledge of maternal milk is needed than that which exist at present, despite the impression to the contrary. Some of the most significant results from these studies, in contrast with what has been reported for decades in the literature, illustrate this point.

Analysis of casein content was carried out in human colostrum samples (1st to 3rd day of lactation) and in mature milk. In both cases similar "casein pellets" were obtained by ultracentrifugation of skimmed samples. Unlike mature milk, colostrum pellets do not contain any protein belonging to the casein group or subunits, but only some precipitated typical whey proteins (Lactoferrin, Igs, α-lactalbumin) (Fig. 1). The absence or the presence of casein micelles in mammary secretions has been proposed as discriminating element for unequivocal definition of colostrum and milk, respectively. The literature reports quantities of casein in human colostrum varying between 10% and 30% of total proteins[1].

In the casein fraction of mature human milk, α-casein has also been found, quantities being about 19% that of β-casein. The literature reports that human milk contains no α-casein

---

[*] Represents person presenting Paper

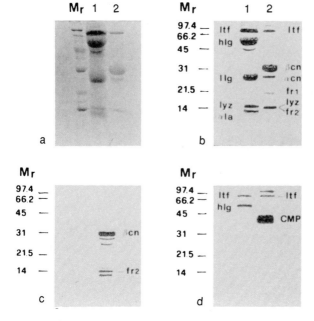

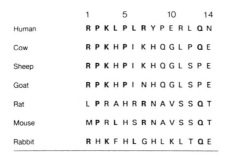

Figure 1. Line 1 = "casein pellet" proteins from colostrum samples; line 2 = "casein pellet" proteins from mature milk samples. a) and b) = Coomassie blue staining; c) = immunostaining with antibodies anti bovine β-lactoglobulin; and d) = PAS staining for glycoproteins. Ltf = lactoferrin; hIg = Igs heavy chains; lIg = Igs light chains; lyz = lysozyme; α-la = α-lactalbumin; β-cn = β-casein; α-cn = α-casein; fr1 and fr2 = β-casein fragments.

whatever[2,3]. This result has been obtained by a modification of the classical Leammli's method for SDS-PAGE. The presence in human milk of bioactive peptides with opioid (exorphins) and antihypertensive activity derived from the α-casein fragmentation, can be then speculated. In Fig. 2 amino acid sequence homology of the human α-casein-like protein (α-CN) and αs₁-caseins already known from other species is shown.

	1			5					10				14
Human	R P K L P L R Y P E R L Q N
Cow	R P K H P I K H Q G L P Q E
Sheep	R P K H P I K H Q G L S P E
Goat	R P K H P I N H Q G L S P E
Rat	L P R A H R R N A V S S Q T
Mouse	M P R L H S R N A V S S Q T
Rabbit	R H K F H L G H L K L T Q E

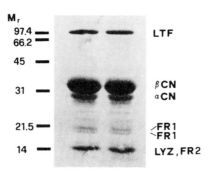

Figure 2. Amino acid sequence homology of the human α-casein-like protein (α-CN) and αs₁-caseins already known from other species.

PROPOSED DYNAMIC MODEL LEADING TO THE GLYCOSYLATION
OF HUMAN α-LACTALBUMIN

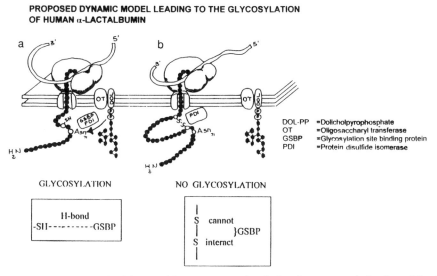

Figure 3. A dynamic model of the possible events occurring during the post-translational modification of human α-lactalbumin. DOL-PP = Dolycholpyrophosphate, OT = Oligosaccharil transferase, GSBP = Glycosylation Site Binding Protein, PDI = Protein Disulfide Isomerase.

Human α-lactalbumin, the main protein component of maternal milk involved in the synthesis of lactose[4], has so far not been described as a glycoprotein, despite the fact that several α-lactalbumins both of ruminant and non-ruminant species are known to be partially[5-7] or fully glycosylated[8,9]; in all these species the glycosylation site is the Asparagine$_{45}$ in the usual triplet Asn$_{45}$-Gly/Gln-Ser$_{47}$, but the functional significance of this glycosylation is still not known. A relatively small portion of human α-lactalbumin (about 1%) has now been found to be glycosylated at the unusual site Asparagine$_{71}$ in the triplet Asn-Ile-Cys, confirming the hypothesis that N-glycosylation can also occur at the consensus site Asn-Xxx-Cys. A dynamic model of the possible events occurring during the post-translational modification of human α-lactalbumin, to explain the low percentage of glycosylation of the protein, is proposed in Figure 3.

Even though no biological function for the post-translational modifications of α-lactalbumin leading to the glycosylation of the molecule has yet been reported, the finding that human α-lactalbumin is partially glycosylated at an unusual glycosylation site, might be of importance in understanding some unexplained physiological properties of this protein (metal binding) and for the production of bioactive molecules by biotechnological processes.

It is normally reported that β-lactoglobulin of cow's milk is the main allergen causing intollerance to this milk in new-born babies; it is also generally believed that cow's milk proteins, in particular β-lactoglobulin, ingested by the mother, can pass into breast milk and thus sensitize predisposed infants[10-12]. However the studies to evaluate bovine β-lactoglobulin in human milk have given conflictual results, with unexplaned intra- and inter-individual variability[10,13]. When bovine β-lactoglobulin has been measured in human milk by immunotechniques (ELISA, RIA), the possibility of cross-reactions between the antibodies employed and proteins different from bovine β-lactoglobulin was never taken into consideration, while cross-reactions between antibodies anti-bovine β-lactoglobulin and some other human milk proteins, such as β-casein and its fragments[14], β2-microglobulin[15] and

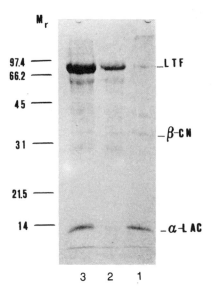

Figure 4. SDS-PAGE of the three fractions separated by affinity chromatography.

lactoferrin,[16] are well documented. A possible explanation of this cross-reactivity based on homologies of the primary structures of these proteins has been proposed[14,15,17].

The aim of this particular study was to analyse the correlation between different, controlled amounts of cow's milk in the mother's diet and the presence of bovine β-lacto-globulin in breast milk.

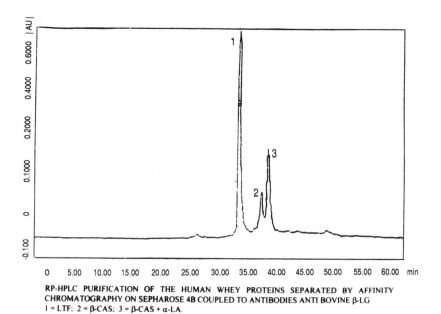

RP-HPLC PURIFICATION OF THE HUMAN WHEY PROTEINS SEPARATED BY AFFINITY
CHROMATOGRAPHY ON SEPHAROSE 4B COUPLED TO ANTIBODIES ANTI BOVINE β-LG
1 = LTF; 2 = β-CAS; 3 = β-CAS + α-LA.

Figure 5. RP-HPLC purification of the human whey priteins separared by affinity chromatography on sepharose 4B coupled to antibodies antibovine β-lg.

The total concentration of bovine β-lactoglobulin or β-lactoglobulin-immuno-like-components was determined by ELISA. To identify the human whey components recognized by the antibodies anti-β-lactoglobulin, an affinity chromatography column, followed by SDS-PAGE and N-terminal sequence determination was set up. Three fractions were collected from the affinity column after starting an acidic gradient to release the protein bounded to the antibody. In Fig. 4 the SDS-PAGE of the three fractions is shown.

After Western blotting on Problot membrane, three components were identified by N-terminal sequencing: human lactoferrin, human β-casein and human α-lactalbumin but no trace of β-lactoglobulin could be detected. Purification of the components of fraction 3 by RP-HPLC has confirmed the result as shown in Fig. 5.

These results would suggest, at least in healthy subjects, that there is no passage of integer bovine β-lactoglobulin ingested by the mother into breast milk, and that false-positive results in ELISA determinations may be due to cross-reaction between polyclonal antibodies and different protein antigens. The biological function of the β-lactoglobulin is yet not known but its direct involvement in causing atopy, still must be proved.

The effectiveness of formulas based on hydrolized cow's milk protein in the prevention of atopic deseases is under discussion, but the starting point to prepare more suitable artificial foods for babies must be a better knowledge of maternal milk.

REFERENCES

1. Kunz, C. and Lonnerdal, B.(1992) *Acta Paediatr.* 81: 107-112.
2. Azuma, N., Kadoya, H. and Yamauchi, K. (1985) *J. Dairy Sci.* 68: 2176-2183.
3. Brignon, G., Chtourou, A. and Ribadeau-Dumas, B. (1985) *FEBS Lett.* 188: 48-54.
4. Ebner, K.E., Denton, W.L. and Brodbeck, U. (1966) *Biochem. Biophys. Res. Commun.* 24: 232-236.
5. Brew, K., Castellino, E.J., Vanaman, T.C. and Hill, R.L. (1970) *J. Biol. Chem.* 245: 4570-4582.
6. Halliday, J.A., Bell, K., McKenzie, H.A. and Shaw, D.C. (1980) *Comp. Biochem. Physiol.* 4: 773-779.
7. MacGiluvray, R.T.A., Brew, K. and Barnes, K. (1979) *Arch. Biochem.Biophys.* 197: 404-414.
8. Prasad, R., Hudson, B.G., Butkowsky, R., Hamilton, J.W. and Ebner, K.E. (1979) *J. Biol. Chem.* 254: 10607-10614.
9. Cantisani, A., Napolitano, L., Giuffrida, M.G. and Conti, A. (1990) *J. Biochem. Biophys. Meth.* 21: 227-236.
10. Axelsson, I., Jakobsson, I., Lindberg, T. and Benediktsson, B. (1986) *Acta Paediatr. Scand.* 75: 702-707.
11. Bjorksten, B. and Kjellman, N-I M. (1990) *Clin. Exp. Allergy* 3: 3-8.
12. Harmatz, P.R. and Bloch, K.J. (1988) *Ann. Allergy* 61: 21-24.
13. Sorva, R., Makinen-Kiljunen, S. and Juntunen-Backman, K. (1994) *J. Allergy Clin. Immunol.* 93: 787-792.
14. Neuteboom, B., Giuffrida, M.G., Cantisani, A., Napolitano, L., Alessandri, A., Fabris, C., Bertino, E. and Conti, A. (1992) *Acta Paediatr.* 81: 469-474.
15. Conti, A. and Godovac-Zimmermann, J. (1990) *Biol. Chem. Hoppe-Seyler* 371: 261-263.
16. Brignon, G., Chtourou, A. and Ribadeau-Dumas, B. (1985) *J. Dairy Res.* 52: 249-254.
17. Monti, J.C., Mermoud, A-F. and Jolles, P. (1989) *Experientia* 45: 178-180.

6

STRUCTURE-FUNCTION OF TWO MEMBRANE PROTEIN SYSTEMS INVOLVED IN TRANSFER OF ELECTRICAL CHARGE

W. A. Cramer,[*] J. B. Heymann, D. Huang, S. E. Martinez, S. L. Schendel, and J. L. Smith

Department of Biological Sciences
Purdue University
West Lafayette, Indiana 47907-1392

ABSTRACT

Structure-function relationships are presented for two membrane proteins or protein complexes that are involved in the transfer of electrical charge through or across biological membranes: (i) the cytochrome b_6f complex and its subunit, cytochrome f, that transfer electrons on a sub-millisecond time scale from the lipophilic quinol to the photosystem I reaction center in oxygenic photosynthetic membranes. Cytochrome f carries out one step in this transfer by oxidizing the membrane-bound iron-sulfur [2Fe-2S] protein in the complex and reducing its soluble electron acceptor, plastocyanin. (ii) Colicin E1 is a toxin-like bactericidal protein that forms a voltage-gated ion channel with a single channel conductance of 10^6-10^7 ions/channel/sec in the *E. coli* cytoplasmic membrane.

The structure of the active lumen-side domain (252 residues) of cytochrome f has been solved at atomic (1.96Å) resolution, the first subunit of the integral polytopic cytochrome bc_1 or b_6f complexes for which an atomic structure has been obtained. The existence of two domains, a predominantly β-strand motif, and the use of the N-terminal amino group as the sixth ligand are unique features of the structure.

The colicin ion channel contrasts with the cytochrome complex in not being permanently embedded in the membrane, but in making a transition, involving a very large structural change, from a water-soluble to an integral membrane-bound state. This kind of structural change is characteristic of the mode of action of many toxins, such as diphtheria toxin, and toxin-like molecules such as the colicins. The voltage-driven gating of the channel, in which the colicin channel is converted from a closed to an open ion-conductive state, involves import into the membrane of at least one *trans*-membrane amphiphilic helical hairpin.

---

[*] To whom correspondence should be addressed.

INTRODUCTION

The cytochrome bc_1 or b_6f complex from respiratory and photosynthetic membranes, and the bactericidal voltage-gated colicin ion channels, both carry out transfer of electrical charge across membranes. They participate, respectively, in the energization and deenergization of membranes. The cytochrome complex transfers electrons and protons across the membrane on a millisecond (10^{-3} s) time scale[1] and the colicin channel translocates monovalent ions (Na^+, K^+, Cl^-) with a single channel conductance $\gtrsim 10^6$ ions/channel/sec[2].

The above-mentioned cytochrome and colicin complexes represent two very different kinds of membrane proteins. The cytochrome complex is an integral membrane protein complex containing four $M_r > 15,000$ polypeptides and three-four small hydrophobic M_r 5,000 polypeptides that span the membrane bilayer with 8-12 *trans*-membrane α-helices per monomeric unit[3,4]. On the other hand, the toxin-like colicin molecules enter the membrane from a water-soluble state. Thus, understanding the fundamental properties of the two protein systems involves not only an analysis of their *structures*. In the case of the colicin molecule, it is necessary to also understand the basis of large scale *structural changes* that accompany the transition of the colicin from the aqueous to the membrane phase.

THE CYTOCHROME b_6f COMPLEX

The cytochrome b_6f electron and proton transfer complex in oxygenic photosynthesis bears many analogies to the bc_1 complex of mitochondrial respiration and bacterial photosynthesis. The cytochrome f [*feille* (Fr.) – leaf] is functionally analogous to cytochrome c_1 of the bc_1 complex. As discussed below, the structure of the active peripheral domain of cytochrome f has been solved by X-ray diffraction to high resolution.

The b_6f complex is one of three integral membrane protein complexes responsible for electron transfer from H_2O to $NADP^+$ and for the generation of the electrochemical H^+ gradient, $\Delta\tilde{\mu}_H^{+1,3-5}$. The four polypeptide subunits of this complex with molecular weight $> 15,000$ are: cytochrome f (285 residues; MW = 32,028 in spinach thylakoid membranes; organelle (o)-encoded); cyt b_6 (215 residues; MW = 24,166; 'o'-encoded); the high potential ($E_{m7} \sim +290$ mV) Rieske iron-sulfur [2Fe-2S] protein (179 residues; MW = 19,116; nuclear-encoded); and "subunit IV" (160 residues, MW = 17,445, 'o'-encoded). The complex contains three-four small ($M_r \sim 4,000$) hydrophobic subunits. Cyt b_6 with four *trans*-membrane helices corresponds to the N-terminal half, and suIV to the C-terminal half of the larger (MW ~ 40-45,000) cytochrome b from mitochondria and photosynthetic bacteria[6]. The cytochrome b family is the most thoroughly sequenced integral membrane protein[7].

The cyt b_6f complex has a central position, in terms of its oxidation-reduction potential, in the electron transport chain, accepting electrons and protons from plastoquinol, and transferring electrons from cytochrome f to its acceptor, the soluble copper protein, plastocyanin. The central positions of the b_6f and bc_1 complexes in the photosynthetic and mitochondrial respiratory chains are shown in figs. 1a,b. As part of this electron transfer process, protons are transferred to the lumen-side aqueous space, contributing to the $\Delta\tilde{\mu}_H^+$. It will be of interest in connection with the structure of cytochrome f, discussed below, that the pathways of H^+ release from, and uptake into, the membrane may utilize chains of bound waters. This hypothesis is based on the finding of an internal water chain in cytochrome f and the precedent of the water chain found in the extrinsic region of the 'H' polypeptide that connects the bulk aqueous phase to the quinone, q_b, binding site in the photosynthetic reaction center complex from Rb. *sphaeroides*[8].

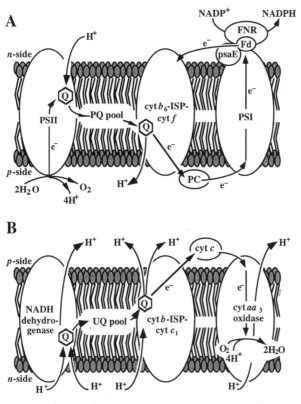

Figure 1. Schematic showing the three integral membrane protein complexes involved in electron transport and proton translocation from the electrochemically negative (n) to the positive (p) side of the membrane in (A) oxygenic photosynthesis, where the terminal electron donor and acceptor are water and NADP$^+$, respectively, and (B) mitochondrial respiration, where the donor and acceptor are NADH and molecular oxygen. Other notation described in ref. 1.

THE "DIVIDE AND CONQUER" APPROACH TO THE STRUCTURE SOLUTION OF MEMBRANE PROTEINS

The structures of only 3-4 fundamentally distinct integral membrane proteins have been solved: (i) the bacterial photosynthetic reaction center[8,9], (ii) the outer membrane porin from gram-negative bacteria[10-12], (iii) the light-harvesting chlorophyll pigment protein complex[13], and (iv) bacteriorhodopsin to high resolution in two dimensions[14]. (v) The surface-bound prostaglandin synthase may also be included in this group[15]. Of these, only the bacterial reaction center is an integral hetero- and oligomeric multisubunit complex like the cyt b_6f complex. The reaction center contains eleven *trans*-membrane helices, and the b_6f complex eight-twelve in the monomeric state, depending on whether the [2Fe-2S] subunit and three hydrophobic subunits each contribute one such helix. Unlike the reaction center, whose light- or redox-active prosthetic groups are contained within the membrane bilayer, the active domains of cytochrome f and the [2Fe-2S] protein are in the membrane-peripheral domain. Thus, one strategy for obtaining partial structural information about such a complex is to isolate and purify its active peripheral domains. This "divide and conquer" approach was taken with cytochrome f after initial attempts to crystallize (a) the entire b_6f complex or (b) the intact 285 residue cyt f polypeptide with its hydrophobic membrane-spanning C-terminus were unsuccessful.

Solution of the Structure of the Lumen-Side Domain of Cytochrome f

Highly diffracting, 2.3Å[16,17] and 1.96Å[18], crystals of a soluble 252 residue lumen-side fragment of the 285 residue cytochrome f were obtained. In addition to the approximately 250 residues that had been designated to the lumen side of the membrane, the fragment also contains two residues of the 20 residue *trans*-membrane segment. In retrospect, it was important that the cleavage not have occurred too far from the membrane interface, because this would have interfered with the protein fold[17].

The cleaved cyt f fragment was obtained from the action of a natural protease activated during the process of organic solvent-extraction of the pigment from thylakoid membranes from turnip leaves. The existence of a somewhat smaller soluble cyt f in cruciferous plants including turnip was first noted by Gray[19], and thus now appears to be due to a unique activable protease near the aqueous interface of the bilayer in thylakoid membranes of the cruciferous plants.

The purified cyt f fragment is highly active in redox reactions with its acceptor, plastocyanin, has a normal spectrum (reduced α–band peak, 554 nm), and normal redox potential, $E_{m7} = +0.365$ V[18].

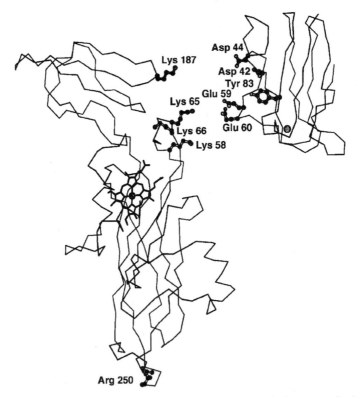

Figure 2. Alpha-carbon traces of cyt f and plastocyanin (poplar) in a 'pre-docking' state. Residues are noted on cytochrome f and plastocyanin that are believed to be involved in docking between the two proteins. The distance from the heme Fe to the alpha-carbon of Arg 250 (last charged residue before membrane-spanning anchor) is 45Å. Sequence information in ref. 20.

The cytochrome structure is elongate, 75Å x 35Å x 24Å, with the heme Fe 28Å and 45Å from the alpha-carbons of Lys187 and Arg250 (Fig. 2). The structure has four unprecedented features for a c-type cytochrome, including two such features for heme proteins: in contrast to the smaller (ca. 100 residues) soluble c cytochromes, cyt f contains (i) two distinguishable domains, "large" and "small," and (ii) a predominantly β-strand secondary structure; (iii) the sixth (axial) ligand is the free N-terminal α-amino group of the amino-terminal residue, Tyr-1. The use of the N-terminal amino group as a ligand is thus far unprecedented in the whole family of heme proteins. (iv) Refinement of the atomic model at a resolution of 1.96Å revealed the presence of five internal water molecules[17,18], four of which define a "vector-like" pathway that extends 12Å from the His25 heme ligand to within approximately 6Å of Lys66 in the basic patch believed to be involved in the binding of plastocyanin[17]. The residues that can provide side- and main-chain H bonds are virtually totally conserved in thirteen cytochrome f sequences, which are compiled in ref[20].

Resemblance of the Cytochrome F Fold to the Fibronectin Fold in Cell Surface Proteins[17]

The overall fold of the large domain of cyt f resembles that of the type III domain of the animal cell surface protein fibronectin (FnIII). The FnIII fold has been observed in four proteins with no functional or evolutionary relation to cyt f. These are (a) two domains of the bacterial chaperone protein PapD, (b) the D2 domain of the human cell surface glycoprotein CD4, (c) two extracellular domains of the human growth hormone receptor, and (d) an FnIII of the human extracellular matrix protein tenascin. Because of the absence of significant sequence identity, there is no proposal for a common origin of cyt f and the above proteins with an FnIII fold. This common fold must have arisen through convergent evolution, (i) because of favorable thermodynamic properties, or (ii) because cell surface proteins and cytochrome f have in common the requirement to escape the jungle of peripheral proteins at the membrane surface in order to capture an agonist, a ligand, or in the case of cytochrome f, its electron acceptor.

Relation to Cytochrome c_1

In spite of the apparent similarity of the function of cytochromes f and c_1 mentioned above and the correspondence in the overall domain structure (i.e., asymmetric distribution of peripheral domain size; one *trans*-membrane helix) implied by their primary sequences, the lack of identity in these sequences implies that cytochromes f and c_1 may also be present as a result of convergent evolutionary processes[21]. Both cytochromes may have evolved to satisfy the general requirement in energy-transducing membranes for a membrane-anchored polypeptide containing an extra-membrane high potential covalently bound redox center (e.g., a c-type heme) on the electrochemically positive side of the membrane.

Electron Transfer to the Acceptor, Plastocyanin

Electron transfer from cyt f ($E_{m7} \approx +0.365$ V) to the ca. 99 residue soluble acceptor plastocyanin ($E_{m7} \approx +0.34-0.37$ V) is approximately isopotential, and occurs with a half-time of approximately 0.1 msec *in situ*. The soluble cyt f fragment can also be utilized to study the *in vitro* rate constant, k_{et}, for its oxidation and electron transfer to plastocyanin. The dependence of k_{et} (2800 s^{-1} at pH 7, ionic strength = 0.004) on ionic strength[22], the inhibitory effect on the electron transfer reaction of modification of plastocyanin carboxylates in the PC acidic clusters 42-45 and 59-61[23], and the cross-linking of cyt f Lys187 to PC Asp44[24] imply that the reaction is governed by electrostatic interactions between a positively charged domain of cyt f near Lys187 and a negative domain(s) of PC near Asp44 and/or Glu59. The dominant region of positive potential on the surface of cyt f includes Lys187 in the small domain and Lys58, 65 and 66 in a neighboring region of its large domain (Fig. 2). The involvement of Lys187 in the docking complex is further suggested by the observation that cyt f of the cyanobacterium, Nostoc PCC 7906, which binds a basic PC, has a pronounced acidic sequence Glu185-Glu186-Asp187-Glu188-Asp189. On the other hand, the electron transfer reaction appears to require conformational flexibility because the cyt f (Lys187) – PC (Asp44) cross-linked complex is not active[22]. Inspection of the position of the respective charges and protein orientations suggest that if the Lys187-Asp44 charge pair participates in the docking between the proteins, then the electron transfer occurs over long (~20Å) distances.

THE VOLTAGE-GATED COLICIN ION CHANNEL[25]

The six well-studied toxin-like plasmid-encoded channel-forming colicins are polypeptides of approximately 500-600 residues that are bactericidal to cells of *E. coli* and related bacterial strains. These colicins are classified (E, I, A) in terms of the outer membrane receptor to which they bind. Translocation from the outer membrane to the cytoplasmic membrane, which is the target for channel formation, utilizes an intermembrane protein network whose general purpose is macromolecule (e.g., DNA) translocation. The major activities associated with the colicin, "translocation," "receptor-binding," and "channel formation" are organized in N-, central, and C-terminal domains, each of which occupies approximately one-third of the polypeptide, approximately 150-200 residues for the channel domain of colicin E1. *In vitro* studies, mostly with colicins E1, A, and Ia, show that these colicins can form an ion channel sufficiently conductive that the electrogenic H^+ pump of the *E. coli* bacterial cell is not able to preserve the cytoplasmic membrane potential, thus causing substantial depolarization and cellular deenergization that precede cell death.

As a channel system, the unique aspects of the colicins are: (i) the intermembrane translocation system described above; (ii) the transition at the charged membrane surface (~30% of the cytoplasmic membrane is anionic lipid) of the ~190 residue channel domain from a compact globular structure[26,27] to one that is unfolded[28,29] on the surface of the membrane (Fig. 3A). The latter conclusion is based on *in vitro* studies of the interaction of the channel domain with anionic liposomes. Such studies also indicate that the channel polypeptide is anchored by surface electrostatic[30] and hydrophobic forces, and by a hydrophobic helical hairpin inserted into the bilayer. At acid pH (~4.0), the channel polypeptide binds very tightly to the surface of anionic (~30% PG) membranes (K_d ~10^{-9} M at 0.15 M ionic strength)[30]. (iii) Channel gating requires protein import. Imposition of a *trans*-negative membrane potential that is known to cause channel opening in planar lipid bilayers was shown by differential (± membrane potential, $\Delta\psi$) photoaffinity labeling experiments to also cause insertion into the membrane bilayer of a segment of the channel polypeptide upstream of the membrane anchor segment that spans residues 471-508 (Fig. 3B). This upstream segment includes the charged and largely amphiphilic region between residues 424 and 460 in the 522 residue colicin E1 molecule[31], and may also include charged segments further upstream in the polypeptide[32].

Channel polypeptide insertion into the membrane, driven by the *trans*-negative $\Delta\psi$, is reversible. Dissipation of the $\Delta\psi$ results in extrusion of the labeled peptide segment from the membrane with a $t_{1/2} \approx 1$ min[31]. It is suggested that this may be a characteristic time for protein import into/export from membrane bilayers.

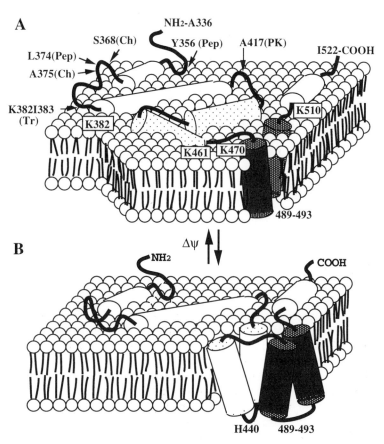

Figure 3. (A) Model for surface-bound-intermediate state showing hydrophobic anchor (darkly hatched), surface-bound helices, protease-accessible sites, and key residues of 187 residue colicin E1 COOH-terminal channel polypeptide. (B) Insertion of amphiphilic helical hairpin (lightly hatched) by membrane potential ($\Delta\psi$) that is associated with channel gating, inferred from differential labeling of channel domain in the presence and absence of a membrane potential.[31]

(iv) The *dynamic* nature of colicin channel formation and the involvement of protein import into the membrane is further illustrated by the measurement of translocation across the membrane of a peptide segment in colicin Ia using a novel trapping technique[32]. Single Cys residues were inserted into the channel domain sequence of colicin Ia by site-directed mutagenesis. The Cys residues were labeled with a covalently bound biotin. The biotin was trapped on the *trans*-side of the membrane, in the presence of a membrane potential, by streptavidin. *Trans*-trapping by streptavidin was demonstrated over an interval > 50 residues. It is not known whether all of these residues are translocated simultaneously or whether different segments are translocated at different times. The latter possibility would be consistent with a view of multiple open, as well as closed, states of the colicin channel, inferred from electrophysiological measurements[33]. It appears that the colicin channel may be a much more flexible and variable structure than has been considered previously for membrane channel proteins. This may be because the colicin channel, unlike eukaryotic channel proteins, does not have any regulatory or metabolic function. Its function is merely to keep the cell deenergized for approximately 30 min, the time required for cell division in *E. coli*. If the cell does not have sufficient energy to divide, it is dead.

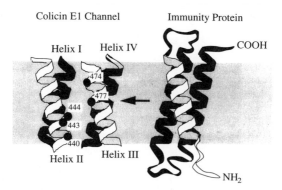

Figure 4. Proposed mechanism of intra-membrane interaction of colicin immunity protein, possessing three *trans*-membrane helices, with *trans*-membrane helices of colicin E1 channel domain. The residues marked in the TM helices of the channel domain are sites of mutations that bypass the immunity phenotype[34].

(v) An additional unique aspect of the colicin channel system is the existence of a specific immunity protein that protects colicin-producing cells from the colicins that they produce[25]. The immunity proteins of the channel-forming colicins are (a) polytopic (the colicin E1 immunity protein appears to span the membrane three times), (b) hydrophobic, (c) located in the cytoplasmic membrane, and (d) appear to exert their inhibitory effect on the colicin channel through specific intramembrane helix-helix interactions[34] (Fig. 4).

In summary, the mechanism of action of the voltage-gated colicin ion channel appears to involve (1) initial electrostatic interaction with the membrane, (2) formation of a hydrophobic hairpin membrane anchor, (3) voltage-dependent import of an amphiphilic hairpin into the membrane resulting in four *trans*-membrane helices that define a channel with a broad spectrum of single channel conductances, and (4) an immunity protein that can act by specific intramembrane helix-helix interactions to inhibit channel function.

Acknowledgment

This research was supported by grants from the NIH, GM-38323, GM-18457 (WAC) and the USDA, 9301586 (JLS and WAC). We thank John J. G. Tesmer for assistance with the cytochrome *f* drawing, and Janet Hollister for her expert help in the completion of this manuscript.

REFERENCES

1. Cramer, W.A. and Knaff, D.B., (1991), *Energy Transduction in Biological Membranes,* Chapt. VII, Springer Study Edition, New York.
2. Bullock, J.O., Cohen, F.S., Dankert, J.R. and Cramer, W.A. (1983) *J. Biol. Chem.*, 258: 9908-9912.
3. Cramer, W.A., Martinez, S.E., Huang, D., Tae, G.-S., Everly, R.M., Heymann, J.B., Cheng, R.H., Baker, T.S. and Smith, J.L. (1994), *J. Bioenerg. Biomem.* 26: 31-47.
4. Cramer, W.A., Martinez, S.E., Furbacher, P.N., Huang, D. and Smith, J.L. (1994) *Curr. Opin. Struct. Biol.* 4/4: 536-544.
5. Szczepaniak, A. and Cramer. W.A. (1990) *J. Biol. Chem.* 265: 17720-17726.
6. Widger, W.R., Cramer, W.A., Herrmann, R. and Trebst, A. (1984) *Proc. Natl. Acad. Sci., U.S.A.* 81: 674-678.
7. Degli Esposti, M., De Vries, S., Crimi, M., Gbelli, A., Patarnello, T. and Mayer, A. (1993) *Biochim. Biophys. Acta* 1143: 243-271.
8. Ermler, U., Fritsch, G., Buchanan, S.K. and Michel, H. (1994) *Structure* 2: 925-936.

9. Deisenhofer, J. and Michel, H. (1989) *EMBO J.* 8: 2149-2169.

10. Weiss, M.S., Abele, U., Weckesser, J., Welte, W., Schiltz, E. and Schulz, G.E. (1991) *Science* 254: 1627-1630.

11. Cowan, S.W., Schirmer, T., Rummel, G., Steiert, M., Ghosh, R., Pauptit, R.A., Jansonius, J.N. and Rosenbusch, J.P. (1992) *Nature* 358: 727-733.

12. Schirmir, T., Keller, T.A., Wang, Y.F. and Rosenbusch, J.P. (1995) *Science* 267: 512-514.

13. Kühlbrandt, W., Wang, D.N. and Fujiyoshi, Y. (1994) *Nature* 367: 614-621.

14. Henderson, R., *et al.* (1990) *J. Mol. Biol.* 213: 899-929.

15. Picot, D., Loll, P.J. and Garavito, R.M. (1994) *Nature* 367: 243-249.

16. Martinez, S.E., Smith, J.L., Huang, D., Szczepaniak, A. and Cramer, W.A. (1992) In: *Research in Photosynthesis*, Vol. II, (Murata, N. ed.), pp. 495-498, Kluwer, Dordrecht.

17. Martinez, S.E., Huang, D., Szczepaniak, A., Cramer, W.A. and Smith, J.L. (1994) *Structure* 2: 95-105.

18. Martinez, S.E., Cramer, W.A. and Smith, J.L., (1995), *Biophys. J.,* 68: 246a.

19. Gray, J.C. (1978) *Eur. J. Biochem.* 82: 133-141.

20. Gray, J.C. (1992) *Photosyn. Res.* 34: 359-374.

21. Doolittle, R.F. (1994) *Trends Biochem. Sci.* 19: 15-18.

22. Qin, L. and Kostic, N.M. (1992) *Biochemistry* 31: 5145-5150.

23. Gross, E.L. (1993) *Photosyn. Res.* 37: 103-116.

24. Morand, L.Z., *et al.* (1989) *Biochemistry* 28: 8039-8047.

25. Cramer, W.A., Heymann, J.B., Schendel, S.L., Deriy, B.N., Cohen, F.S., Elkins, P.S. and Stauffacher, C.V. (1995) *Ann. Rev. Biophys. Biomolec. Structure* 24: 611-641.

26. Parker, M.W., Postma, J.P.M., Tucker, A.D. and Tsernoglou, D. (1992) *J. Mol. Biol.* 224: 639-657.

27. Elkins, P., Bunker, A., Cramer, W.A. and Stauffacher, C.V. (1995) *Biophys. J.* 68: 369a.

28. Zhang, Y.-L. and Cramer, W.A. (1992) *Protein Science* 1: 1666-1676.

29. Schendel, S.L. and Cramer, W.A. (1994) *Protein Science* 3: 2272-2279.

30. Zakharov, S.D., Zhang, Y.-L., Heymann, J.B. and Cramer, W.A. (1995) *Biophys. J.* 68: 368a.

31. Merrill, A.R. and Cramer, W.A. (1990) *Biochemistry* 29: 8529-8534.

32. Slatin, S.L., Qui, X.-Q., Jakes, K.S. and Finkelstein, A. (1994) *Nature* 371: 158-161.

33. Deriy, B.N., Cramer, W.A. and Cohen, F.S. (1995) *Biophys. J.* 68: 368a.

34. Zhang, Y.-L. and Cramer, W.A. (1993) *J. Biol. Chem.* 268: 10176-10184.

β,γ-CRYSTALLINS

Modular Building Blocks of the Eye Lens

Huub Driessen

ICRF Unit of Structural Molecular Biology and Laboratory of Molecular
  Biology
Department of Crystallography
Birkbeck College, London, United Kingdom

The β,γ-crystallins are major components of the structural proteins of the mammalian eye lens. Their structures are highly symmetrical being composed of repeating 'Greek key' motifs. The protein subunits have two similar N- and C-terminal domains, linked by a short connecting peptide. While in monomeric γ-crystallins domain interactions are intramolecular, the analogous contacts in oligomeric β-crystallins are intermolecular. In both β- and γ-crystallins the domains associate through topologically equivalent surface hydrophobic patches although in γ-crystallins the domains are also covalently connected by a linker. Thus the β,γ-superfamily demonstrates how modification of an existing interface rather than evolution of a new one can give rise economically to novel assemblies during evolution. The role in domain swapping of the linker peptide versus the β-crystallin specific N- and C-terminal extensions is discussed.

THE LENS

The eye has evolved in such a way that an unobstructed, clear view has been achieved. It is composed of live tissues, which do not have particles in the light path, but require continuous nourishment. In the eye lens throughout life the anterior epithelial cells continue to migrate laterally, elongate near the equator and differentiate into fibre cells, which overlay the older cells in concentric layers. During this process cell nuclei and other organelles are lost, and as a consequence there is little turnover and renewal of proteins in the central nucleus of the lens. Lens cells do not die; therefore, the proteins in the nucleus are as old as the organism itself. The transparency of the lens therefore depends on an intact energy metabolism and on the integrity of the cellular organisation. Moreover, the transparency and refractive power of the lens depend on a smooth gradient of refractive index for visible light. This is achieved not only by the regular arrangement of the fibre cells, but also by a smoothly changing concentration gradient of the structural lens proteins, the crystallins. The regions of high and low protein concentrations are characterised by different proportions of component proteins of different sizes in a species-specific, development-specific way, also contributing to the particular optical requirements. The proteins must be maintained in a native state with controlled aggregation for a lifetime. A change in the oxidation-reduction state of the cells, the osmolarity of the cells, excessively increased concentrations of metabolites, and various physical insults such as ultraviolet radiation, may cause disturbances to this organisation, leading to uncontrolled aggregation and crosslinking, giving rise to opacities leading to the disease cataract. It is therefore important to understand how the lens proteins interact with each other and with water, and how they can remain stable for the lifetime of the organism.

THE CRYSTALLINS

The crystallins[1,2] which make up more than 90% of the proteins in the lens, can be defined as water-soluble structural proteins that occur in high concentrations in the cytoplasm of the fibre cells. Classically four major groups of crystallins have been distinguished. The α-,β- and γ-crystallins occur in all vertebrate classes, although γ-crystallins are low or absent in avian lenses. δ-crystallin is found exclusively in reptiles and birds. Additional types of crystallins have been reported, often occurring only in restricted taxonomic groups. They are examples of hijacked genes. α-crystallins are small heat-shock proteins, stress-related proteins. While no structures for α-crystallins are known, the structure of turkey δ-crystallin has recently been reported by our laboratory[3]. The β- and γ-crystallin families are members of a wider superfamily, and have approximately 30% identity with each other [4,5]. While γ-crystallins are monomeric, β-crystallin subunits associate to form a wide range of oligomers, from dimers to octamers. Over the last decade great progress has been made in structural research of the members of the β,γ-superfamily using single-crystal X-ray crystallography, a technique highly suitable, because the proteins in the crystal lattice are at concentrations comparable to those in the nucleus of the lens.

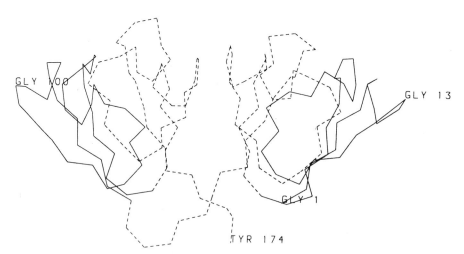

Figure 1. Cα backbone trace of β-crystallin showing the tertiary structure. The γB chain consists of two similar domains, closely interacting around a pseudo-twofold axis and linked by a short connecting peptide. In each domain two 'Greek key' motifs form a pair of four-stranded antiparallel β-pleated sheets, each sheet composed of three strands from one motif and one from the other. The sheets pack together in a wedge shape,The N- and C-terminal amino acid residues, and topologically equivalent glycines in motifs 1 and 3 are indicated.

STRUCTURE OF γ-CRYSTALLINS

The γ-crystallins comprise four topologically equivalent 'Greek key' motifs, with pairs of motifs organized symmetrically around a local dyad to give globular domains[6-10]. The two domains pack together with a short, single connection and are related by a further pseudo 2-fold axis that bisects the angle between the intradomain dyads (Fig. 1). In each domain the two 'Greek key' motifs form a pair of fourstranded antiparallel β-pleated sheets, each sheet composed of three strands from one motif and one from the other. The sheets pack together in a wedge shape, closed at the top by the loops connecting the third and fourth strands of each motif. Each domain has a tightly packed hydrophobic core, while the domains associate through topologically equivalent surface hydrophobic patches. Unusually, at the surface of the molecule over half the ionic side chains are closely paired, which is thought to stabilise the tertiary structure and to order the water around the molecule, a feature that will be highly important in the dense nucleus of the lens.

STRUCTURE OF β-CRYSTALLINS

Based on the sequence identity the structure of β-crystallin subunits was predicted to be similar with two globular domains, but with N- and C-terminal extensions or arms, that have little homology between them[4,11,12]. It was thought that these extensions would be playing a critical role in the oligomerisation of β-crystallins. However, when the structure of the simplest β-crystallin, the βB2 homodimer, was determined (called I-form because it crystallised in space group I222)[13,14], the prediction about the globular domains was confirmed, but unexpectedly the connecting peptide was extended and the two domains separated in a way quite unlike γ-crystallins (Fig. 2). Domain interactions analogous to those within monomeric γ-crystallins are intermolecular and related by a crystallographic dyad in the βB2 dimer. In the crystal lattice two dimers associate around further dyads to give a tetramer with 222 symmetry. Thus the β,γ-superfamily demonstrates how modification of an existing interface rather than evolution of a new one can give rise economically to novel assemblies during evolution. This was recognised as a novel way of exploiting symmetry for assembly of oligomers[13].

DOMAIN SWAPPING AS AN EVOLUTIONARY TOOL

This phenomenon has recently been generalised by Eisenberg and coworkers[15] into a hypothesis, which predicts that domain swapping between subunits of oligomeric proteins is a mechanism that favours the evolutionary transition from monomeric to oligomeric proteins. In this hypothesis it is proposed that the monomeric form of the protein has two domains with a primary interdomain interface between them (Fig. 3). The domain-swapped dimer has two identical primary interdomain interfaces, each formed between a domain from one polypeptide chain and a domain from the other chain. An additional interface, which does not exist in the monomer, is a secondary interdomain interface, which can mutate, allowing the formation of a stable dimer. Proteins that have been suggested as examples of domain swapping include RNase A, bovine seminal RNase, diphteria toxin, β-crystallin/γ-crystallin, γ-interferon/β-interferon, interleukin-5/granulocyte macrophage-colony stimulating factor. However, the relevance of domain swapping to the Rnases is disputed[16]. Slingsby et al.,[17] have shown that motif sharing between protein domains and subunits is also a common tool for oligomerisation. The β,γ-superfamily is a good example of domain

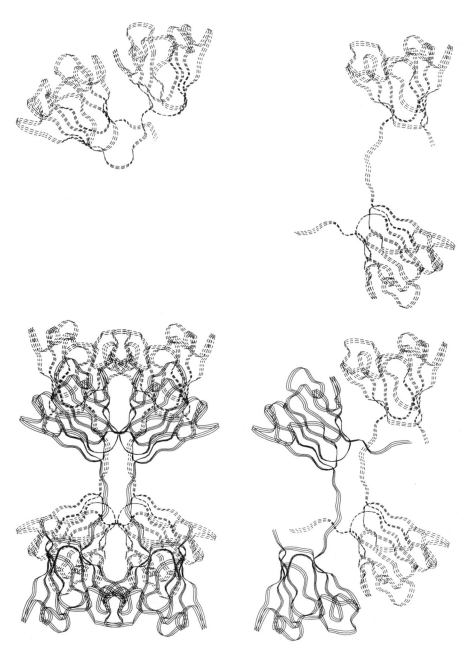

Figure 2. Ribbon diagrams of γB-crystallin and βBp-crystallin showing the domain organisation. The linker peptide of the βB2 chain is extended (top right) rather than bent as in γ-crystallins (top left). In the βB2 dimer (bottom right) the domain-domain interactions are similar to those found in monomeric γ-crystallins. However, these interactions are intersubunit rather than intrasubunit. Subunit 1 is in broken line and subunit 2 is in solid line. Note how the C-terminal extension interacts with the N-terminal domain of the partner subunit. Two βBp-crystallin dimers combine into a tetramer with 222 symmetry (bottom left).

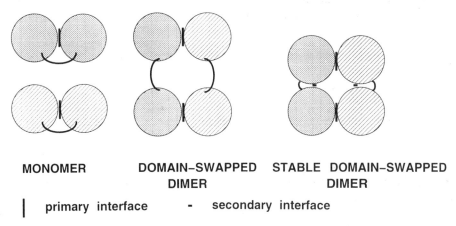

MONOMER   **DOMAIN–SWAPPED**   **STABLE DOMAIN–SWAPPED**
                              DIMER   **DIMER**

| primary interface   -   secondary interface

Figure 3. A model for protein domain swapping and dimer evolution (Bennet *et al.*, 1994), showing how the monomeric and dimeric form of the protein share the same primary interdomain interface, of which there are two in the dimer. Further mutations in the secondary interdomain interfaces specific to the dimer give further stabilisation.

	80	81	82	83	84	85	85A	86	87	88
bovine βB1	P	I	K	M	D	A	-	Q	E	H
rat	P	I	R	M	D	S	-	Q	E	H
chick	P	I	R	M	E	A	-	E	D	H
bovine βB2	P	I	K	V	D	S	-	Q	E	H
rat	P	I	K	V	D	S	-	Q	E	H
bovine βB3	P	I	K	I	D	G	H	P	D	H
rat	P	L	H	I	D	G	-	P	D	H
human	P	L	N	I	D	S	-	P	D	H
bovine βA2	P	V	L	C	A	N	H	S	D	S
bovine βA3	P	I	C	S	A	N	H	K	E	S
mouse	P	I	C	S	A	N	H	K	E	S
human	P	F	C	S	A	N	H	K	E	S
chick	P	V	C	S	A	N	H	K	E	S
frog	P	I	C	S	A	N	H	K	E	S
bovine βA4	P	V	A	C	A	N	H	R	D	S
rat γA	S	I	P	Y	T	S	-	-	S	H
bovine γB	L	I	P	Q	H	T	-	G	T	F
rat	L	I	P	Q	H	S	-	G	T	Y
human	L	I	P	P	H	S	-	G	A	Y
rat γC	L	I	P	H	T	G	-	-	S	H
human	L	I	P	Q	T	V	-	-	S	H
bovine γD	L	I	P	H	A	G	-	-	S	H
rat	L	I	P	H	A	G	-	-	S	H
human	L	I	P	H	S	G	-	-	S	H
bovine γE	L	I	P	H	T	S	-	-	S	H
rat	L	I	P	H	S	S	-	-	S	H
carp γm1	M	I	P	P	Y	R	-	G	S	Y
carp γm2	M	I	P	M	H	R	-	G	S	Y
frog γ-1	V	I	P	Q	H	R	-	G	S	F
frog γ-2	V	I	P	Q	Q	K	-	G	P	H

Figure 4. Alignment of sequences of the connecting peptide region of β- and γ-crystallins. Residues 83-87 define the linker proper.

swapping, in which γ-crystallins do not form dimers in solution, while the current β-crystallins are not found as monomers in solution.

THE LINKER PEPTIDE IN βB2

The critical question is, which factors determine the different type of assembly in γ- and β -crystallins, as they share a similar hydrophobic interface. Whatever dissimilarities there are in the shared interface would be present both for an extended and a bent linker structure. What relative role is played by the linker peptide and the extensions? In I-form βB2 the N- and C-terminal arms do not have any density, and therefore unexpectedly do not appear to be involved in specific interactions in the dimer. The connecting peptide in βB2 is extended but takes a sharp turn in γ-crystallins so that the two domains interact. Whereas all β-crystallins have a proline at position 80, none of the γ-crystallins does, and whereas all βγ-crystallins have a proline residue at position 82, this is not the case for any of the β-crystallins (Fig. 4). In γB most of the residues between positions 80 and 87 are in the polyproline conformation with Gly86 allowing a sharp turn, whereas in βB2 many of the equivalent residues are in β extended conformation (Fig. 5). Presumably the presence of Pro82 in γ-crystallins ensures the polyproline conformation is maintained, whereas the extended conformation of Lys82 in βB2 helps to position Val83 into the domain interface region compared with Gln83 in γB, which sticks into the solvent. In βB2 N82 hydrogen

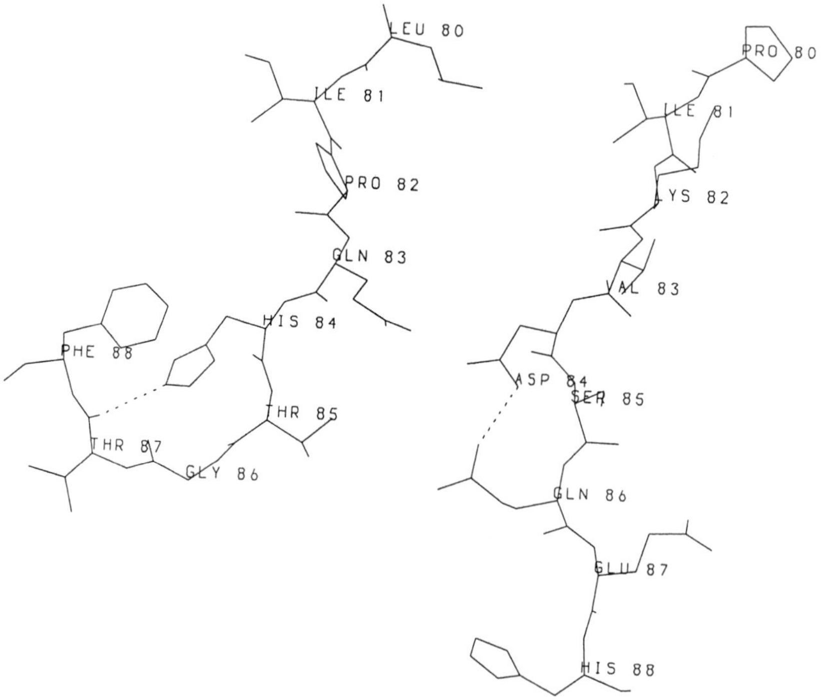

Figure 5. Comparison of conformation of the connecting peptide regions of γB- and βB2crystallins (left and right respectively). Residues 83-87 define the linker proper. While βB2 displays an extended conformation, γB initially is in the polyproline conformation followed by a sharp turn and change of direction. Although the chains start to diverge at residue 82, the sharp bend in γB is behind residue 86.

bonds with a water molecule that in turn bonds with the carboxylate group of the Glu147, which is conserved in β-crystallins, whereas it is a conserved Arg in γ-crystallins. Although the conformation of β- and γ-linker peptides start to diverge at the ψ torsion angle of residue 82, the extended connecting peptide deviates most markedly after position 86. In γ-crystallins the sharp turn is achieved in one of two ways. Either residue 86 is a glycine with positive Φ torsion angle allowing a change in direction that in γB is stabilised by the side chain of His84 hydrogen bonding to the backbone carbonyl of residue 87; or sometimes the connecting peptide is shorter by one residue and in this case the three central residues of the turn (84, 85, 87) always comprise those amino acids that most frequently occur in linker peptides between domains, allowing some stabilisation of local polar backbone atoms. On the other hand, there are no β-crystallins with short connecting peptides and none has a glycine at position 86. Instead several of the residues in the connecting peptide have bulky, polar side chains with a conserved acidic residue at position 87. These comparisons suggest a critical role for the linker peptide in the β,γ-superfamily in directing dimerisation.

THE N-TERMINAL EXTENSIONS IN βB2

At this stage we solved a crystal form of βB2 in space group C222 (called C-form), which contains less water in the lattice[18,19]. This structure is also a tetramer with 222 symmetry but differs in that part of the N-terminal extensions have become visible. The asymmetric unit contains two dimers, which combine around crystallographic twofold axes into two of the 222 tetramers seen in the I-form. These two dimers are usually described as the AB- and CD-dimers, as they are distinguished by small differences. The tetramers in the C-form are more tightly packed than in the I-form. Analysis of the tetramer contacts shows

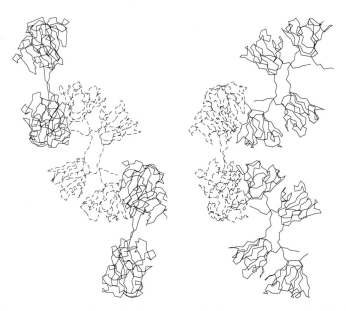

Figure 6. Two perpendicular views of the interactions of the N-terminal arms in the C222-lattice βB2 tetramer. The N-terminal arm of the A subunit of the AB dimer at the bottom (solid line) interacts with the N-terminal domain of the D subunit of the CD dimer in the middle (broken line). The N-terminal arm of the C subunit of the CD dimer in the middle (broken line) interacts with the N-terminal domain of the B subunit of the AB dimer at the top (broken line). This leads to an endless helical array of interactions in the z direction.

that the sites of interaction break the 222 symmetry of the tetramers. The N-terminal extensions play a major role in directing interactions between tetramers. One of the N-terminal arms interacts with a β-specific hydrophobic patch on the N-terminal domain of another tetramer (Fig. 6). These crystallographic observations for the I- and C-form obtained over a physiological concentration range show how in β-crystallin oligomers, the N-terminal extensions of βB2 can switch from interacting with water to interacting with protein depending on their relative concentrations. While the sequence extensions of βB2-crystallin may be used as spacers in the liquid-like outer regions of the lens, in the core regions they provide a framework for longer-range protein interactions. This could be useful in maintaining a gradient of refractive index.

CONCLUSIONS

Although a comparison of three γ-crystallin structures and one basic β-crystallin structure, suggests simple solutions for what is the driving force in oligomerisation in the β,γ-superfamily, further analysis is clearly necessary. Recent protein engineering experiments in which the linker of βB2 has been transplanted to γB[20] and vice versa[21] have shown that in both cases monomers are formed. By contrast switching the connecting peptide of γB into the sequence of βA3 does not prevent oligomer formation[22]. This suggests that the interactions of the connecting peptide with the domain structure are quite important. Also, differences between the basic and acidic β-crystallins become apparent. When both the N- and C-terminal arms of βB2[21], or either the N-terminal or the C-terminal arm of βB2[23] are removed, the protein still forms dimers, confirming our hypothesis that the arms are not required for dimerisation. However, removal of the N-terminal arm of βA3 does prevent oligomer formation[24]. These differences between the acidic and basic β-crystallins need further experimentation. Furthermore NMR spectroscopy has shown how the C-terminal three residues in γB have enhanced conformational flexibility compared with the rest of the molecule and that this extension is involved in interactions with α-crystallins[25]. The role of the C-terminal extensions in βB2 and βA3 needs to be examined. In the β,γ-superfamily individual domains have been shown to fold independently[26] and so they are currently being employed in further crystallographic work to determine the relative contributions of the interface residues, the linker peptide and the extensions.

While the γ-crystallins make a large contribution to tight packing in the nucleus of the lens, the β-crystallin oligomers with their variety of terminal extensions are probably more involved in less dehydrated areas of the lens. The effect of this is that there is a gradient of protein packing from the watery edge of the lens towards a hard to very hard nucleus where water is virtually excluded. The phenomenon of domain swapping therefore may be thought to have had a direct effect on the evolution of the lens by using β,γ-domains as flexible building blocks in forming a taylor-made protein gradient involved in providing the required gradient of refractive index for the particular organism.

REFERENCES

1. Wistow, G. J. and Piatigorsky, J. (1988) *Ann. Rev. Biochem.* 57: 479-504.
2. Lubsen, N.H., Aarts, H.J.M. and Schoenmakers, J.G.G. (1988) *Molec. Biol.* 51: 47-76.
3. Simpson, A., Bateman, O., Driessen, H., Lindley, P., Moss, D., Mylvaganam, S., Narebor, E. and Slingsby, C. (1994) *Nature Structural Biology* 1: 724-734. For Table 1 see also *Nature Structural Biology* 1: 831 (1994).
4. Driessen, H.P.C., Herbrink, P., Bloemendal, H. and De Jong, W.W. (1980) *Exp. Eye Res.* 31: 243-246.

5. Berbers, G.A.M., Hoekman, W.A., Bloemendal, H., de Jong, W.W., Kleinschmidt, T. and Braunitzer, G. (1984) *Eur.J.Biochem.* 139: 467-479.
6. Blundell, T., Lindley, P., Miller, L., Moss, D., Slingsby, C., Tickle, I., Turnell, B. and Wistow, G. (1981) *Nature* 289: 771-777.
7. Wistow, G., Turnell, B., Summers, L., Slingsby, C., Moss, D., Miller, L., Lindley, P. and Blundell, T. (1983) *J. Mol. Biol.* 170: 175-202.
8. Sergeev, Y.V., Chirgadze, Y.N., Mylvaganam, S.E., Driessen, H., Slingsby, C. and Blundell, T.L. (1988) *Proteins* 4: 137-147.
9. White, H.E., Driessen, H.P.C., Slingsby, C., Moss, D.S. and Lindley, P.F. (1989) *J. Mol. Biol.* 207: 217-235.
10. Najmudin, S., Nalini, V., Driessen, H.P.C, Slingsby, C., Blundell, T.L., Moss, D.S. and Lindley, P.F. (1993) *Acta Cryst.* D49: 223-233.
11. Wistow, G., Slingsby, C., Blundell, T., Driessen, H., de Jong, W. and Bloemendal, H. (1981) *FEBS Lett.* 133: 9-16.
12. Slingsby, C., Driessen, H.P.C., Mahadevan, D., Bax, B. and Blundell, T.L. (1988) *Exp. Eye Res.* 46: 375-403.
13. Bax, B., Lapatto, R., Nalini, V., Driessen, H., Lindley, P.F., Mahadevan, D., Blundell, T.L. and Slingsby, C. (1990) *Nature* 347: 776-780.
14. Lapatto, R., Nalini, V., Bax, B., Driessen, H., Lindley, P.F., Blundell, T.L. and Slingsby, C. (1991) *J. Mol. Biol.* 222: 1067-1083.
15. Bennett, M.J., Choe, S. and Eisenberg, D. (1994) *Proc Natl. Acad. Sci. USA* 91: 3127-3131.
16. D'Allessio, D. (1995) *Nature Structural Biology* 2: 11-13.
17. Slingsby, C., Bateman, O.A. and Simpson, A. (1993) *Mol. Biol. Rep.* 17: 185-195.
18. Driessen, H.P.C, Bax, B., Slingsby, C., Lindley, P.F., Mahadevan, D., Moss, D.S. and Tickle, I.J. (1991) *Acta Cryst.* B47: 987-997.
19. Nalini, V., Bax, B., Driessen, H., Moss., D.S., Lindley, P.F. and Slingsby, C. (1994) *J. Mol. Biol.* 236: 1250-1258.
20. Mayr, E-M., Jaenicke, R. & Glockshuber, R. (1994) *J. Mol. Biol.* 235: 84-88.
21. Trinkl, S., Glockshuber, R. and Jaenicke, R. (1994) *Protein Science* 3: 1392-1400.
22. Hope, J.N., Chen, H-C. and Hejtmancik, J.F. (1994a) *J. Biol. Chem.* 269: 21141-21145.
23. Kroone, R.C., Elliott, G.S., Ferszt, A., Slingsby, C., Lubsen, N.H. and Schoenmakers, J.G.G. (1994) *Prot. Engineering* 7: 1395-1399.
24. Hope, J .N., Chen, H-C. and Hejtmancik, J. F. (1994b) *Prot. Engineering* 7: 445-451.
25. Cooper, P. G., Carver, J. A., Aquilina, J.A., Ralston, G. B. and Truscott, R. J. W. (1994) *Exp. Eye Res.* 59: 211-220.
26. Rudolph, R., Siebendritt, R., Nesslauer, G., Sharma, A.K. and Jaenicke, R. (1990) *Proc. Natl. Acad. Sci. USA* 87: 4625-4629.

8

THE USE OF METAL IONS AS SPECTROSCOPIC PROBES IN THE ELUCIDATION OF pKa's OF SMALL MOLECULES

Bijan Farzami

Department of Biochemistry
Tehran Medical Science University
P.O. Box 14155-5399, Tehran, Iran

The apparent pKas' of the functional groups interacted with metal ions in solutions were determined using a technique devised in this laboratory (Farzami 1992). In these studies several groups of molecules such as coenzyme thiamine diphosphate and some of its derivatives, several dipeptides, amino acid and nucleotides and their bases were the subject of our studies. The results obtained were correlated with that obtained by other techniques such as NMR. In instances where the metal ion could indirectly alter the ionization properties of a group, a conformational scheme could be drawn. This method proved applicable to larger molecules such as proteins, enzymes and nucleic acids (Farzami 1990-1992).

The determination of pKa's of biological molecules are important in that they involve complex biological processes that are dependent on the pH of the reaction and hence the pKa's of the groups that are directly or indirectly involved in such reactions.

Many other applications of pKa knowledge includes its use in structural studies, separation techniques and thermodynamic studies that rely on the nature of the ionizing species.

The use of different techniques such as direct or potentiometric titration, spectro-scopic, nuclear magnetic resonance (NMR), have been used extensively for such determi-nations but all such methods have their own limitations and short comings. The most recent technique evolved during the last two decades has been NMR used for pKa determination of functional groups in small biological molecules.[1-4] It involves the use of large quantities of pure sample. In present study a novel spectrophotometric method is employed to determine the pKa of a group that is attached directly to a chromophor or located at the position where it can induce changes in the electronic state of a chromophor. The use of metal ion as a probe could relay changes due to ionization of the group to a nearby chromophor and act as an intermediary agent. Theoretically a relationship exists between the degree of electronic transition and the electron density in the exited state.[5,6] The role of the metal ion is to transfer the changes due to ionization of a group to an adjacent chromophor. We have tested these effects in several small molecules as well as macromolecules such as proteins and nucleic acids.[7-12]

In present study we concern ourselves with the use of the method for small molecules. The assumption made for these studies are as follows:

I) If an ionizable group is attached to an aromatic ring in such a way that the ionization of that group induced changes in the pi energy structure of the ring the effect could induce changes in the total absorbance with a trend similar to the trend observed from Henderson-Hasselback equation. It is suggested that if some quantity "L" (an absorption coefficient or rate constant, etc.) exists such that the corresponding property of a solution, the absorbance or the reaction rate, etc., is the product of that quantity and the concentration, the dependence of "L" for a molecule HA could be related to the proton concentration through the following equations for a single ionizing species.[13]

$$LH[A]o = LHA\,[HA] + LA-\,[A-]$$

$$LH[A]o = \frac{LHA\,[A]o\,[H+]}{Ka + [H+]} + \frac{La-\,[A]o\,Ka}{Ka + [H+]}$$

$$LH = \frac{LHA\,[H+] + LA-\,Ka}{Ka + [H+]}$$

where LA- and LHA is the contributions from the conjugate base and the acid respectively and LH is their total quantity.

The LH quantity, in our studies were taken as the total absorption of the chromophor which was affected by the change of energy in the course of ionization. These changes, when plotted against the incremental changes of pH, a minimum was observed which corresponded to the pKa of the ionizing group under the study.

The point correlated closely with the pKa obtained by microtitration where the measurement was feasible. Using this method 0.1 unit of pKa difference between ionizing groups in a compound could be differentiated.

II) In molecules where the changes in energy during the process of ionization could not be directly relayed to a chromophor, a metal ion as a meditor could be employed that specifically bind to the functional group under study. The metal ion could relay the change in energy due to the ionization to a nearby chromophor and thus similar effect as the direct transition of energy could be observed.

Several different types of molecules with ionizing groups were the subject of our studies. These include coenzyme molecules such as thiamine pyrophosphate and several of its derivatives, amino acids, dipeptides and nucleotides as well as proteins and enzymes that are discussed elsewhere[7-11].

EXPERIMENTAL

Materials

All the reagents used were analytical grade obtained from Sigma and May and Baker. Thiamine monotitrate was the gift of the department of Biochemistry and Biophysics, Biozentrum, Basel.

A. Preparation of thiamine derivatives.
1. 4-amino-2, 6- dimethyl pyridinium sulfonate was prepared by sulfate ion cleavage of thiamine hydrochloride (THC), compound (II).[14]

2. Hydroxymethyl thiazole(HMT), compound (III) was prepared by the addition of dried chloroform by potassium carbonate to the remaining solution of part (I), at neutral pH chloroform layer containing compound (III) was evaporated off and was eluted on a silica gel column. To quaternarized the product one volume of HMT and two volumes of methyl iodide in three volumes of ETOH is used. The mixture is refluxed at 40-50° C for 24 hours.[15]

3. Reduction of thiamine and the preparation of tetrahydro-thiamine; The reduction is performed by using sodium borohydride. Care should be taken not to allow temperature rise above 10° C. The precipitate thus collected was dried at 40° C melt.p. 129-131 compound (IV).[16]

4. N'1-methylthiamine perchlorate was prepared by methylating thiamine chloride hydrochloride by dimethyl sulfonate at strict pH of 6.5[17]. The perchlorate salt was then obtained by treating the final product with sodium perchlorate. The filtered product was dried under vacuum, recrystalized from 0.1 M perchloric acid.

B. Synthesis of Dansyl-Glycine-Aspartate.[18] Dansylation was carried out by dissolving 20 mg. of Gly-Asp in 15 ml of 0.1 M solution of sodium bicarbonate followed by the addition of 15 ml. of dansyl chloride in acetone (1mg/ml). The resulting solution was evaporated to dryness, heated at 100° C in a sealed tube with 6N HCl for 10 hours. The product was evaporated in vacuum to dryness and purified on TLC of silica gel with a mixture of 20/10/10 volume of N-butanol, glacial acetic acid and acetone respectively. The band identified by its fluorescence emission, extracted by a 1:1 mixture of acetone/water. The dansylated product was then evaporated to dryness.

C. Preparation of Solutions for Spectral Studies. Crystalline thiamine hydrochloride (THC), thiamine pyrophosphate (TPP obtained from Sigma and used without further purification. 10^{-3} M solutions of each compound was prepared with 0.1 N NaCl. 21.2 mg of the crystalline final product of N'1-methylthiamine was weighed and made up to 50 ml. with 0.1 N NaCl solution (10^{-3}) M. ADPS (compound II), tetrahydrothiamine (compound IV) and hydroxyethylthiazol were weighed separately and dissolved in 0.1 N solutions of NaCl to obtain concentrations of 10^{-3} M in each of the compound Magnesium sulfate (10^{-2} M) solution was made in 0.1 N solution of NaCl. Acetate and Tris buffer were used to prepare solutions with pH's ranging from 3 to 10 (.01 unit interval) with constant ionic strength maintained at 0.1 N. The final concentrations of compounds were 10^{-4} M. All samples were read against the blank in a thermostated Cary 118 spectrophotometer. The total spectra were monitored between 200-400 nm. The wavelength of study for THC and TPP were chosen to be 230, 248 and 265 nm. For N'1-methyl THC, the 250 nm. wavelength was chosen. Tryptophanyl- glutamic acid solutions of 10^{-3} M in 0.1 M NaCl solutions, were used. Similarly the solutions of ATP, ADP and GTP were prepared.

RESULTS AND DISCUSSION

Results obtained from direct spectrophotometric analysis; pKa determinations were carried out by two different methods; a: A direct measurement of absorbance change in specific wavelength as a function of incremental changes of pH (0.1 to 0.15). The plots of dOD vs. dpH reaches a minimum point on the dpH scale where it signifies the pka of the ionizing group. This method was applied to molecules such as THC, TMN, TDP, TMP, THT, HET, ADPS and N'1-methyl-THC all the above compounds were studied at wavelengths 230-300 nm. The selection of thiamine and its derivations for our pKa studies stems from the fact that the ionizing groups on the pyrimidyl moiety and thiazol are known to be essential

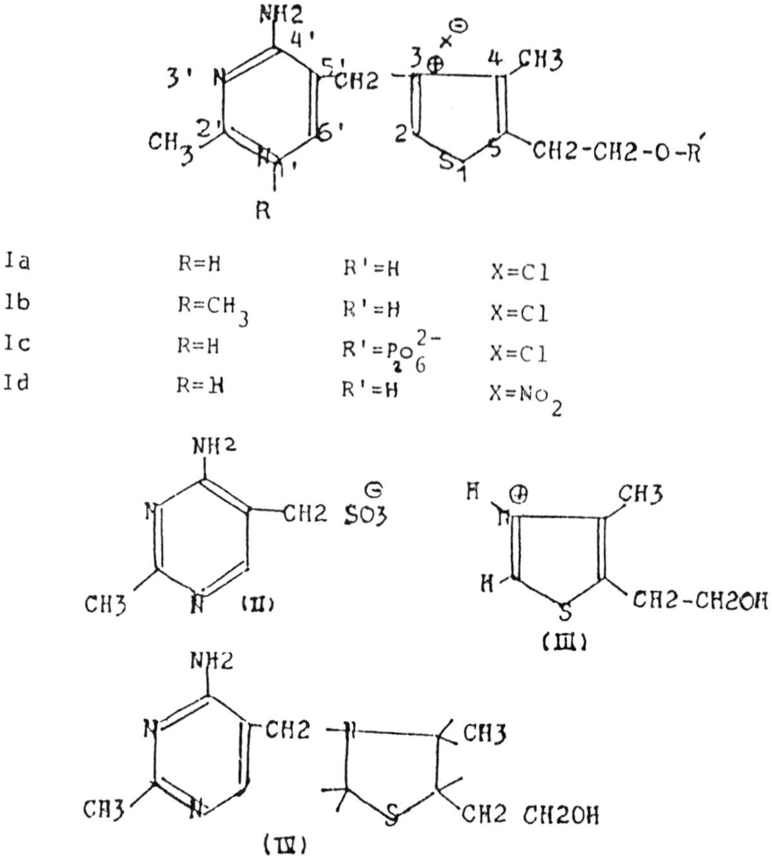

Figure 1. Thiamine derivatives used in present studies.

in thiamin dependant enzyme reactions.[19] The importance of pyrimidyl ring and more specifically the (C4-NH₃) in TDP enzymatic reactions was first proposed by Schellenberger and co-workers for enzyme pyruvate decarboxylase.[20] The ionization properties of this group in stabilization of the intermediates and the release of product acetaldehyde was considered important (scheme 1). Some of the results obtained were substantiated by microtitration where the determination was feasible.

The pKa estimations were either obtained from the point of reversal of the plot of dOD vs. dpH or by plotting log {(A+)-(A+)}/{(A+)-(A)} or log {(A-)-(A)}/{(A-)-(A+)} against the pH where A+ is the total absorption of full protonated species, A- is the total absorption of the anionic and A+- is the total absorbance of the both forms in equal proportions. Each of the logarithmic quantities gave a linear segment of the plot on the two sides of pH scale, with a common intercept at pH=pK for the ionizing group under the study.

The spectral studies with (1C) (Fig. 2) showed that there are two consecutive pKa's of 6.1 and 6.8 that are specific for TDP among thiamine derivatives. In THC and TMN the two pKa's are replaced by the pKa of 6.25. Another pKa of 3.48 to 3.75 was obtained for all the derivatives of THC except the N'1-methyl derivative. The absence of pKa in this range points to the fact that the replacement of hydrogen by a methyl group has abolished the pKa at this range. Therefore, the pKa could be assigned to N'1 functional group of the pyrimidyl ring. Furthermore the evolution of plus charge on the ring causes a decrease in the pKa

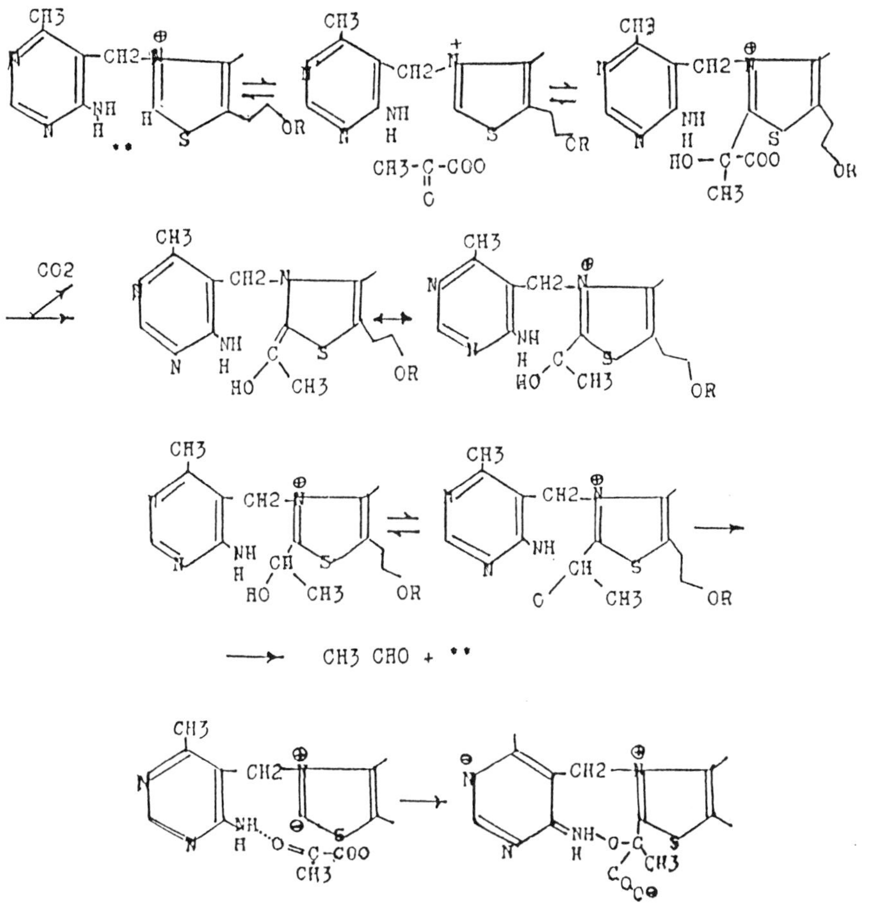

Scheme 1. a)Mechanism of Pyruvate decarboxylation by the enzyme pyruvate decarboxylase. b) The role of 4' NH_2 in stabilizing the initial substrate binding and the subsequent intermediate.

assigned to the 4'NH_3 while splitting of this pKa to pKa= 5.54 and 5.8 (Table 1) could be assigned to the production of two canonical structures that are produced in the methylated form of the thiamine. In TDP the 4-amino group pKa's could be similarly produced by the presence of two canonical structures induced and stabilized by diphosphate moiety of thiamine (Fig. 3). Compound (III) gave a pKa = 3.3 that was assigned to (N3) of the thiazole ring. This assignment was based on the fact that when the compound with the pirimidyl moiety as a substituent to thiazol was used, the respective spectral changes due to its ionisation was vanished. The effect of other side groups on the pKa's under the study was worth considering. For instance when the derivative (II) was used, the pKa of (4'-NH_3) in thimamine was raised from 6.34 in (Ia) to 6.5 and the N'1 pKa from 3.48 to 3.75.

This increase was also evidenced in IV. The increase in N'1 pKa to 3.6 in IV compared to that in Ia signifies the role of quaternary amino group with the positive charge that is eliminated in IV. In thiamine diphosphate, the two pKa's 6.1 and 6.8 were assigned to C4-NH_2. These pKa's are collectively seen as one in microtitration and was found to be equal to 6.47. The effect of pyrophosphate on C4-NH_2 may be through the induction and stabilization of two structures that are produced in the process of ionization (Fig. 3). Ib

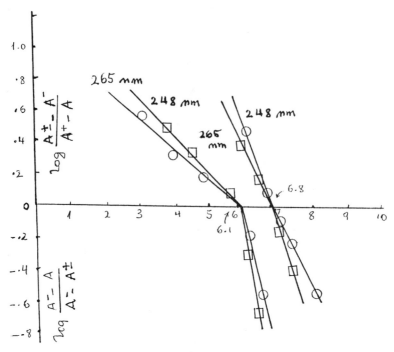

Figure 2. Logarithmic plots of TPP absorbance changes due to incremental changes of pH at constant ionic strength of 0.1 M., T = 25°C.

showed two C2-NH$_2$ pKa's that were the result of positive charge induced by N'1 methylation. pKa/'s of C4-NH$_2$ and N'1 has been reported previously using NMR techniques; (pKa C4-NH$_2$) = 7.5-7.8 and pKa (N'1) = 3.54-4.3[21,22]. The use of Mg^{2+} with the reaction mixture of several thiamine analogues signified the effect of metal ion as a probe to investigate the environment of groups that are affected by metal bindings. Among the thiamine derivatives only IC showed this effect. The pKa's of C4 amino group were enhanced to 6.25 and 7.05

Table 1. pKa's of some small molecules as detected by microtitration and by spectrophotometry with or without metal ions as probes.

amine derivatives pKa as determined
by microtitration and the proposed spectrophotometry

Compound	pK1(N'1)	pK2(4'NH2)	pK1(N'1)	pK2
	microtitration		spectroscopic	
T P P	3.46	6.47	6.1	6.8
T P P + Mg	3.50	6.36	6.25	7.07
T H C	3.48	6.34	6.25	-
T H C + Mg	3.48	6.34	6.25	-
T M N	3.40	6.30	6.25	-
T M N + Mg	3.40	6.31	6.25	
N'1-CH3	-	5.75	5.45	5.80
A D P S	3.75	6.5	N.M.	N.M.
T H T	3.60	6.2	N.M.	N.M.
5- hyd-ethyl -methyl thiaz.	-	-	3.3	

N.M.=Not measured
-troscopic method

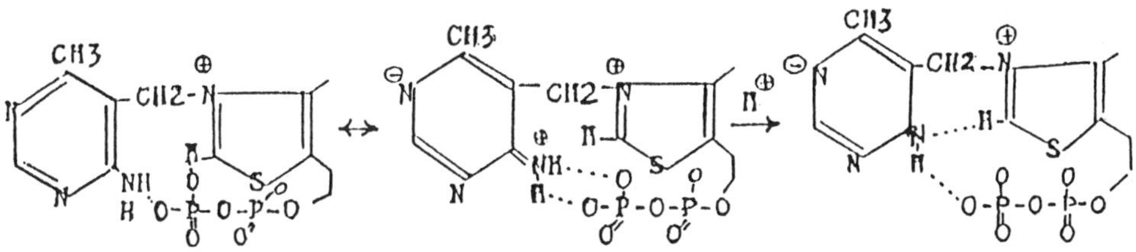

Figure 3. Possible stabilization in conformation of pyrimidyl group due to pyrophosphate.

from 6.1 and 6.80 due to the binding to Mg ion. It could be assumed that Mg ion may bind to pyrophosphate destabilizing the ionization structure. It has been previously reported that the hydrogen bond between the C4-NH$_2$ and the intermediate acetaldehyde in enzymatic reaction could lead to the stability of intermediate as well as the access to the release of the product. Therefore, the conformation that facilitates this mechanism was proposed to be when C4-NH$_2$ is adjacent to C$_2$ carbon of thiazolium ring, where the attack on the carbonyl group of substrate takes place. We suggest that the conformation presented above when stabilized by H bonding of pyrophosphate moiety of thiamine diphosphate with the adjacent (C4-NH$_2$) of pyrimidine ring is in accordance with the conformation that is most suitable for the catalysis in the enzyme reaction.

Further studies were carried out with compounds such as imidazole in which the ionization of imidazole could have a direct effect on the degree of electronic transition associated with the ring. The wavelength of 235 nm was chosen for these studies. One set of experiment was carried out through a direct spectrophotometric titration which used the assumption previously used in thiamine derivatives. The pKas of 6.17 and 6.02 was obtained for imidazole. In a similar system the use of an intermediate ion was proved to produce similar results. It is noteworthy to point that the dimensions of changes due to the changes in pH was more pronounced in all the cases where the metal ions were used. In such a system the pKas of 6.17 and 5.96 were obtained. The imidazole pKa was next studied in dipeptide

Table 2. pKa's of some small molecules measured directly or by using metal ions as probes and the effect of ETOH solvent concentration

no acids, dipeptides , nucleotides

Compound	%ETOH	pKa1	pKa2	Remarks
Imidazol	0	6.18	-	No metal ion
His-Ala	0	5.95	-	- - -
His- Ala	5	6.16	-	- - -
His- Ala	15	6.32	-	- - -
His- Ala	30	6.50	-	- - -
Trp- Glu	0	4.30	-	Ca2+
Trp- Glu	2.5	4.45	-	Ca2+
Trp- Glu	5	5.2	-	Ca2+
ATP	0	7.5	8.3	No metal ion
GTP	0	7.4	8.4	- - -
DNS-Gly-Asp	0	3.4	-	- - -

His-Ala. It may be expected that the alanyl substituent would effect the pKa of imidazole to a lower value due to the presence of carboxylate moiety. In fact this effect was seen when imidazole pKa in the dipeptide compared to that of single imidazole (Table II). As the results indicate, the pKa from the direct titration correlated closely with the pKa obtained using the intermediate ion within .02 units of pKa. The thermodynamic validity of all the pKa determinations were further established using solvent media of varying dielectric properties. The expected change due to the change in sovent polarity was observed.

This was performed by making all the solutions in some mole fraction of an organic solvent such as ethyl alcohol. Thus ETOH concentrations were fixed at 5, 15, and 30% v/v ETOH/H$_2$O, from which the bulk dielectric constant of the solutions were estimated. As it is shown in (Table II), the pKa of the dipeptide His-Ala was increased to 6.16, 6.33 and 6.51 in 5, 15 and 30% ETOH concentrations respectively. Another pKa of 5.64 was observed for imidazole as it was detected from a shoulder of the plot of dOD vs. pH in all the plots. This pKa remained constant at 5.93 in all the solvent concentration. The increase in the second pKa of imidazole as a function of concentration stems from the fact that according to the ionization scheme presented below, in the second ionization step charges are evolved in the process of ionization with more association in higher solvent polarity. In the first step, no charges are evolved or dessipated during the process of ionization and therefore no effect could be seen from the polarity of the solvent.

$$IMID.+ \; \longrightarrow \; IMID. \; + \; H+$$
$$\longleftarrow$$

$$IMID. \; \longrightarrow IMID.- \; + \; H+$$
$$\longleftarrow$$

In a second set of experiment the dipeptide Trp-Glu was used as a model dipeptide. Ca^{2+} ion was used as an intermediary ion. The measurements were carried out at 280 nm, the maximum wavelength of Trp where the effect from the ionization could be more pronounced. For Trp-Glu a sharp minimum at pKa = 4.32 was obtained when d(abs.)/dpH was plotted against pH. The solvent effect test was similarly carried out for Trp-Glu system.

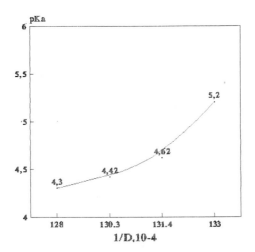

Figure 4. Trp-Glu pKa as a function of the inverse of dielectric constant of solutions.

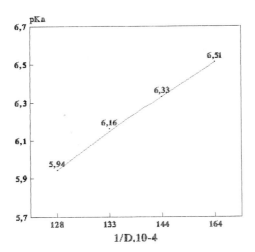

Figure 5. His-Ala pKa as a function of the inverse of dielectric constant of solutions.

ETOH solutions of 2.5, 3.5 and 5% v/v were used for solutions under study. pKas, of 4.40, 4.62 and 5.20 were obtained for above the solvent concentrations respectively.

The sharp increase in pKa due to the effect of solvent polarity signifies that the process of dissociation when the charges are evolved are strongly affected by the change in the dielectric properties of the solvent more so than when the dissociation is not accompanied by the production of charges such as in the previous example of His-Ala. The dependency of pKa on the inverse of dielectric constant is depicted in Fig. 4,5 for Trp-Glu and His-Ala. The non-linearity from the expected trend based on classical Fouss equation could be assigned to the structure of the bulk solution in lower alcohol concentrations. The upward trend in the case of His-Ala could be assigned to the change in the effective distance between the generated charged ions as it is depicted from Fouss equation.

$$Ka = \frac{4p\ N.\ a3}{3000}\ exp.\ (\frac{[ZI].[Z2].\ e2}{a.kT}\ .\ \frac{1}{D})$$

$$-Ln\ Ka = pKa = Ln\ \frac{4p\ N.\ a3}{3000} + \frac{[ZI].[Z2].\ e2}{a.kT}\ \frac{1}{D}$$

Where N is the Avagadro's number, k is the Boltzman constant, e is the electronic charges, a is the distance of closest approach, Z is the number of charges on cation or anion, T is the absolute temperature and Ka is the acid dissociation constant.

Another step in these studies was the determination of ionization constant in molecules that are not chromophor but could be modified covalently with a chromophor. In this respect a peptide such as Gly-Asp was dansylated. The dansyl-Glyc-Asp thus prepared was used for these studies. The wavelength of 247 was chosen for pKa determinations. A pKa of 3.37 was obtained which could be assigned to aspartate group in the molecule (Table II).

CONCLUSIONS

Applications of the method could be delineated as follows:

1. pKa determination of functional groups in small molecules that contain chromophoric groups adjacent to the ionizing species within 0.1 pH unit, its use in evaluation of possible canonical structures.

2. Metal binding site studies in small molecules.

3. Conformational studies of small molecules in solutions.

4. Metal binding site elucidation in large molecules, i.e. protein enzymes and nucleic acids.

5. pKa of the groups that bind specifically to metal ions in large molecules within 0.1 unit of pH.

6. Estimation of the relative polarities of the active site of enzyme by observing pK shift.

7. Configuration of the metal coenzyme complex.

REFERENCES

1. Bovey, F.A. (1972) *High resolution Nuclear Magnetic Resonance of Macromolecules*. Academic Press.

2. Dwek, R.A. (1973) *Nuclear Magnetic Resonance in Biochemistry*, Clarendon Press. Mc Donald, C.C. and W.D. Phillips, 1970, *Proton Magnetic Resonance Spectroscopy in Fine structure of proteins and Nucleic acids,* Vol. 4 (Fasman, G.D. and Timashef, S.N. ed.), pp. 1-48, Dekker.

3. Roberts, G.C.K. and Jardetsky, O. (1970) *Advan. Protein Chem.* 24: 448-545.

4. Sheard, B. and Bradbury, E.M. (1970) *Progr. Biophys. Mol. Biol.* 20: 187-246.

5. Pavia, D.L., Lapman, G.M. and Kriz, G.S. (1976) *Introduction to Spectroscopy*, W.B. Saunders Company, New York.

6. Dyer, R.T. (1965) *Application of Absorption Spectroscopy of Organic Compounds*, Prentice Hall Inc., Englewood Cliffs, N.J.

7. Farzami, B. and Jordan, F. (1990) *Metal Ions in Biology and Medicine* (Collery, P, Poirier, L.A., Manfait, M., Etienne, J.C., eds.) John Libbey Eurotext, Pairs.

8. Farzami, B., Kuimov, A.N. and Kochetove, G.A. (1992) *Metal Ions in Biology and Medicine* (Collery, P, Poirier, L.A., Manfait, M., Etienne, J.C., eds.) John Libbey Eurotext, Paris.

9. Farzami, B., Kuimov, A.N. and Kochetov, G.A. (1992) *J.Sci. I.R. Iran* V3 No. 3,4, pp. 81.

10. Farzami, B. and Zamani, D. (1994) *Metal Ions in Biology and Medicine* (Collery, P, Poirier, L.A., Littlefield, J.C., Etienne, J.C., eds.) pp. 399-404, John Libbey Eurotext, Paris.

11. Farzami, B., Moosavi-Movahedi, A.A. and Naderi, G.A. (1994) *Int. J. Biol. Macromol.* V 16, No. 4, pp. 181.

12. Farzami, B. (1992) *Metal Ions in Biology and Medicine* (Anastassopoulpou, J. Collery, Ph., Etienne, J.C., eds.) pp. 102-107, John Libbey Eurotext, Paris.

13. Fersht, A.R. (1985) *Enzyme Structure and Mechanism* 2nd. Ed. pp. 143-146, W.H. Freeman, Co., New York.

14. Williams, R.R., Waterman, R.E., Keresztesy, J.C and Buchman, E.R. (1935) *J. Am. Chem. Soc.* 57: 536-537.

15. Williams, R.R., Keresztesy, J.C. and Buchman, E.R. (1935) *J. Am. Chem. Soc.* 57: 1850-1851.

16. Clark, C.M. and Sykes, P. (1967) *J. Chem. Soc.(C)* 1411-1414.

17. Jordan, F. Yitbarek, H.M. (1978) *J. Am. Chem. Soc.* 100: 2534-2541.

18. Royt, J.F. and White, B.J. (1987) *Biochemical Techniques, Theory and Practice*, John Wiley.

19. Shellenberger, A. (1967) *Angew. Chem. Int.* Ed. 6: 1024.

20. Shellenberger, A. (1982) *Annals of the New York Academy of Science* Vol. 378: pp. 51.

21. Hopmann, R.F. and Brugnoni, G.P. (1973) *Nature New Biology* Vol. 246: No. 153.

22. Jordan, F. (1982) *J. Org. Chem.* 2748: 47.

9

TRYPSIN COMPLEXED WITH α1-PROTEINASE INHIBITOR HAS AN INCREASED STRUCTURAL FLEXIBILITY

Gyula Kaslik,[1] András Patthy,[2] Miklós Bálint,[1] and László Gráf[1]*

[1] Department of Biochemistry
Eötvös University
Puskin u.3., H-1088 Budapest, Hungary
[2] Agricultural Biotechnology Center
H-2100 Gödöllő
P.O. Box 170, Hungary

ABSTRACT

Mutant rat trypsin Asp[189] Ser was prepared and complexed with highly purified human α_1 -proteinase inhibitor. The complex formed was purified to homogeneity and studied by N-terminal amino acid sequence analysis and limited proteolysis with bovine trypsin. As compared to uncomplexed mutant trypsin the mutant enzyme complexed with α_1 -proteinase inhibitor showed a highly increased susceptibility to enzymatic digestion. The peptide bond selectively attacked by bovine trypsin was identified as the Arg[117] - Val[118] one of trypsin. The structural and mechanistic relevance of this observation to serine proteinase-substrate and serine proteinase-serpin reactions are discussed.

INTRODUCTION

According to Pauling's transition state theory[1] , enzymes provide templates which are complementary to the reactants in their activated transition state rather than to the substrates in their ground state. Despite 50-year extensive research along these lines the molecular mechanism by which enzymes convert substrates from the ground state to the transition state is not clear at all. It is generally accepted, however, that the structure of the enzyme must undergo some changes to complement changes of the substrate throughout catalytic reactions[2-4]. The extent of these changes may depend on the enzyme-substrate system studied, varying from an extensive distortion of the enzyme structure to complement transition state (induced fit) to relatively minor, local conformational changes in the course of catalysis[2-4]. In the case of seine proteinases the latter mechanism seems to be likely, since the structures of crystalline trypsin and chymotrypsin are indistinguishable from those of the same enzymes complexed with "substrate-like" canonical proteinase inhibitors[3-5]. The question, however, is whether the binding of these inhibitors does properly mimic substrate binding. Dufton[6] argued that the modes of interaction of inhibitors and substrates with the same serine proteinase might be markedly different, and that binding of true substrates to the proteinase, unlike that of inhibitors, might induce a relative movement would focus stress on the scissile bond in the substrate, thereby facilitating its hydrolysis[6]. Our own enzyme kinetic studies on trypsin, chymotrypsin and mutant trypsins with modified substrate-binding pockets further support the view that serine proteinase action may be accompanied by some conformational changes of the enzyme[7,8]. Our interpretation of the poor catalytic activity of mutant trypsins with chymotrypsin-like substrate-binding pockets has been that though the pockets properly accommodate the substrate, this interaction does not induce the conformational change crucial for optimal transition state stabilisation[7-10]. Furthermore, we hypothesised that different conformational flexibilities of the substrate-binding sites of trypsin and chymotrypsin, rather than evident differences between the crystalline structures of the binding sites, represent the structural basis for the different substrate specificities of these proteinases[9,10]. Recently, this possibility has also been entertained by Perona and co-workers[11] when interpreting crystallographic data on two mutant trypsins with chymotrypsin-like specificity.

To directly explore the molecular mechanism of serine proteinase action time-resolved diffraction studies will be required. Until such studies will be technically feasible to be performed structural investigation of any kinetically stable reaction intermediates of serine proteinases may provide us with useful information about the extent and nature of conformational changes that the enzymes may undergo throughout catalysis. In this paper we present data on a dramatic conformational change of mutant trypsin Asp[189] Ser upon its interaction with α_1-proteinase inhibitor.

MATERIALS AND METHODS

Enzymes, Inhibitors and Chemicals

Bovine pancreatic trypsin (TPCK) and porcine pancreatic elastase were purchased from Sigma. Mutant rat trypsinogens, Asp[189] Ser and His[57] Ala, Asp[102] Asn, were expressed in an *E.coli* expression-secretion system, purified to homogeneity and activated by enterokinase, as described previously[7,12]. Human α_1-PI was obtained from Serva and purified to homogeneity on a MONO Q column (Pharmacia FPLC system, Uppsala, Sweden) by using a linear gradient from 0 to 0.5 M NaCl in 10 mM sodium phosphate buffer, pH 7.0. Fractions were monitored by SDS-PAGE[13]. The purified inhibitor was stored at -20° C. All chemicals used were of reagent grade.

Polyacrylamide Gel Electrophoresis and Protein Content Determination

SDS-PAGE was carried out in 12% or 15% (w/v) slab gels according to Laemmli[13]. Native non-denatureing PAGE was performed under the same conditions except that SDS and 2-mercaptoethanol were absent from the buffers and that the polyacrylamide gel concentration was 10% (w/v). Protein content was determined by the method of Bradford[14].

Preparation and Isolation of the Mutant Trypsin-α_1-PI Complex

α_1-PI in a slight molar excess was mixed with mutant trypsin Asp[189] Ser in 10 mM sodium phosphate buffer, at pH 7.0. The mixture was loaded onto a MONO Q column and eluted with a linear pH and salt gradient from 7.0 to pH 5.0 and from 0 to 0.2 M NaCl in 10 mM sodium citrate-phosphate buffer. The separation was followed by SDS-PAGE[13]. The purified complex was stored at -20° C.

Limited Proteolysis

Mutant trypsin Asp[189] Ser and its complex with α_1-PI, in a final concentration of 4 μM, in 0.2 mM sodium phosphate buffer of pH 7.0 containing 0.1 M NaCl were incubated with a 5 to 1 molar ratio of bovine trypsin. The digestion was performed at room temperature for different time periods from 0 to 16 hours. Aliquots taken were boiled for 3 minutes and then PMSF was added to them at a final concentration of 1 mM. The samples were analysed by SDS-PAGE applying 15% and 12 % (w/v) polyacrylamide gels for the mutant trypsin and the complex, respectively, and by N-terminal amino acid sequence analysis.

Separation by HPLC

Reversed-phase HPLC was used for the separation of fragments in the intact and enzymatically degraded complexes. Samples were loaded onto an Aquapore OD300 C18 column (Applied Biosystems, 4.6 mm x 220 mm) equilibrated with water containing 0.1% (v/v) TFA (solvent A), and a 30 minute linear gradient was applied by using a mixture of 80% (v/v) acetonitrile, 20% (v/v) water and 0.08% (v/v) TFA, as solvent B. The flow rate was 1 ml/min and detection was done at 220 nm.

N-Terminal Amino Acid Sequence and Amino Acid Analyses

N-terminal amino acid sequences were determined in an Applied Biosystems 471 A pulsed liquid-phase sequencer by using a program adopted from Hunkapiller and co-workers[15]. Amino acid analysis was performed by Pico-Tag method[16].

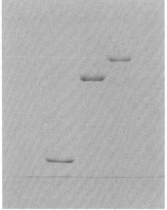

Figure 1. 12% SDS PAGE of (a) trypsin mutant Asp[189] Ser, (b) α_1-PI and (c) purified trypsin mutant Asp[189] Ser-α_1-PI complex.

a   b   c   d   e

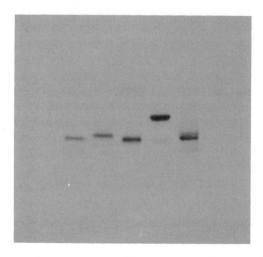

Figure 2. 10% native PAGE of trypsin mutnts (a) Asp^{189} Ser and (b) His^{57} Ala, Asp^{102} Asn, (c) α_1-PI, the equimolar mixtures of α_1-PI with (d) trypsin mutant Asp^{189} Ser, and (e) trypsin mutant His^{57} Ala, Asp^{102} Asn.

RESULTS

Human α_1-proteinase inhibitor and mutant rat trypsin Asp^{189} Ser were mixed and the mixture was subjected to chromatography on a MONO Q column as described in the Materials and Methods. 85% of the protein content of the mixture was eluted in one peak and found to be homogeneous by both SDS-PAGE and native non denaturing PAGE (Figs. 1 and 2). The molecular weight of the mutant trypsin Asp^{189} Ser-α_1-PI complex as estimated by SDS-PAGE (Fig. 1) is about 73 kDa, somewhat less than the sum of those of trypsin (24 kDa) and α_1-PI (53 kDa). Parallel experiments with wild-type rat trypsin and α_1-PI resulted in similar results, except that the complex isolated from their mixture was less homogeneous and stable than that of mutant trypsin Asp^{189} Ser-α_1-PI (data are not shown).

A rat trypsin mutant with destroyed catalytic triad, mutant His^{57}Ala, Asp^{102} Asn,[12] was not able to form stable complex with α_1-PI as shown by native PAGE (Fig. 2).

To compare the structural flexibilities of mutant trypsin Asp^{189} Ser in complex with α_1-PI and in its uncomplexed form, both preparations were digested with bovine trypsin. SDS-PAGE was used to follow the time-course of the digestion (Fig. 3). As seen in the Figure, the complex is much more susceptible to proteolysis than the trypsin mutant is in its uncomplexed form. In fact, the protein band of mutant trypsin-α_1-PI complex was converted to a new electrophoretic component within 15 minutes, while mutant trypsin showed resistance to tryptic digestion. It has to be noted that the presence of the cleaved form of α_1-PI appeared in these gels as a consequence of the 3 minute boiling of samples.

Sequences of the first six N-terminal residues of α_1-PI trypsin mutant Asp^{189} Ser, the isolated mutant trypsin-α_1-PI complex and a 1-hour tryptic hydrolysate of the latter one were determined, and the results are shown in Table I. The N-terminal of α_1-PI is probably blocked since no amino acids could be detected in the first six cycles of Edman degradation. Sequencing the isolated complex, in addition to the N-terminal sequence of the mutant trypsin a new amino acid sequence appeared that corresponds to residues 359-364 of α1-PI. PTH-amino acid signals for the two sequences were found to be

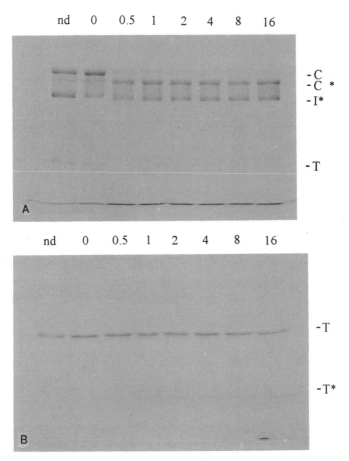

Figure 3. Limited proteolysis with bovine trypsin of trypsin mutant Asp[189] Ser-α_1-PI complex (12% SDS-PAGE; Fig. 3A) and of rat trypsin mutant Asp[189] Ser (15% SDS-PAGE; Fig. 3B). 4 μM complex and 4 μM mutant trypsin were incubated with 0.8 μM bovine trypsin at pH 7 (0.1 M NaCl, 0.2 M sodium phosphate buffer), at room temperature for different time periods (0-16 h). Aliquots removed at the indicated times (h) were boiled for 3 min. and then PMSF was added at 1 mM final concentration to inactivate trypsin. nd (not digested) is the control sample which was incubated for 16 h. under similar conditions but without bovine trypsin. T: trypsin mutant Asp[189] Ser (Mw 24 kDa); T*: C-terminal fragment of cleaved trypsin mutant (Mw 13 kDa); C: trypsin mutnat Asp[189] Ser-α_1-PI complex without the C-terminal fragment of α_1-PI (Mw 73 kDa); C*: digested complex (Mw 62 kDa); I*: cleaved α_1-PI (without its C-terminal fragment; Mw 49 kDa).

comparable in each cycle indicating that α_1-PI was completely cleaved at peptide bond Met[358]-Ser[359]. The separation and identification of the C-terminal fragment, residues 359-394, of α_1-PI (Peak I in Fig. 4A, Table I) by reversed-phase chromatography of the mutant trypsin-α_1-PI complex confirmed this finding. N-terminal sequence analysis of a 1-hour tryptic digest of the trypsin mutant-α_1-PI complex revealed the presence, in equimolar amounts, of three N-terminal sequences (Table I). The new N-terminal sequence released by limited tryptic hydrolysis of the complex is Val-Ala-Thr-Val-ala-Leu-, representing the 118-123 sequence region of rat trypsin. When the 1-hour tryptic digest of the mutant trypsin-α_1-PI complex was subjected to reversed-phase HPLC, the newly released N-terminal fragment of the trypsin mutant coeluted with the C-terminal fragment of α_1-PI (Peak I in Fig. 4B, Table I).

Table 1. Results of the N-terminal amino acid sequence analysis of 1. α_1-PI, 2. trypsin mutant Asp[189] Ser, 3. trypsin mutant Asp189 Ser- α_1-PI complex, 4. complex digested with bovine trypsin for 1 h and 5. HPLC peaks in Fig. 4. (for details see text). The amino acid identified in the first six cycles are shown

Sample	No. of cycles					
	1	2	3	4	5	6
1. α_1-PI	-	-	-	-	-	-
2. trypsin mutant	I	V	G	G	Y	T
3. complex	I	V	G	G	Y	T
	S	I	P	P	E	V
4. digested complex	I	V	G	G	Y	T
	S	I	P	P	E	V
	V	A	T	V	A	L
5. HPLC peaks:						
peak I / Fig. 4A	S	I	P	P	E	V
peak II / Fig. 4A	I	V	G	G	Y	T
peak I / Fig. 4B	I	V	G	G	Y	T
	S	I	P	P	E	V
peak II / Fig. 4B	V	A	T	V	A	L

Schematic representation of the structures of trypsin mutant-α_1-PI complex and its trypsin-nicked form is based on the protein analytical work described above (Fig. 5).

DISCUSSION

The exact molecular mechanism by which serpins including α_1-proteinase inhibitor interact with serine proteinases has not yet been established[17,18]. Since the reactive sites of serpins are susceptible to their target proteinases just like those of true substrates and "substrate-like" canonical proteinase inhibitors, the general kinetic mechanism of their reaction with the enzymes can be written as follows:

$$E + I \Leftrightarrow E{\cdot}I \Leftrightarrow E{\equiv}I \rightarrow E{-}I^* \rightarrow E + I^*$$

where E-I is the initially formed Michaelis complex, E≡I is the tetrahedral complex (TI) and E-I* is the acyl-enzyme intermediate that dissociates into the active enzyme (E) and the cleaved inhibitor (I*). The inhibitory mechanism of serpins differs, however, in many respects from that of canonical "standard mechanism" inhibitors like kunins and kazals[17,18]. One striking difference is that the structure of serpins unlike that of canonical inhibitors,

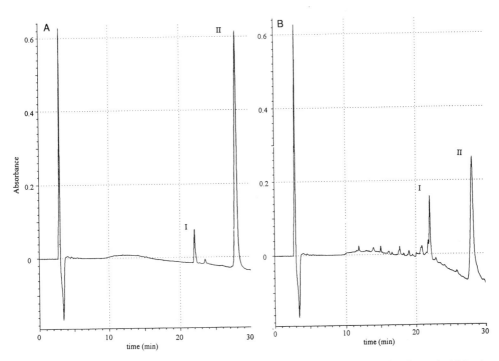

Figure 4. Reversed-phase HPLC analysis of (Fig. 4A) intact complex, (Fig. 4B) complex digested with bovine trypsin for 1 h, injected under reducing conditions (2% v/v 2-mercaptoethanol). HPLC analysis was performed on a C_{18} reversed-phase column using a linear gradient of 0-80% acetonitrile in 0.1% (v/v) aqueous TFA over 30 min. at a flow rate of 1 ml/min. Peaks were identified by N-terminal amino acid sequence and amino acid analyses (Table I). Peak I in Fig. 4A is the C-terminal fragment of α_1-PI (Ser^{359} -Lys^{394}); peak II in Fig. 4A is the trypsin mutant Asp^{189} Ser-α_1-PI complex without the C-terminal fragment of α_1-PI; peak I in Fig. 4B contains the C-terminal fragment of α_1-PI (Ser^{359} -Lys^{394}); and the N-terminal fragment of trypsin mutant Asp^{189} Ser (Ile^{16} -Arg^{117}); peak II in Fig. 4B is the trypsin mutant Asp^{189} Ser-α_1-PI complex without the N-terminal fragment of trypsin (Ile^{16} -Arg^{117}) and the C-terminal fragment of α_1-PI (Ser^{359} -Lys^{394}).

while reacting with the proteinases, undergo a dramatic change from a stressed, labile conformation to a relatively ordered, heat-stable form[17,18]. Furthermore, serpins unlike canonical proteinase inhibitors form kinetically stable complexes with proteinases[19,20]. The question if the general structure of such complexes corresponds to a tetrahedral[20] or an acyl-enzyme intermediate[18,19] is still debated.

Our own sequencing data on the mutant trypsin Asp^{189} Ser α_1-PI complex showed that the P_1-P_1' peptide bond was present in a cleaved form (Table I, Fig. 4), thus supporting the view that the acyl-enzyme intermediate rather than the tetrahedral one may be the kinetically stable form of the complex. There may be some faint doubt, however, that the acyl intermediate was formed from the tetrahedral one under the conditions of phenyl-thio-carbamylation of the protein (first reaction of the Edman-degradation). Model experiments to exclude or confirm this possibility are now being performed in our laboratory. Our failure to form a stable complex between α_1-PI and mutant trypsin His^{57} Ala, Asp^{102} Asn further

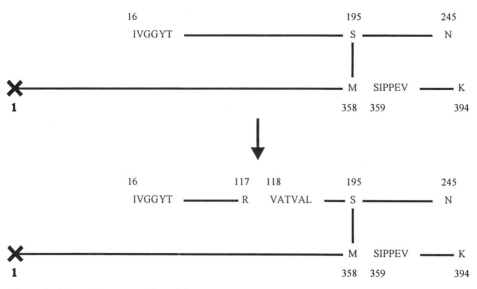

Figure 5. Schematic representation of the structures of mutant trypsin-α_1-PI and its trypsin-nicked form.

confirms the general notion that the catalytic apparatus of the proteinase is essential for complex formation with serpins[17-20].

Instead of native trypsin, a mutant trypsin with a single amino acid replacement in its substrate-binding pocket, Asp[189] to Ser[7], was used in this work. The low (auto) catalytic activity, 5 orders of magnitude smaller than that of wild-type trypsin, made this mutant an apparently ideal model to study the biochemical properties of the enzyme-α_1-PI complex. Since the X-ray structure of mutant rat trypsin Asp[189] Ser complexed with bovine pancreatic trypsin inhibitor was shown to be identical with that of bovine trypsin complexed with the same inhibitor[21] there is no reason to doubt that structural features of the mutant trypsin Asp[189] Ser-α_1-PI complex are relevant to naturally formed proteinase-serpin complexes.

Though a relatively increased sensitivity of different proteinase-serpin complexes to proteolytic attack has already been noted in the literature[22,23], the cleavage sites within the complexes have not yet been identified. For the first time, our present studies provide evidence that the proteinase component of a proteinase-serpin complex underwent a dramatic conformational change upon complex formation with the serpin, and that this increased molecular flexibility of the enzyme serves as the structural basis for the marked susceptibility of the complex to proteolytic attack. The tryptic cleavage site identified as peptide bond Arg[117]-Val[118] in trypsin is of particular interest. This site is located within the hinge region connecting the two large domain of trypsin, the relative movement of which was postulated to play a crucial role in the catalytic process[6]. Considering the rational possibility that our mutant trypsin-α_1-PI represents either the acyl-enzyme or the tetrahedral intermediate of the proteinase-serpin reaction we would be tempted to propose that the proteinase might undergo a similar conformational transition when reacting with its true substrate. However, other possibilities have to be also entertained. Such possibilities are that only serpins can induce such a conformational distortion of target proteinases or that they just reinforce catalytically relevant structural changes of proteinases. Such changes of the enzyme structure would complement, at certain stages of the proteinase-serpin interaction, the remarkable structural transition of serpins from a stressed conformation to a relaxed one.

REFERENCES

1. Pauling, L. (1964) *Chem. Eng. News* 24: 1375.
2. Fersht, A. (1985) *Enzyme structure and mechanism.* Freeman, New York.
3. Kraut, J. (1988) *Science* 242: 533-540.
4. Polgár, L. (1989) *Mechanism of protease action.* CRC Press, Boca Raton.
5. Bode, W. and Huber, R. (1992) *Eur. J. Biochem.* 204: 433-451.
6. Dufton, M.J. (1990) *FEBS Lett.* 271: 9-13.
7. Gráf, L., Jancsö, A., Szilágyi, L., Hegyi, G., Pintér, K., Náray-Szabó, G., Hepp, J., Medzihradszky, K., Rutter, W.J. (1988) *Proc. Nail. Acad. Sci.* USA 85: 4961-4965.
8. Gráf, L, Boldogh, I., Szilágyi, L., Rutter, W.J. (1990) In: *Protein Structure Function,* (Zaidi, Z.H., Abbasi, A. and Smith, D.L. eds.) p.p. 49-55, TWEL Publishers, Karachi.
9. Gráf, L., (1992) *Farady Discuss Chem. Soc.* 93: 135.
10. Gráf, L., (1995) In: *Natural Sciences and Human Though* (Zwilling, R. ed) pp. 139-147, Springer-Verlag, Berlin, Heidelberg.
11. Perona, J.J., Hedstrom, L., Rutter, W.J., Fletterick, R. (1995) *Biochemistry* 34: 1489-1499.
12. Corey, D.R., Craik, C.S. (1992) *J. Am. Chem. Soc.* 114: 1784-1790.
13. Laemmli, U.K. (1970) *Nature* 227: 680-685.
14. Bradford, M.M. (1976) *Anal. Biochem.* 72: 248-254.
15. Hunkapiller, M.W., Hewick, R.M., Dreyer, W.J. and Hood, L.E. (1983) *Methods Enzymol.* 91: 399-413.
16. Bidlingmeyer, B.A., Cohen, S.A., Tarvin, T.L. (1984) *J. Chromatogr.* 336: 93-104.
17. Shulze, A.J., Huber, R., Bode, W. and Engh, R. (1994) *FEBS Lett.* 344: 117-124.
18. Potempa, J., Korzus, E. and Travis, J. (1994) *J. Biol. Chem.* 269: 15957-15960.
19. Cohen, A.B., Geczy, D. and James, H.L. (1978) *Biochemistry* 17: 392-400.
20. Matheson, N.R., van Halbeek, H. and Travis, J. (1991) *J. Biol. Chem.* 266: 13489-13491.
21. Perona, J.J., Hedstrom, L., Wagner, R., Rutter, W.J., Craik, C.S. and Fletterick, R.J. (1994) *Biochemistry* 33: 919-933.
22. Oda, K., Laskowski, M., Sr., Kress, L.P. and Kowalski, D. (1977) *Biochem. Biophys. Res. Commun.* 76: 1062-1070.
23. Cooperman, B.S., Stavridi, E., Nickbarg, E., Rescorla, E., Schechter, N.M. and Rubin, H. (1993) *J. Biol. Chem.* 268: 23616-23625.

10

STRUCTURES AND FUNCTIONS OF LEGINSULIN AND LEGINSULIN-BINDING PROTEIN

Hisashi Hirano

National Institute of Agrobiological Resources
Kannondai 2-1-2, Tsukuba, Ibaraki, 305, Japan

The involvement of insulin-like proteins in regulatory metabolic mechanism in plants and microorganisms has yet to be fully determined. The authors isolated a soybean seed glycoprotein, basic 7S globulin (Bg), which is capable of binding to bovine insulin and insulin-like growth factors(IGF)-I and II[1]. Bg showed similarity to the animal insulin receptor in protein structure[2], subcellular localization[3] and protein kinase activity[4]. Bg would thus appear to have insulin receptor-like functions.

Plant peptides like animal insulin which are capable of binding Bg were attempted to purify and a 4-kDa peptide, leginsulin could be isolated from radicles of germinated soybean seeds by affinity chromatography[5]. Leginsulin was found to bind to Bg and compete with insulin for this binding. Leginsulin is localized in cell walls and has stimulatory effect on the phosphorylation activity of Bg and thus may be involved in cellular signal transduction as animal insulin, though there is no amino acid sequence similarity between leginsulin and insulin.

The structures and functions of leginsulin and Bg as determined in this study are presented in the following.

STRUCTURE OF BASIC 7S GLOBULIN

Bg isolated from the soybean seed has a high isoelectric point (pH 9.05-9.26) and molecular mass of about 168-kDa[6]. It is comprised of four pairs of α (27-kDa) and β (16-kDa) subunits linked together by a disulfide bond. Bg separated from crude extracts of soybean seed by two-dimensional polyacrylamide gel electrophoresis (2D-PAGE) was transferred from a gel onto a polyvinylidene difluoride (PVDF) membrane. The electroblotted protein was sequenced directly with a gas-phase sequencer[7]. Bg separated by 2D-PAGE was partially digested with endopeptidases such as *Staphylococcus aureus* V8 protease and α-chymotrypsin by the method in[8]. The digests were separated by SDS gel electrophoresis, electroblotted onto PVDF membranes and sequenced[7]. Amino acid sequences suitable for construction of oligonucleotide probes to clone the DNA encoding Bg were obtained and used to identify a 1.4 kb cDNA encoding Bg from a soybean seed cDNA library[9].

A comparison of the amino acid sequence deduced from the cDNA sequence with the actual sequence determined showed that the cDNA contained open reading frames for a putative signal peptide followed by sequences for the Bg α and β subunits. It was thus considered from the results of DNA and protein sequencing that: (1) Bg is synthesized as a precursor polypeptide containing a putative signal peptide and Bg α and β subunits in ribosomes bound to the endoplasmic reticulum; (2) the putative signal peptide is co-translationally removed; (3) disulfide bonding post-translationally occured and (4) the precursor polypeptide is cleaved to generate α and β subunits and finally (5) Bg is co- or post-translationally glycosylated[7]. These considerations were confirmed by *in vivo* pulse chase labeling, using the developing soybean seeds[10].

Amino acid sequences of Bg subunits were compared with those of more than 16,000 proteins, recorded in the amino acid sequence database PRF-SEQDB (Protein Research Foundation). Conglutin γ, a lupin seed protein[11] was shown to possess a sequence homologous to the β subunit of Bg. The amino acid sequence of a carrot glycoprotein, GP-57 deduced from cDNA sequence has been shown highly homologous with that of Bg by sequence homology search using the SWISS-PROT database[12].

HEAT SHOCK INDUCTION OF BASIC 7S GLOBULIN

When immersed in water (50-60 °C), dry mature soybean seeds initially swell rapidly and thereafter release Bg protein enters into the surrounding water[13,14]. There is thus the question as to whether the released protein represents stored Bg or is a consequence of *de novo* synthesis. Although this question cannot be fully answered at present, it was suggested from the following findings that Bg released during hot water treatment, may possibly be produced *de novo*: (1) a polypeptide with the same molecular mass as the Bg precursor polypeptide and which cross-reacts with anti-Bg antibody has been detected and small increase in the amount of mature Bg protein has been observed during hot water treatment of soybean seeds; (2) by Northern hybridizations, Bg mRNA was found present in the seeds following immersion in hot water and (3) at three sites in the 5' non-coding region in Bg genomic DNA, the heat shock element consensus sequence was identified[15].

BASIC 7S GLOBULIN-LIKE PROTEINS IN PLANTS

Seed protein extracts from azuki-bean, cowpea, common-bean, lupin, mung-bean, pea and winged-bean were analyzed by Western blotting, using antiserum raised against purified soybean Bg α and β subunits[16]. Strong reaction was observed with the extracts from azuki-bean, cowpea, common-bean and mung-bean, though reaction with extracts prepared from winged-bean was much weaker. No cross-reaction could be detected with proteins isolated from pea. These results indicate proteins similar to Bg to be widely distributed in legume species[16].

When immersed in hot water, azuki-bean, cowpea, common-bean, mung-bean and winged-bean seeds also released proteins[14]. Analysis of the N-terminal amino acid sequences of the proteins indicated the major released proteins to have amino acid sequences highly homologous to those of soybean Bg subunits. All Bg-like proteins released during the hot water treatment reacted strongly with antiserum raised against soybean Bg protein[14].

The carrot Bg-like protein, GP-57 is released from cultured cells, into the medium during somatic embryo formation. GP57 mRNA in storage roots increases rapidly in response to wounding[12]. Thus, this protein may also present a stress-inducible protein.

Eight-kDa proteins highly homologous in the N-terminal sequence with ubiquitin are released from the common-bean, cowpea and mung-bean seeds and carrot cells, following stress treatment[14]. Ubiquitin is induced by stress and in eukaryotes, interaction of ubiquitins with proteins induced by stress may possibly serve to make them targets for degradation. Three species of genomic DNAs encoding soybean ubiquitin were cloned and the heat shock element-like sequences at three sites of all 5' non-coding regions were identified[17]. The release of ubiquitin from legume seeds and carrot cells may thus possibly be related to that of Bg-like proteins.

INSULIN-BINDING ACTIVITY OF BASIC 7S GLOBULIN

Following immobilization on PVDF membranes, soybean Bg protein released into hot water, was shown to bind [^{125}I]insulin and [^{125}I] IGF-I and -II[1]. Preliminary competitive binding assay, indicated the dissociation constants of Bg-insulin and Bg-IGF-I complexes to be about 15 nM and 60 nM, respectively[1]. Bg-like proteins released by hot water treatment of other legume species (azuki-bean, cowpea, lupin and mung-bean) as well as GP-57 protein from cultured callus cells have been shown capable of binding to insulin and IGF. Thus, all Bg-like proteins in plants may be concluded to possess insulin binding activity as well as IGF-I and -II binding activity.

A comparison of the structural characteristics of Bg protein with those of the insulin receptor indicated the following: (1) There are structural similarities in glycosylation, the presence of a cysteine-rich domain and a putative transmembrane domain and a disulfide-bound α-β subunit structure; (2) the insulin receptor is synthesized as a precursor polypeptide which is post-translationally cleaved on the N-terminal side of the serine residue to generate α and β subunits. These findings are consistent with the proposed post-translational processing scheme for Bg precursor polypeptide and (3) a consensus sequence of ATP-binding site indispensable for protein phosphorylation is present in Bg[15].

Bg has a protein kinase activity about two thirds of tyrosine kinase activity of rat insulin receptor[4]. Immunocytochemistry has indicated Bg to be localized in plasma membranes as is also the insulin receptor and middle lamellae of cell walls[3]. It follows then that Bg may thus have insulin receptor-like functions. Bg may thus possibly be involved in an insulin-like regulatory mechanism in plants.

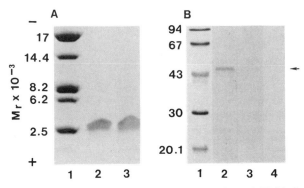

Figure 1. (A) SDS gel electrophoresis analysis of the purified leginsulin and (B) binding of radioiodinated leginsulin to the soybean insulin-binding protein in the presence or absence of bovine insulin probed by ligand blotting [5]. (A) Lane 1, molecular marker peptides; lane 2, non-reduced leginsulin; lane 3, leginsulin reduced by 2-mercaptoethanol. (B) Lane 1, molecular marker proteins; lane 2, no insulin added to the incubation buffer; lane 3, bovine insulin present at a concentration of 0.1 μM; lane 4, bovine insulin is present at a concentration of 10 μM. Arrow indicates the position of the blotted Bg.

ISOLATION AND STRUCTURE OF LEGINSULIN

The presence of insulin receptor-like protein suggests that hormonal peptides such as insulin which are capable of binding to Bg should be present in plants. To isolate insulin-like peptides, affinity chromatography using Sepharose CL-4B column immobilized Bg was conducted. A 4-kDa peptide named leginsulin was purified from the fractionated extract of soybean radicles[18]. Ligand blotting experiments using radioiodinated leginsulin confirmed the leginsulin to be capable of binding to Bg and compete with insulin for this binding (Fig. 1).

Leginsulin consists of 37 amino acid residues with half-cystines in three disulfide bridges. Electrospray ionization mass spectrometric analysis revealed that a portion of the peptide is processed to delete the C-terminal glycine. Among animal peptide hormones, about 50 % possess a C-terminal amide produced from the C-terminal glycine in non-activated form of the peptides. It was considered possible that the C-terminus of leginsulin was amidated like a number of animal peptide hormones. However, a C-terminal α-amide could not be detected in leginsulin by mass spectrometric analysis. A possibility exists that although the C-terminus is not amidated, the C-terminal processing is essential for producing the active form of leginsulin.

Based on amino acid sequences, cDNA clones encoding leginsulin were screened from a soybean seed cDNA library. Out of ten clones selected, one cDNA consisted of 545 bp and included an open reading frame of 357 bp containing regions encoding the putative signal peptide and mature leginsulin. The open reading frame from the first initiation codon to the termination codon encoded a polypeptide of 112 amino acids. Leginsulin may thus be considered to be translated as a larger precursor polypeptide in ribosomes bound to the endoplasmic reticulum.

FUNCTIONS OF LEGINSULIN

In the signal transduction of animal insulin, which is mediated by the insulin receptor, insulin first binds to the extracellular domain of the insulin receptor and then stimulates tyrosine kinase activity in the intracellular domain of the receptor for the transmission of the insulin signals into cells. Since Bg has protein kinase activity, should leginsulin have animal insulin-like functions, the protein kinase activity of Bg ought to be stimulated in the presence of leginsulin.

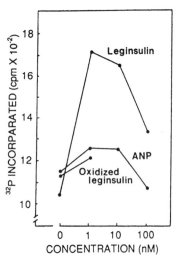

Figure 2. Stimulation of Bg phosphorylation activity with leginsulin. Phosphorylation activity was measured in a reaction mixture containing $[\gamma\text{-}^{32}P]ATP$, Bg and different concentrations of leginsulin, oxidized leginsulin or atrial natriuretic peptide (ANP). Results are the mean of three experiments.

Assessment was made of the ability of leginsulin to stimulate the phosphorylation activity of Bg *in vitro*[17]. Leginsulin was shown to significantly stimulate phosphorylation (Fig. 2). Maximum stimulatory effect was observed at relatively low leginsulin concentration (1nM), indicating the possible involvement of leginsulin and Bg in the cellular signal transduction mechanism in a manner similar to that of insulin and insulin receptor. When disulfide bridges of leginsulin were cleaved, no stimulatory effect on the phosphorylation of Bg was observed. The intradisulfide bonded-structure of leginsulin may possibly be required for such stimulation.

In general, amino acid sequences are conserved in proteins having similar functions. However, there are several reports describing that even though proteins have similar functions, their primary structures are not similar, but higher order structures of them are highly conserved, as noted for hexokinase, an ATPase fragment of heat shock protein HSP70 and actin[18]. Although there is no sequence identity between leginsulin and insulin, or between Bg and insulin receptor, these proteins have similar functions and possibly similar three-dimensional structures.

In preliminary experiment using cultured carrot callus, leginsulin significantly affected the re-differentiation of the callus; leginsulin induced shoot differentiation from the callus when present at relatively low concentration and the specific formation of embryos at relatively high concentration (unpublished data). Leginsulin thus appears likely involved in the regulation of growth and differentiation of plants, as in the case of animals.

In Western-blotting experiments using the anti-leginsulin antibody, leginsulin-like peptides were detected in plant species such as azuki-bean, mung-bean and carrot. The leginsulin/ leginsulin-binding protein (Bg) signal transduction system may thus be widely distributed in plant species.

ACKNOWLEDGMENTS

The author would like to thank S. Komatsu, H. Kagawa, Y. Watanabe, H. Kajiwara, N.-K. Nishizawa and S. Tsunasawa for their collaboration in this study.

REFERENCES

1. Komatsu, S. and Hirano, H. (1991) *FEBS Lett.* **294:** 210-212.
2. Hirano, H., Komatsu, S., Kajiwara, H. and Watanabe, Y. (1992) In: *Plant Tissue Culture and Gene Manipulation for Breeding and Formation of Phytochemicals* (Oono, K., ed.) pp.143-146, NIAR, Tsukuba.
3. Nishizawa, N,-K., Mori, S., Watanabe, Y. and Hirano, H. (1994) *Plant Cell Physiol.* **35:** 1079-1085.
4. Komatsu, S., Koshio, O. and Hirano, H. (1994) *Biosci. Biotech. Biochem.* **58:** 1705-1706.
5. Watanabe, Y., Barbashov, S. F., Komatsu, S., Hemmings, A. M., Miyagi, M., Tsunasawa, S. and Hirano, H. (1994) *Eur. J. Biochem.* **224:** 167-172.
6. Yamauchi, F., Sato,K. and Yamagishi, T. (1994) *Agric. Biol. Chem.* 645-650.
7. Hirano, H. and Watanabe, T. (1990) *Electrophoresis* **11:** 573-580.
8. Cleveland, D.W., Fischer, S.G., Kirschner, M.W. and Laemmli, U.K. (1977) *J. Biol. Chem.* **252:** 1102-1106.
9. Kagawa, H. and Hirano, H. (1989) *Nucl. Acids Res.* **17:** 8868.
10. Hirano, H. (1990) *Tissue Culture* **16:** 464-467.
11. Blagrove, R. J. and Gillespie, L. M. (1975) *Aust. J. Plant Physiol.* **2:** 13-27.
12. Satoh, S., Sturm, A., Fujii, T. and Chrispeels, M. J.(1992) *Planta* **188:** 432-438.
13. Asano, M., Okubo, K. and Yamauchi, F. (1989) *Nippon Shokuhin Kogaku Kaishi* **36:** 19-25.
14. Hirano, H., Kagawa, H. and Okubo, K. (1992) *Phytochemistry* **31:** 731-735.
15. Watanabe, Y. and Hirano, H. (1994) *Plant Physiol.* 1019-1020.

16. Kagawa, H., Yamauchi, F. and Hirano, H. (1987) *FEBS Lett.* **226:** 145-149.
17. Xa, B.-S., Waterhouse, R. N., Watanabe, Y., Kajiwara, H., Komatsu, S. and Hirano, H. (1994) *Plant Physiol.* **104:** 805-806.
18. Flaherty, K. M., Deluca-Flaherty, C. and McKay, D. B. (1990) *Nature* **346:** 623-628.

11

FROM STRUCTURE TO THERAPY?

L. N. Johnson

Laboratory of Molecular Biophysics
University of Oxford
Oxford, OX1 3QU, United Kingdom

ABSTRACT

Protein crystal structure determinations have made important contributions to understanding the molecular basis of recognition and function. In some selected examples, protein structures have been used to design new therapeutic agents. This paper summarises recent achievements and reports on progress in our long term study to provide potential agents for the treatment of diabetes. Structural studies with glycogen phosphorylase have led to the design of a series of glucose analogue inhibitors that can act as powerful regulators of liver glycogen metabolism. The best compound to date has involved some novel carbohydrate chemistry and exhibits a K_i that is 3 orders of magnitude lower than that of glucose. Physiological studies have shown that inhibitors more potent than glucose produce dramatic effects in inhibition of glycogen degradation and promotion of increased glucose utilisation and glycogen deposits in isolated hepatocytes.

INTRODUCTION

The earliest known drugs were those derived from natural products whose beneficial effects were detected from observation. The study of such natural products has long been a focus of research at the HEJ Research Institute of Chemistry, Karachi where many new and interesting molecules have been isolated from a varitey of plants and marine organisms. An example of a medicinal natural product is aspirin, from willow bark, which was used as an analgesic long before its pharmacological properties or molecular basis of action was understood. The conceptual breakthrough, developed in the nineteenth century, that a drug may have specific target at the cellular level led to the notion of receptor based drug design that has profoundly benefited the drug discovery process. The strategy in the design of new agents has been to modify the structure of a lead compound by systematic chemistry in conjunction with biological and physiological evaluation to produce a compound of the required potency. The most recent phase in drug design has utilised knowledge of the three dimensional structures of target molecules or of related compounds. From experimental observations at the atomic level of how inhibitors bind to their macromolecules, specific interactions that are important in molecular recognition can be inferred. This knowledge can

Table 1. A selection of proteins where structural knowledge has contributed to the design of new therapeutic agents and/or an understanding of the action of existing agents

Protein	Disease	Inhibitor / Drug
Haemoglobin	Sickle cell aneamia	Salicylaldehyde compounds, clofibric acid, bezafibrate
Dihydrofolate Reductase	Cancer	Methotextrate
Thymidylate Synthase	Cancer	Fluorouracil and 6,7-imidazotetrahydroquinoline compounds
Nucleoside Phosphorylase	Cancer	9-(arylmethyl) derivatives of 9-deazaguanine
Angiotensin Converting Enzyme	Hypertension	Captopril and enalopril
Renin	Hypertension	Peptide based transition state analogues
Human Leucocyte Elastase	Degenerative Lung Disease	α1-antitrypsin, fluoromethyl ketones and cephalosphorin analogues
Thrombin	Anticoagulant therapy	Hirudin and non-cleavable hirudin analogues; argatroban
Tissue Plasminogen Activator Inhibitor	Anticoagulant therapy	Engineered recombinant protein to prevent inhibition
Carbonic anhydrase	Occular hypertension, Glaucoma	Thienothiopyran-2-sulfonamides (MK507)
DNA Gyrase	Bacterial infection	Quinolones (Nalidixic acid), Coumarins (Novobiocin)
Adenosine Deaminase	Sever Combined Immune deficiency disease	Gene therapy
HIV Protease	AIDS	C2 symmetric diols based on 7 membered cyckic urea scaffold; Peptide based transition state analogues
HIV Reverse Transcriptase	AIDS	AZT, nevirapine
Human Rhino Virus	Common Cold	Sterling-Winthrop Pharmaceutical Research Division and Janssen Research Foundation compounds
Influenza Virus Neuraminidase	Influenza	Modified neuraminic acid analogue

be applied to lead compounds in *de novo* design as well as to improvement in existing leads and can provide insights into mechanisms of existing drugs. However although much attention has been focused on the potential of structure based drug design the number of compounds that have neared the market place is limited. The structure of prostaglandin synthase, the target for aspirin, is now known and this has provided a structural explanation for one aspect of aspirin's anti-inflammatory properties[1]. Can we utilise protein structures to design better drugs?

The design of tight inhibitors has been impressively accomplished but the more stringent requirements that a drug must fulfil in order to treat patients such as bioavailability, toxicity, metabolism, half life and cost effectiveness are more difficult to address. Structure based drug design has a long way to go before it can rival the products such as cortisones, anti-inflammatory agents, antibiotics, antidepressants and hypoglycaemic agents that were produced by more conventional methods. But it provides an alternative approach. It has been estimated[2] that on average in a trial and error procedure some 10000 compounds will be screened, 10 will go forward to trials and 1 may become a prescription medicine. That 7 out of 10 medicines do not recoup their expenses and that fewer than 5 earn more than $1B per year.

CURRENT ACHIEVEMENTS IN STRUCTURE BASED DRUG DESIGN

There has been a host of structural results on proteins of medical importance[3]. They include studies that contribute to understanding how existing agents work and which may lead to new compounds with improved properties (Table 1). Examples include work with acetyl cholinesterase for agents that block neurotransmission, with bacterial DNA gyrase for understanding the action of some antibiotics, with HIV reverse transcriptase for understanding the mechanism of AZT and other potential drugs, with thrombin and tissue plasminogen activator for regulation of blood coagulation, and with prostaglandin synthase for understanding the action of aspirin and other anti-inflammatory agents. Structural data have led to clues for improvement in properties of the anti-viral compounds against human rhino virus. The structure of the cyclophilin A-cyclosporin complex is relevant for understanding the immunosupressant activity of the drug cyclosporin. Knowledge of protein structure has led to new proteins produced by recombinant DNA technology such as a fast acting insulin for treatment of diabetes or "humanised" antibodies for the treatment of leukaemia and arthritis. Results on the human histocompatibility complex and associated bound peptides, of cytokines and a cytokine receptor complex, super antigens, cell surface adhesion molecules, and the outstanding work on the human growth hormone receptor complexed with the hormone and the tumour suppressor gene product p53 complexed with DNA have provided new insights into basic recognition events in a variety of biological responses.

Figure 1 shows 4 examples of compounds that have been designed on the basis of structure. Each of these compounds are in clinical trials. They include:

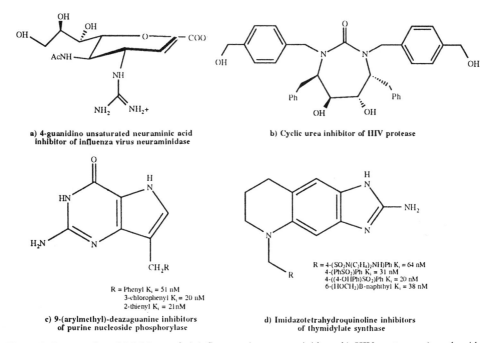

a) 4-guanidino unsaturated neuraminic acid inhibitor of influenza virus neuraminidase

b) Cyclic urea inhibitor of HIV protease

c) 9-(arylmethyl)-deazaguanine inhibitors of purine nucleoside phosphorylase

R = Phenyl K_i = 51 nM
3-chlorophenyl K_i = 20 nM
2-thienyl K_i = 21 nM

d) Imidazotetrahydroquinoline inhibitors of thymidylate synthase

R = 4-(SO$_2$N(C$_3$H$_6$)$_2$NH)Ph K_i = 64 nM
4-(PhSO$_2$)Ph K_i = 31 nM
4-((4-OHPh)SO$_2$)Ph K_i = 20 nM
6-(HOCH$_2$)B-naphthyl K_i = 38 nM

Figure 1. Structure based inhibitors of a) influenza virus neuraminidase; b) HIV protease; c) nucleoside phosphorylase and d) thymidylate synthase.

A) Influenza Virus Neuraminidase: inhibitors where knowledge of the structure of the neuraminidase complexed with a transition state compound and the computer programme GRID[4] were used to design a new compound[5]. The addition of a guanidino group to interact with 2 carboxylates resulted in a compound (Fig. 1a) that exhibited a $K_i = 2 \times 10^{-10}$ M , 4 orders of magnitude better than the starting compound and significant *in vivo* activity against influenza virus replication.

B) HIV Protease Inhibitors. In the search for new therapeutic compounds to combat AIDS the HIV protease is probably the most widely studied protein by X-ray methods with over 150 structure determinations of inhibited complexes reported. Recent work has concentrated on non-peptide based inhibitors that have better oral bioavailability. Mechanistic information was used in the design of an agent that exploited the 2-fold symmetry of the dimeric enzyme with a C2-symmetric diol in order to target the 2 aspartates, included the use of a cyclic urea carbonyl oxygen to mimic the hydrogen bond features of a key structural water molecule, and in which selectivity was engineered and restrained by a preorganised scaffold from knowledge of the binding site geometry[6]. The result was a series of relatively low molecular weight compounds with high oral bioavailability and good potency against the virus. Compound DMP 323 (Fig. 1b) has a $K_i = 2.7 \times 10^{-10}$ M.

C) Purine Nucleoside Phosphorylase. Interest in this enzyme as a drug target arises from two rather different properties. The enzyme metabolises purine nucleosides, including anticancer agents and anti-AIDS drugs and hence its inhibition might allow these drugs longer action. Secondly inhibitors may have application in the treatment of T-cell proliferative diseases. Starting with the observation that 5'-deoxy-5'-iodo-9-deazainosine was the most potent available inhibitor and that the N-9 position allowed substitution of groups. The design and synthesis of a number of 9-(arylmethyl) derivatives of 9-deazaguanine has been reported[7] (Fig. 1c). The improved potency of the 9 substituted compounds has been rationalised from X-ray analysis where subtle changes in hydrogen bonding patterns and location of the phenyl group between 2 phenylalanine side chains were observed.

D) Thymidylate Synthase. Thymidylate synthase is the limiting enzyme in the metabolic pathway for the *de novo* synthesis of thymidylate and as such is a target for drugs against cancer and other proliferative cell diseases. Using the E. coli structure of thymidylate synthase as a model for the human enzyme (70% identity in sequence), a series of 6,7-imidazotetrahydroquinoline inhibitors have been developed[8] (Fig 1d). Some of the resulting compounds were shown to effectively inhibit the growth of 3 tumour cell lines *in vitro* . Curiously it appeared that cell growth inhibition may not have been the primary consequence of thymidylate synthase inhibition. As the authors discuss, the information from a protein crystal structure has the potential of increasing selectivity for a given target but having a tight inhibitor does not necessarily result in the specific targeting of that enzyme in cells. Factors which are difficult to predict such as binding to other cellular proteins, metabolism or transport properties can render a potent inhibitor ineffective or possibly more effective in cell culture and *in vivo*.

GLYCOGEN PHOSPHORYLASE AND DIABETES

The structure of glycogen phosphorylase has been used to design glucose-analogue inhibitors that have increased potency compared to the parent compound glucose and which may prove beneficial in the regulation of glycogen metabolism in Type II diabetes. Diabetes mellitus is characterised by chronic elevated blood glucose levels. The disorder affects 2%

of the population in the Western world and 75% of this total is accounted for by the non-insulin dependent form of the disease (NIDDM or Type II diabetes). NIDDM is managed by diet, exercise, hypoglycaemic drugs, which are based on 3rd generation sulphonylureas that were originally developed in the 1930s as antibiotics, and, if these fail, by insulin therapy. The current drugs are not entirely satisfactory and there is a continued interest in new agents that can control blood glucose levels. Hyperglycaemia in NIDDM patients is a result of diminished insulin release and or insulin resistance that leads to impaired tissue glucose uptake and impaired suppression of the output of glucose from liver, even when blood glucose levels are already high[9].

A simplified scheme for the regulation of glycogen in liver is shown in Figure 2. Glycogen concentrations are regulated by the activities of glycogen phosphorylase (GP) and glycogen synthase (GS). Activation of GP by phosphorylation (GPb to GPa) is achieved by the action of a single enzyme, glycogen phosphorylase kinase, at a single site, Ser14, but the reciprocal inhibition of glycogen synthase (GSI to GSD) through phosphorylation is effected by at least 5 different kinases acting on multiple different serine sites. The reverse reactions of inactivation of phosphorylase and activation of glycogen synthase are achieved *in vivo* largely through the action of a single enzyme, protein phosphatase-1G (PP-1G). It is mainly through this enzyme that the co-ordination of glycogen breakdown and synthesis is achieved in response to glucose in liver[10,11] and in response to insulin in muscle[12]. In liver PP-1G is targeted to glycogen by a glycogen binding subunit and the PP1 catalysed dephosphorylation of glycogen synthase is inhibited allosterically by extremely low concentrations (2-20 nM) of the active form of phosphorylase, GPa[13,14]. GPa does not inhibit its own dephosphorylation by PP-1G and hence activation of GS is achieved after a lag period during which hepatic GPa is converted to its inactive form GPb by PP-1G and inhibition by GPa of PP-1G activity against GS is relieved.

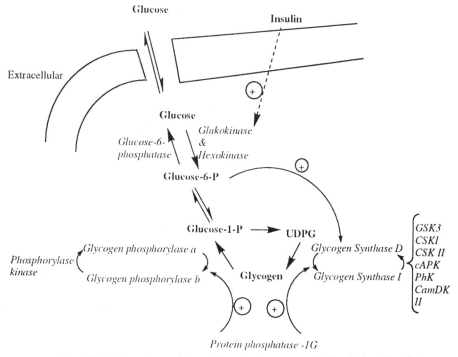

Figure 2. Simplified scheme for regulation of hepatic glycogen metabolism by insulin and glucose.

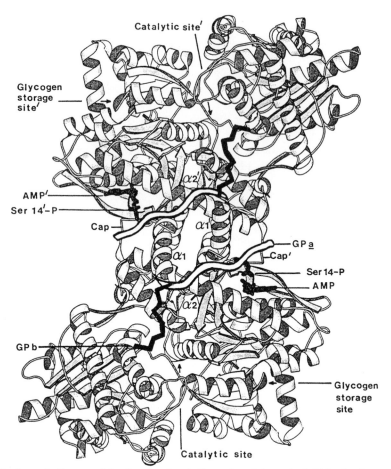

Figure 3. A schematic diagram of the phosphorylase b dimer viewed down the 2 fold axis of symmetry. The position of the N-terminal 20 residues in the non-phosphorylated phosphorylase b (GPb) are shown in black and their positions in the phosphorylated Phosphorylase a (GPa) are shown in white. The catalytic site is over 30 Å away from the Ser14-P site.

Glucose is able to augment these effects by competitive inhibition of GPa and by the promotion of the less active T state of GPa which is a better substrate for PP-1G than the active R state GPa. The demonstration that insulin augmented the glucose stimulation of GS and inhibition of GP[15] provided an important link between the roles of glucose and insulin in hepatic glycogen metabolism.

A rationalisation for the effects of glucose on GP has come from the X-ray crystal structure of the active and inactive forms of the rabbit muscle GP[16,17]. Glucose is an inhibitor that binds to the catalytic site in competition with substrate but also stabilises the T state (less active) form of the enzyme (in the nomenclature of Monod, Wyman and Changeux) by making specific interactions with a loop of chain (the 280s loop) that blocks access to the catalytic site.

In Figure 3 we show a schematic diagram of the phosphorylase dimer structure determined by X-ray diffraction studies in which the essential conformational change observed on the phosphorylation of serine residue 14 is illustrated[18-23]. n the non-phosphorylated T (inactive state) the N-terminal 20 residues are poorly ordered and make contacts

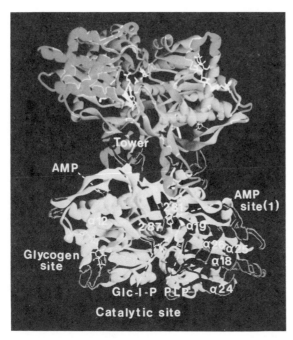

Figure 4. A colour diagram of T state GPb viewed down the 2-fold axis of the dimer (180° from the view shown in Figure 3). One subunit is coloured blue and the other purple. Regions of the molecule that changes in conformation between the T and R states are shown in red (lower subunit) and yellow (upper subunit). The binding sites for substrate glucose-1-phosphate (Glc-1-P) at the catalytic site, AMP at the allosteric site and maltopentose at the glycogen binding site are shown.

within their own subunit. On phosphorylation to the R (active state) there is a dramatic change and the N-terminal 20 residues swing up to contact the other subunit of the dimer. The Ser-phosphate is stabilised by interactions with 2 arginine residues one from each of the subunits. These changes lead via the subunit-subunit interactions to changes at the catalytic site which is over 30 Å away from the Ser-14-P site.

In Figure 4 we show a view of the phosphorylase b molecule in the T state where the view is 180° from that in Figure 3. Here we can look into the catalytic site, which is marked by the binding of the substrate glucose-1-phosphate (Glc-1-P), and which is buried about 15 Å below the surface. Access to the site is blocked by the loop of chain termed the 280s loop, residues 281-287. On the T to R transition this loop is displaced and becomes disordered.

Movement of the loop appears to be correlated with shifts in the packing of the tower-tower helices and these changes at the subunit-subunit interface are communicated to the Ser-14-P site which is also at the subunit interface. Hence there is indirect allosteric communication between the Ser-P site and the catalytic site.

These results have suggested that a more powerful T state inhibitor of GPa than glucose itself, might be of interest in regulation of glycogen metabolism and may provide leads for compounds that could alleviate hyperglycaemia for treatment of Type II diabetes[24]. A systematic analysis of glucose analogue inhibitors has been carried out using knowledge of the T state structure of rabbit muscle GPb as a model and involving the design and organic synthesis of novel carbohydrate compounds[24-28]. A summary of some of these compounds and their K_i values is given in Figure 5.

One of the most effective compounds discovered early on is an N-linked C1 derivative of β-D-glucose (N-acetyl-β-D-glucopyranosylamine (1-GlcNAc)) compound 6

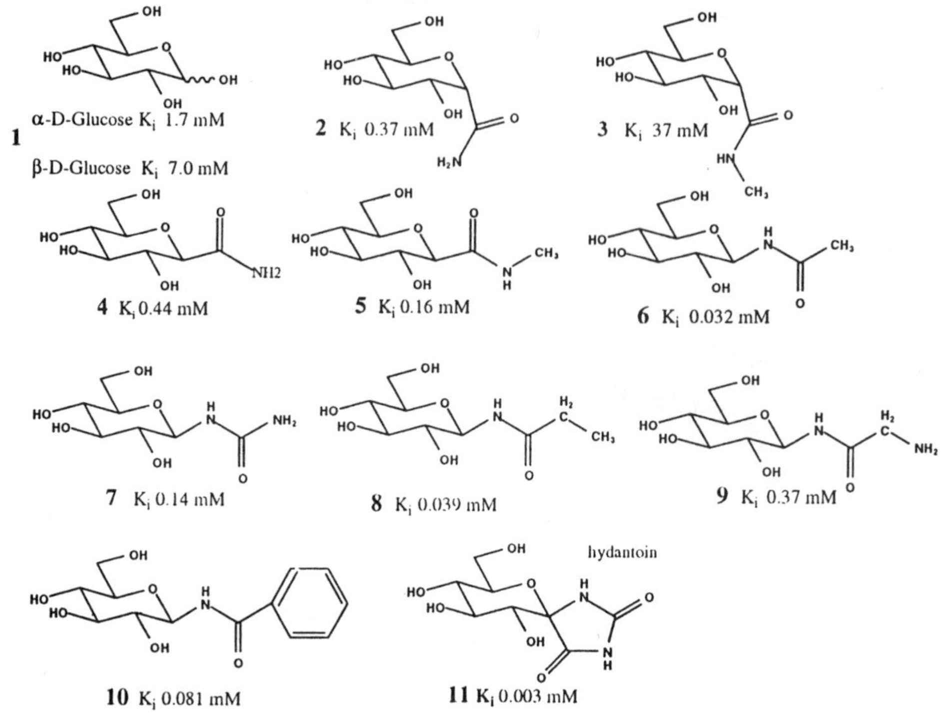

Figure 5. Glucose analogue inhibitors of glycogen phosphorylase and their K_i values.

(Fig. 5). 6 exhibits a K_i for rabbit muscle GPb of 32 μM, a value which is 200 fold lower than the corresponding K_i of 7 mM for β-D-glucose. In recent work the glucopyranose analogue of hydantocidin, compound 11, has been found to be an even better inhibitor with a K_i 3 μM, 10^3 times better than the parent compound. This compound was discovered following interest in hydantocidin, a furanose based spirohydantoin. This naturally occurring compound has promise as a herbicide with very little evidence of toxicity to mammals. The corresponding glucopyranose analogue of hydantocidin was modelled into the catalytic site of GPb and found to exploit additional hydrogen bonds to the protein to those made by the parent compound glucose. Accordingly the 2 epimeric spirohydantoins of glucopyranose were synthesised and tested with glycogen phosphorylase to reveal the first specific enzyme inhibition by a spirohydantoin at the anomeric position of the sugar[27]. Here we provide a summary of the design and rational for the biochemical activities of the different glucose analogues that led to the synthesis of this potent GP inhibitor.

Crystallographic Studies

In the structural studies, crystal of T state GPb were soaked in solutions containing 100 mM or 50 mM of the compound of interest for 1-2 h and X-ray diffraction data to 2.4 Å resolution collected with typical merging R values of 7-8 %. The structures of the complexes were refined with XPLOR to crystallographic R values of 17-20 % and an estimated error in coordinates of about 0.2 Å, rms deviation from ideal bond lengths of 0.017 Å and angles 3.5°. Examination of the structures with reference to the glucose complex revealed the following observations on the correlation between structure and activity.

(i) Binding of the Lead Compound. α-D-glucose (1) binds with each of its peripheral hydroxyl groups involved in hydrogen bonds both as donors and as acceptors and there is little scope for modification at these sites (Fig. 6a). There is however a deep pocket adjacent to the C1 atom in the β configuration and a smaller pocket partially blocked by water molecules adjacent to the α configuration. The hydrogen bonds from O2 and a α-O1 through water to Asn 284 and Asp 283, respectively, are important for stabilisation of the T state structure[17,24]. Despite these specific polar interactions, glucose binding is relatively weak probably because of few strong van der Waals interactions (there are no interactions with aromatic groups) and the energy cost of transferring a polar molecule from bulk solvent to the catalytic site.

(ii) Additional Hydrogen Bonds through Water Molecules from Ligand to Protein Are Important. The α-heptonamide 2 (K_i 0.37 mM) bound 5 times better than glucose and exploited hydrogen bonds through water molecules from its CO and NH groups to Asp283[26].

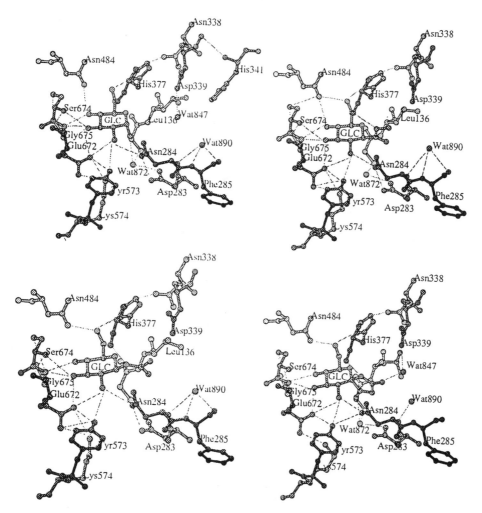

Figure 6. Details of the interactions of glucose analogue inhibitors at the catalytic site of T state glycogen phosphorylase b. a) 1 α-D-glucose; b) 5 N-methyl-β-heptonamide; c) 6 β-N-acetyl glucopyranosylamine; d) 11 the spirohydantoin glucopyranose.

(iii) The Conformation of the Ligand Is Important. Attempts to improve on 2 with an additional methyl group (compound 3) led to a substantial reduction in affinity (K_i 37 mM). Single crystal studies revealed 3 adopted a skew boat conformation which, when bound to GP, made fewer favourable hydrogen bonds than the corresponding glucopyranose chair. Additional atoms can cause unexpected effects[29].

(iv) Similar K_i Values Can Be Generated by Quite Dissimilar Contacts. Compound 4, the β-C-amide compound of the β-glucoheptonic acid series had a K_i 0.44 mM similar to 2, the α-C-amide. But in the complex with 4 the stabilising interactions were not to water but from the amide N to the main chain carbonyl oxygen of His377 and from the CO to Asn 284 side chain[26].

(v) Displacement of Water Molecules Can Prove Advantageous. Compound 5, the methyl analogue of 4, bound with a K_i 0.12 mM in a similar position to compound 4. There was no hydrogen bond to Asn 284 but the water molecule, Wat OH4 847, was displaced by the methyl group (Fig. 6b). The additional van der Waals contacts to the methyl and the displacement of the water contribute about 0.6 kcal/mol to the binding energy[26].

(vi) Linear Hydrogen Bonds and Displacement Of More Water Molecules Are Even Better. Compound 6, the β-N-acetyl glucopyranosylamine, (1-GlcNAc), had a K_i 0.032 mM, two orders of magnitude tighter than the corresponding parent β-D-glucose compound and one order of magnitude tighter than the corresponding glucoheptonic acid 5. The reversal of the amide functionality (6 vs 5) led to a shorter, more linear hydrogen bond from the amide nitrogen to the main chain carbonyl oxygen of His 377 (Fig. 6c). In addition 2 waters, OH4 847 and OH8 872, were displaced and their displacement appears to be correlated with the overall slight tightening of the site in which shortening of the hydrogen bonds of the peripheral hydroxyl group on the sugar to the enzyme are noted[25,28].

(vii) Replacement of a Non-Polar Group with a Polar Group Is Not Always Advantageous Even in a Polar Environment. The methyl group of 6 is 2.8 Å from the position of the water OH4 847 which it displaces. This suggested that a polar group in this position might hydrogen bond to this water and create additional binding energy. Compound 7 has an NH_2 group in place of the methyl but exhibited a poorer K_i of 0.14 mM. The structure of 7 complexed with GPb showed that in order to make the hydrogen bond to the water, the main hydrogen bonding contact to CO His 377 had been lengthened beyond 3.3 Å. The poorer K_i could be accounted for by loss of this hydrogen bond which is not compensated by the hydrogen bond to water. Evidently it is more favourable to displace this water with a non-polar group in the vicinity than to exploit its hydrogen bonding potential[28].

(viii) The Change from Methyl to Ethyl Can Be Neutral. Compound 8 has an ethyl group in place of the methyl of 6 but exhibited a comparable K_i. Although there may be some favourable energy change from partially shielding the ethyl group on transfer from solvent to the protein, the van der Waals interactions do not provide any substantial gain over those made by the methyl group[28].

(ix) Desolvation Effects Can Be Important for Polar Substituents. Compound 9, the N-glycinyl glucopyranosylamine, has an NH_2 group in place of the terminal methyl of 8, the N-ethyl acetyl glucopyranosylamine. 9 exhibits a higher K_i (0.37 mM vs 0.039 mM). The decrease in affinity may be attributed to a poorer hydrogen bond contact to CO His 377 which is partially compensated by a hydrogen bond OH0 Wat 887 from the carbonyl oxygen

and to the energy needed to desolvate 9 on to transfer to the catalytic site. OH4 Wat 847 is displaced but the NH2 group is not able to hydrogen bond directly to Asp339[28].

(x) Larger Groups Are Not Always Advantageous. The complex with N-benzylacetyl glucopyranosylamine 10 shows that the phenyl ring is accommodated in the same region as the CH_2-CH_3 moiety of compound 8 but in order to accommodate the ring the contact to CO His 377 has slightly lengthened[28]. This compound exhibited a poorer Ki 0.081 mM than the corresponding methyl 6 or ethyl 8 compounds.

(xi) Rigid Groups That Are Able to Exploit Several Hydrogen Bonds Are Favourable. Compound 11, the spirohydantoin, is the best inhibitor to date (Ki 0.003 mM[27]. The NH is able to make the hydrogen bond to CO His377 without distortion (Fig. 6d). The CO group hydrogen bonds through water to Asp339 and the α-CO group hydrogen bonds through Wat 872 to Asp 283 (as in the complex with 2). The other NH group does not make a direct hydrogen bond but is in a favourable electrostatic environment just 4 Å from Asp 283. These extra hydrogen bonds through water appear to compensate the otherwise favourable entropy gain when they are displaced as observed in the complex with 6. Further we suspect there may be a contribution from the rigid substituent groups, as has been observed in other complexes[24] although this has not yet been tested.

Physiology

We have recently explored whether inhibitors, that have been developed on the basis of the rabbit muscle GPb structure, are effective regulators of liver glycogen metabolism. At the time only 6 was available in sufficient quantities. Previous work has shown that 6 (1-GlcNAc) is a competitive inhibitor of both liver and muscle isozymes of GP and is indeed considerably more effective than glucose[30]. In intact hepatocytes 1-GlcNAc has been shown to be an effective regulator of liver GP producing substantial inhibition. At 1 mM concentration 1-GlcNAc promotes activation of liver protein phosphatase by 600 % whereas glucose at 50 mM concentration produces only a 200% enhancement. These results are fully in accord with the regulatory role of glucose, as discussed above and indicate that the glucose analogue inhibitor is considerably more effective than glucose (Board *et al.*, in preparation). The effects on GS are more complex. In gel-filtered liver extracts (where ATP has been removed) 1-GlcNAc leads to activation of GS but in intact hepatocytes there was no direct activation of GS. There is evidence to support the notion that in intact hepatocytes, 1-GlcNAc is metabolised by glucokinase to 1-GlcNAc-6-phosphate and that this compound interferes with GS activation. Despite the lack of activation of GS in intact hepatocytes, subsequent work has shown that glycogen deposition is enhanced and glucose uptake stimulated by the 1-GlcNAc inhibition of glycogen degradation in a dose dependent manner (Board & Johnson, in preparation). These effects on isolated hepatocytes indicate a potential hypoglycaemic action and a positive role for the analogues in the treatment of Type II diabetes. Experiments are underway to assess the effects of 1-GlcNAc on glycogen metabolism *in vivo*.

ACKNOWLEDGMENTS

This work has been supported by the MRC, the Oxford Centre for Molecular Sciences and by the EC contract number B 102 CT94 3025. I wish to acknowledge contributions by my colleagues whose names appear on the original papers.

REFERENCES

1. Picot, D., Loll, P. J. and Garavito, M. (1994) *Nature* 367: 243-249.
2. Vagelos, P. R. (1991) *Science* 252: 1080-1084.
3. Perutz, M. F. (1992). *Protein Structure: New approaches to disease and therapy*. W. H. Freeman, New York.
4. Goodford, P. J. (1985) *J. Med. Chem.* 28: 849-857.
5. Von Itzstein, M., Wu, W.-Y., Kok, G. B., Pegg, M. S., Dyasson, J. C., Jin, B., Phan, T. V., Smythe, M. L., White, H. F., Oliver, S. W., Colman, P. M., Varghese, J. N., Ryan, D. M., Woods, J. M., Bethell, R. C., Hothmam, V. J., Cameron, J. M. and Penn, C. R. (1993) *Nature* 363: 418-423.
6. Lam, P. Y. S., Jadhav, P. K., Eyermann, C. J., Hodge, C. N., Ru, Y., Bacheler, L. Y., Meek, J. L., Otto, M. J., Rayner, M. M.]., Wong, Y., N,, Chang, C.-W., Weber, P. C., Jackson, D. A., Sharpe, T. R. and Erickson-Viitanen, S. (1994) *Science* 263: 380-384.
7. Montgomery, J. A., Niwas, S., Rose, J. D., Secrist, J. A., Babu, Y. S., Bugg, C. E., Erion, M. D., Guida, W. C. and Ealick, S. E. (1993) *J. Med. Chem.* 36: 55-69.
8. Reich, S. H., Fuhry, M. A. M., Nguyen, D., Pino, M. J., Welsh, K. M., Webber, S., Janson, C. A., Jordan, S. R., Matthews, D. A., Smith, W. W., Bartlett, C. A., Booth, C. L. J., Herrmann, S. M., Howland, E. F., Morse, C. A., Ward, R. W. and White, J. (1992) *J. Med. Chem.* 35: 847-858.
9. DeFronzo, R. A. (1988) *Diabetes* 37: 667-687.
10. Hers, H. G. (1976) *Ann. Rev. Biochem.* 45: 167-189.
11. Stalmans, W. (1976) *Curr. Topics Cell Reg.* 11: 51-97.
12. Dent, P., Lavoinne, A., Nakielny, S., Caudwell, F. B., Watt, P. and Cohen, P. (1990) *Nature* 348: 302-308.
13. Allemany, S. and Cohen, P. (1986) *FEBS Lett.* 198: 194-.
14. Wera, S., Bollen, M. and Stalmans, W. (1991) *J. Biol. Chem.* 266: 339.
15. Witters, L. A. and Avruch, J. (1978) *Biochemistry* 17: 406-410.
16. Sprang, S. R., Goldsmith, E. J., Fletterick, R. J., Withers, S. G. and Madsen, N. B. (1982) *Biochemistry* 21: 5364-5371.
17. Martin, J. L., Withers, S. G. and Johnson, L. N. (1990) *Biochemistry* 29: 10745-10757.
18. Barford, D. and Johnson, L. N. (1989) *Nature* 340: 609-614.
19. Acharya, K. R., Stuart, D. I., Varvill, K. M. and Johnson, L. N. (1991). *Glycogen phosphorylase: Description of the protein structure*. World Scientific, London and Singapore.
20. Barford, D., Hu, S.-H. and Johnson, L. N. (1991) *J. Mol. Biol.* 218: 233-260.
21. Johnson, L. N. (1992) *FASEB J.* 6: 2274-2282.
22. Johnson, L. N. and Barford, D. (1993) *Annu. Rev. Biophys. Biomol. Struct.* 22: 199-232.
23. Johnson, L. N. and Barford, D. (1994) *Protein Sci.* 3: 1726-1730.
24. Martin, J. L., Veluraja, K., Ross, K., Johnson, L. N., Fleet, G. W. J., Ramsden, N. G., Bruce, I., Orchard, M. G., Oikonomakos, N. G., Papageorgiou, A. C., Leonidas, D. D. and Tsitoura, H. S. (1991) *Biochemistry* 30: 10101-10116.
25. Johnson, L. N., Watson, K. A., Mitchell, E. P., Fleet, G. W. J., Son, J. C., Bichard, C. J. F., Oikonomakos, N. G., Papageorgiou, A. C. and Leonidas, D. D. (1994) *Complex carbohydrates in drug research*: (Bock, K. and Claussen, H. eds.) pp. 214-226, Copenhagen, Munksgaard.
26. Watson, K. A., Mitchell, E. P., Johnson, L. N., Son, J. C., Bichard, C. J. F., Orchard, M. G., Fleet, G. W. J., Oikonomakos, N. G., Leonidas, D. D., Kontou, M. and Papageorgiou, A. C. (1994) *Biochemistry* 33: 5745-5758.
27. Bichard, C. J. F., Mitchell, E. P., Wormald, M. R., Watson, K. A., Johnson, L. N., Zographos, S. E., Koutra, D. D., Oikonomakos, N. G. and Fleet, G. W. J. (1995) *Tetrahedron Letts*. In press,
28. Watson, K. A., Mitchell, E. P., Johnson, L. N., Crucianni, G., Son, J. C., Bichard, C. J. F., Fleet, G. W. J., Oikonomkos, N. G., Leonidas, D. D. and Kontou, M. (1995) *Acta Cryst*. D In press,
29. Watson, K. A., Mitchell, E. P., Johnson, L. N., Son, Y., Fleet, G. W. J. and Oikonomakos, N. G. (1993) *J. Chem. Soc. Chem. Commun.* 654-656.
30. Board, M., Hadwen, M. and Johnson, L. N. (1995). *Eur. J. Biochem*. In press.

12

NEW INSIGHTS INTO THE MOLECULAR STRUCTURE OF THE AGONISTS BINDING SITE OF PROTEIN KINASE C BY PSEUDORECEPTOR MODELING

G. Krauter,[1] C. W. von der Lieth,[2] and E. Hecker[*1]

[1] Research Program 3: Risk Factors of Cancer and Cancer Prevention
[2] Central Spektroscopy Institute
German Cancer Research Centre
Im Neuenheimer Feld 280, W-69120 Heidelberg, Germany

ABSTRACT

Protein kinase C (PKC) comprises a family of isoenzymes with serine/threonine kinase activity similar in molecular size, structure and mechanism of activation. This enzyme family is an important part of one of the major signal transduction pathways regulating many intracellular processes such as modulation of gene expression, cell proliferation and differentiation. PKC was identified as the major cellular receptor for skin tumor promoting phorbol esters and certain other compounds exhibiting skin tumor promoting bioactivity. Moreover, PKC is a target for molecules with antineoplastic activity like bryostatin. Thus by various investigators it is considered as a potential target for discovery and development of new anticancer drugs. On the other hand for PKC so far no potent and isoenzyme selective activators or inhibitors are known, and a 3-D structure of the enzyme is not available. Therefore a new technique of molecular modeling was investigated to design the agonist binding site of PKC: 'Pseudoreceptor modeling'. It is a comprehensive strategy in the design of potent, selective and novel ligands for unknown receptors (Vedani *et al.,* 1993). The new approach is focused on the binding site of the receptor and allows the construction of hypothetical 3-D-models of binding pockets using the directionality of receptor-ligand interactions (hydrogen bonds, metal-ligand interaction, hydrophobic interactions). In the investigation described an ensemble of six ligands, binding all specifically to the regulatory domain of the enzyme and representing together a pharmacophore model for activation of PKC, was successfully used. The pseudoreceptor model of PKC will be presented. It may be used as a surrogate of the agonist binding site of PKC to guide the synthesis of new and selective PKC-agonists allowing further investigation of PKC-isoforms in the signal

---

[*] Represents person presenting Paper

transduction pathway. In addition it may be of assistance in the development for new antineoplastic drugs.

INTRODUCTION

In the search for novel cellular targets for the treatment of cancer it appears a sophisticated novel strategy to intervene with the components of cellular signalling systems that are known or assumed to be altered in malignant cells. An ultimate goal of such intervention is to restore the normal control mechanisms rather than to simply kill the proliferating cancer cells by trying to block more or less selectively their DNA synthesis. Mostly the essential components of cellular signal transduction pathways are proteins with receptor functions. The first step of an innovative medicinal chemistry therefore will be to understand the role of these proteins in signalling cascades and more specifically to become knowledgeable of the molecular 3-D-structure of the receptor proteins and/or their ligands. An obvious approach to new classes of cancer therapeutic agents may be to identify agonists or antagonists of appropriate receptor proteins.

In recent years, an attractive target for investigations of this kind has become the Protein Kinase C (PKC) family[1-3]. PKC is one of the key elements in signal transduction pathways involved in regulation of cell growth, cell differentiation and tumor promotion[4-6]. The molecular heterogeneity of the family, the distinct biochemical properties of individual isoenzymes and their anticipated functional divergence makes them to appear attractive for anticancer drug development. Yet no X-ray crystallographic data have been reported for anyone member of this enzyme family. In the absence of such data studies of the molecular mechanism of their activation, a prerequisite for a rational design of modulators of their activity to be used as new drugs, have been impeded.

In this situation an entirely new approach of molecular modeling called 'pseudoreceptor modeling' was applied. It aims at unravelling the molecular structures or features of PKC that are involved in agonist binding of the receptor. In this way it might be possible to gain at the molecular level new insights into what may be called the agonist binding site of PKC as well as into the structure of new agonists and antagonists.

PROTEIN KINASE C (PKC)

Activation of Protein Kinase C

PKC was originally identified by Nishizuka and co-workers as a cytoplasmic calcium-activated, phospholipid-dependent serine/threonine kinase from rat brain[7-8]. Somewhat later the previously postulated receptor of diterpene ester (DTE) type tumor promoters such as TPA[9] was detected to be identical with PKC[6,10].

Today PKC is known as a ubiquitous protein family of ten isoenzymes. All family members are similar in size, structure and mechanism of activation. Physiologically the enzyme is activated upon stimulation of cells by various external messengers including hormones, neurotransmitters and growth factors. The external signals induce the phospholipase catalysed hydrolysis of membrane inositol phospholipids, generating among other products, phosphatidylinositol-4,5-bisphosphate [PtdIns(4,5)P$_2$] and sn-1,2-diacylglycerol (DAG, 11). Sn-1,2-diacylglycerol (DAG) activates PKC directly as an agonist and [PtdIns(4,5)P$_2$] mediates calcium mobilisation, which in a synergistic fashion may also activate PKC. DAG may be generated also alternatively from phosphatidylcholine via activation of phospholipase C or phospholipase D to yield phosphatic acid which is cleaved to sn-1,2-diacylglycerol by

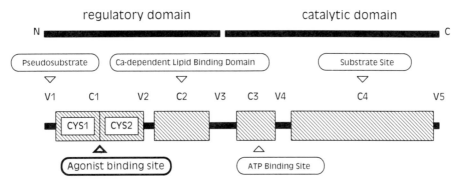

Figure 1. Structural elements of Protein Kinase C (PKC) and their functionality C1-C4: conserved regions of PKC, V1-V5: variable regions of PKC.

phosphatic acid phosphohydrolase[12]. The different ways of mobilisation of endogenous agonists (DAG + Ca^{2+} or DAG without Ca^{2+}) in conjunction with the different molecular species of PKC may allow to account a variety of cellular effects.[4]

Structural Aspects of the PKC Family

Molecular cloning of PKC so far has releaved the existence of a multigene family coding for ten isoenzymes.[13] The PKC isoenzymes may be categorised in three subgroups, the 'conventional' PKC(cPKC), the 'novel' PKC (nPKC), and the 'atypical' PKC (aPKC, Table I). All of them bind DAG, whereas binding of TPA and requirement of CA2+ is typical for most of the isoenzymes. Generally the structure of PKC is considered to be composed of two main domains with four conserved (C1-C4) and five variable regions (V1-V5; see Fig. 1). The approximately 30-kD N-terminal domain contains a tandem repeat of a cystein-rich zinc finger like motif (C1-region) responsible for binding PKC-agonists. In cPKC but not in nPKC this domain has in addition a Ca^{2+} binding domain (C2-region). The approximately 50-kD C-terminal contains the catalytic subunit which binds ATP (C3-region) and the substrate (C4-region, Fig. 1).

Sequence Alignment of the Cystein Rich Zinc Finger-Like Motif (C1-Region) of Various PKC Isoenzymes

The conserved cystein rich zinc finger like motif in the C1-region of PKC is known as the binding domain of the endogenous activator DAG, of the exogenous skin tumor promoting DTE and of several structures of non-tumor promoters with antineoplastic

Table 1. The ten presently known isoenzymes of PKC, in the three subgroups and some of their characteristics

PKC- subgroups	Isoenzymes	Characteristics		
		DGA-binding capability	TPA-binding capability	Ca^{2+} requirement
'conventional' PKC (c-PKC)	α, βI, βII, γ	+	+	+
'novel' PKC (n-PKC)	δ, ϵ, η, θ	+	+	-
'atypical' PKC (a-PKC)	λ, ζ	+	-	-

```
PKC-α  (Ra)    37 HKFIARFFKQPTFCSHCTDFI-WG-FGKQGFQCQVCCFVVHKRCHEFVTFSCPGA  89
PKC-βI (Ra)    37 HKFTARFFKQPTFCSHCTDFI-WG-FGKQGFQCQVCCFVVHKRCHEFVTFSCPGA  89
PKC-γ  (Ra)    36 HKFTARFFKQPTFCSHCTDFI-WG-IGKQGLQCQVCSFVVHRRCHEFVTFECPGA  88
PKC-δ  (Ra)   159 HEFIATFFGQPTFCSVCKEFV-WG-LNKQGYKCRQCNFAIHKKCIDKIIGRCTGT 211
PKC-ε  (Ra)   170 HKFMATYLRQPTYCSHCRDFI-WGVIGKQGYQCQVCTCVVHKRCHELIITKCAGL 223
PKC-ζ  (Ra)    31 HLFQAKRFNRRAYCGQCSERI-WG-LARQGYRCINCKLLVHKRCHVLVPLTCRHH  83
PKC-η  (Ma)   172 HKFMATYLRQPTYCSHCREFI-WGVFGKQGYQCQVCTCVVHKRCHHLIVTACTCQ 225
PKC-α  (Ha)    37 HKFIARFFKQPTFCSHCTDFI-WG-FGKQGFQCQVCCFVVHKRCHEFVTFSCPGA  89
PKC-L  (Me)   171 HKFMATYLRQPTYCSHCREFI-WGVFGKQGYQCQVCTCVVHKRCHHLIVTACTCQ 224
PKC-β  (Me)    37 HKFTARFFKQPTFCSHCTDFI-WG-FGKQGFQCQVCCFVVHKRCHEFVTFSCPGA  89
PKC-γ  (Me)    36 HKFTARFFKQPTFCSHCTDFI-WG-IGKQGLQCQVCSFVVHRRCHEFVTFECPGA  88
tpa-1          19 HQFVATFFRQPHFCSLCSDFM-WG-LNKQGYQCQLCSAAVHKKCHEKVIMQCPGS  71
X-PKCII        34 HKFTARFFKQPTFCSHCTDFI-WG-FGKQGFQCQVCCFVVHKRCHEFVTFSCPGA  86
dPKC 98f       72 HKFMATFLRQPTFCSHCREFI-WG-IGKGYQCQVCTLVVHKKCHLSVVSKCPGM 124
UNC-13        615 HNFATTTFQTPTFCYECEGLL-WG-LARQGLRCTQCQVKVHDKCRELLSADCLQR 667
n-Chimaerin    46 HNFKVHTFRGPHWCEYCANFM-WG-LIAQGVKCADCGINVHKQCSKMVPNDCKPD  98
                * *        * * *      *      *   *   *    * *     *
```

```
Konsensus      H-F-------P--C--C------G----QG--C--C----H--C-------C---
Sequenz
```

```
                * *        * * *      *      *   *   *    * *        *
PKC-α  (Ra)   102 HKFKIHTYGSPTFCDHCGSLL-YG-LIHQGMKCDTCDMNVHKQCVINVPSLCGMD 154
PKC-βI (Ra)   102 HKFKIHTYSSPTFCDHCGSLL-YG-LIHQGMKCDTCMMNVHKRCVMNVPSLCGTD 154
PKC-γ  (Ra)   101 HKFRLHSYSSPTFCDHCGSLL-YG-LVHQGMKCSCCEMNVHRRCVRSVPSLCGVD 153
PKC-δ  (Ra)   231 HRFKVYNYMSPTFCDHCGTLL-WG-LVKQGLKCEDCGMNVHHKCREKVANLCGIN 283
PKC-ε  (Ra)   243 HKFGIHNYKVPTFCDHCGSLL-WG-LLRQGLQCKVCKMNVHRRCETNVAPNCGVD 295
PKC-ζ  (Ra)     -----------------------------------------------------
PKC-η  (Ma)   246 HKFNVHNYKVPTFCDHCGSLL-WG-IMRQGLQCKICKMNVHIRCQANVAPNCGVN 298
PKC-α  (Ha)   102 HKFKIHTYGSPTFCDHCGSLL-YG-LIHQGMKCDTCDMNVHKQCVINVPSLCGMD 154
PKC-L  (Me)   245 HKFSIHNYKVPTFCDHCGSLL-WG-IMRQGLQCKYVNECAYSMSSERG-PNCGVM 296
PKC-β  (Me)   102 HKFKIHTYSSPTFCDHCGSLL-YG-LVHQGMKCSCCEMNVHRRCVRSPSLCGVD 154
PKC-γ  (Me)   101 HKFRLHSYSSPTFCDHCGSLL-YG-LVHQGMKCSCCEMNVHRRCVRSVPSLCGVD 153
tpa-1          91 HRFKTYNFKSPTFCDHCGSML-YG-LFKQGLRCEVCNVACHHKCERLMSNLCGVN 143
X-PKCII       100 HKFRIHTYSSPTFCDHCGSLL-YG-LIHQGMKCETCMMNVHKRCVMNVPSLCGTD 152
dPKC 98f      147 HRFVVHSYKRFTFCDHCGSLL-YG-LIKQGLQCETCGMNVHKRCQKNVANTCGIN 199
UNC-13          -----------------------------------------------------
N-Chimaerin     -----------------------------------------------------
```

Chart 1. Sequence alignment of the two cystein-rich domains of PKC. The numbering to the left and right of the sequences correspond to the amino acid number predicted from cDNA sequences. The following Alignment data are used[34-43]: PKC-Isoenzyme of rat (Ra) Ono *et al.*, 1987; mouse-PKC-η (Ma) Osada *et al.*, 1990; rabbit-PKC-α (Ha) Ohno *et al.*, 1987; -human-PKC-L (Me) Bacher *et al.*, 1991; human-PKC-β,γ (Me) Coussen *et al.*, 1986; tpa-1 Caenorhabditis elegans-PKC Tabuse *et al.*,1989; Xenopus laevis-X-PKCII Chen *et al*, 1989; drosophila-dPKC 98f Schaeffer *et al.* 1989; unc-13 (Protein of Caenorhabditis elegans) Maruyama *et al.*, 1991; n-Chimaerin (Protein of the human brain) Hall *et al.*, 1989.

activity, the best investigated being bryostatin.[14] These structurally diverse molecules all specifically activate PKC by binding to the same agonist binding site.[15] These findings were specified further in that, it was shown that of the C1-region both cystein-rich zinc finger like motifs (Cys1 and Cys2, see Fig. 1) are able to bind separately (independently) phorbolesters (PKC-agonists of DTE type).[16] Therefore the assumption was made that only those amino acids which occur in all isoenzymes in both zinc finger like motifs are regarded as essential for the PKC-agonist binding (or PKC activation). To find out these special amino acid residues in PKC a sequence alignment of the zinc finger like region of various PKC isoenzymes was carried out. Also the corresponding sequence of tpa-1, X-PKCII and dPKC

98f (PKC from lower eukaryotes) indicating the developmental conservation of the sequence together with the phorbolester binding proteins n-chimaerin and unc-13 protein where included (Chart 1). This alignment shows that in most PKC-agonist binding proteins 13 amino acids are conserved, namely 6 cysteines, 2 histidines, 2 glycines, 1 glutamine, 1 proline, 1 phenylalanine yielding a 'consensus sequence'. The alignment of rat PKCζ with only one zinc finger like motif shows that only 12 of the 13 residues are present.[17] As the only difference proline is replaced by a arginine. But PKCζ is the only isoform that lacks both phorbol ester binding and phorbolester dependent kinase activity (Table I). Therefore the exchange proline - arginine may be responsible for the lack of activity.[4]

CONSTRUCTION OF A 3-D-SURROGATE OF THE AGONIST BINDING SITE, A PSEUDORECEPTOR MODEL OF PKC

To obtain additional new insights in the structural elements of the agonist binding site of PKC and more information of which functional groups of the amino acids in the binding site interact with functional groups of the PKC-agonists the computer assisted method 'pseudoreceptor modeling' appears a most appropriate new approach. As an alternative the well known approach of 'Modeling by Homology' was excluded because (i) it is impossible to find a similar experimentally, well-defined, protein structure (zink finger protein) that may be used as a template of the zink finger like motif of PKC (ligand/agonist binding site); (ii) from the data available PKC appears not to bind to DNA thus excluding use of DNA binding zink finger proteins as a starting template.

Method

The pseudoreceptor modeling technique attempts to close the gap between a known 3-D structure of an agonist and an unavailable 3-D-structure of the agonist binding site of

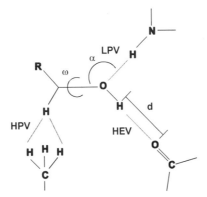

Figure 2. Directionality of the non-bonded interactions. A schematic representation of the possible interaction types: H-extension vectors (HEV, H-bond donor) generated for corresponding possible H-bond acceptor sites, lone-pair vector (LPV, H-bond acceptor) generated for corresponding possible H-bond donor sites and hydrophobicity vectors (HPV) generated for vectors of a possible hydrophobic interaction. The orientation, length and distribution of the vectors are assigned by geometric parameters for each specific interaction, e.g. α, ω and d. These data are stored in a database and activated when constructing the interaction vectors. For the vector cluster analysis normally idealised interaction vector are generated. This means that only one vector with mean values for α, ω and is constructed. The distribution of the vectors is taken into account by special algorithms.[18]

the receptor. Based upon the known 3-D structure of certain ligand molecules the new approach allows construction of a hypothetical surrogate of the agonist binding site for a structurally uncharacterised bioregulator. The technique involved was developed mainly at SIAT Biographics Laboratory by A. Vedani[18] and is restricted to the binding site of the receptor. The first software available to implement the concept of pseudoreceptor modeling which was used in this study, is the program YAK© [18].

The basic idea of the approach is to use information on the directionality of non-bonded interactions of the ligands (hydrogen-bonding, hydrophobic-, ion-pair - and metal-ligand interactions) to mimic the essential agonist-macromolecule interactions in the true biological receptor. The more or less directional nature of non-bonded interactions (Fig. 2) are intensively analysed by statistical methods using experimentally determined 3-D structures of small organic molecules as well as of macromolecules.[19-23]

The analyses of the directionality of all possible non-bonded interactions of a set of overlaid bioactive ligands, which bind to the same receptor site representing a pharmaco- phore model, may allow to identify type and approximate position of possible important sites of interaction in the receptor protein. These interaction sites are used as a starting point to orient suitable entities of the receptor (amino acid, metal ions, solvent) in 3-D space and to construct a 3-D pseudoreceptor model. For more detailed information on the concept, the algorithms of that concept and the program yak see loc. cit[18, 24].

The Pharmacophore Model

As mentioned above the first step in the process of pseudoreceptor modeling is the definition of a pharmacophore model. Therefore the different classes of known PKC-agonists are examined using experimental data such as structure activity relations and molecular modeling techniques (2-D-QSAR, 3-D-QSAR, molecular dynamics). For each class of PKC-agonists a prototype is defined and the bioactive conformations are evaluated (for detailed discussion see[25]).

In the ligands used to represent the prototypes the lipophilic residue of the latter are missing (mostly aliphatic ester chains). By Rippmann[26] it was demonstrated that the lipophilic parts of the agonists are responsible for (unspecifical partition or transport between biological phases. Thus, as long as special hydrophilic groups remain unchanged, a medium hydrophobicity is a sufficient condition for biological e.g. high promoting acitivity. Fig. 3 shows the structures of the six ligands representing the corresponding six prototypes of PKC-agonists selected for the definition of the pharmacophore - the ensemble of this trial. In the pharmacophore model elaborated certain functional groups of the ligands match with each other. In contrast to other published models in the matching of oxygens of ligands (prototypes, Fig. 3) in a special spatial orientation for the present pharmacophore (see also below Fig. 6) only two functional groups were identified to be necessary for PKC activation (see QSAR); one primary (or secondary) OH-group (e.g. OH-20 using the phorbol number- ing) and one carbonyl oxygen (carbonyl of ester group in position 13 using the phorbol numbering). However, to meet the result of structure activity studies an additional geometric assumption has to be made: any substitution of the hydrogen in the OH-4 group of TPA causes a complete loss of its biological activity, although this oxygen is not necessary for activity as best illustrated by 4-deoxy-TPA. The latter is highly active as an irritant and tumor promoter[6]. Therefore the spatial volume above this functional group and it's matching counterparts in the ensemble was defined as a 'negative steric hindrance point'. It means that this 3-D-volume is a relevant part of the receptor in that it is not allowed to be occupied by structural elements of agonists.

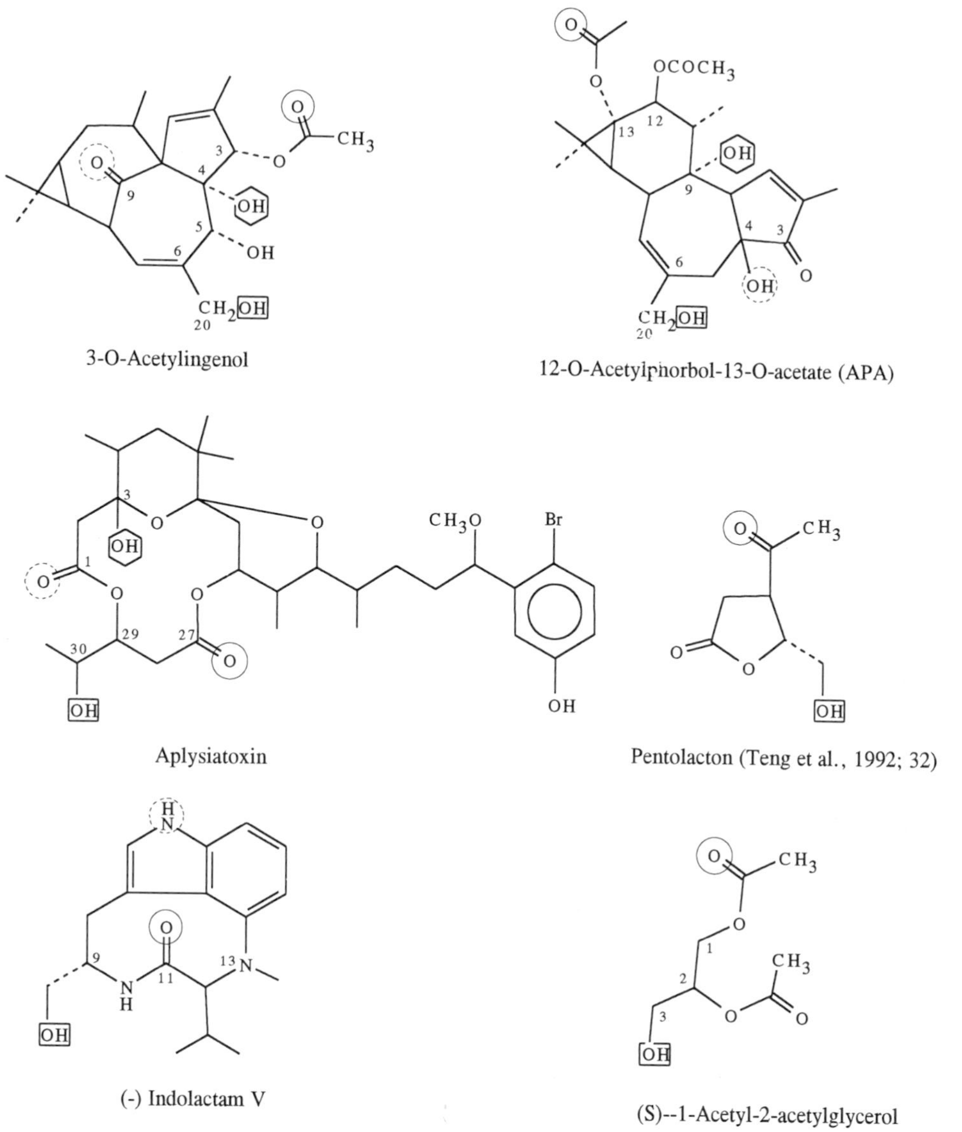

3-O-Acetylingenol

12-O-Acetylphorbol-13-O-acetate (APA)

Aplysiatoxin

Pentolacton (Teng et al., 1992; 32)

(-) Indolactam V

(S)--1-Acetyl-2-acetylglycerol

Figure 3. Structures of the six prototypes of agonists selected for the definition of the pharmacophore. The functional groups matched are marked with identical symbols in individual molecules. In the matching functional groups have a maximum distance of 1 Å is not exceeded.

Identification of Functional Groups of the Pharmacophore Relevant for Receptor Binding

To obtain information on the relevance of functional groups in the pharmacophore model defined above the directionality of all possible non-bonded interactions are evaluated for the ensemble with the program YAK$^{©}$. Regions in 3-D-space with a high density of similar non-bonded interactions indicate as important interaction sites at the receptor. In Fig. 4 an idealised representation of such an analysis is shown. For the pharmacophore

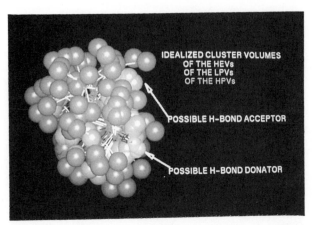

Figure 4. Vector cluster analysis of the idealised interaction vectors of all six selected prototypes: H-extension vectors (HEV, H-bond donor) generated for corresponding possible H-bond acceptor sites (cyan), lone-pair vector (LPV, H-bond acceptor) generated for corresponding possible H-bond donor sites (yellow) and hydrophobicity vectors (HPV) generated for vectors of a possible hydrophobic interaction (magenta). The pronounced clusters of the two dominant interactions - one H-bond donor site and one H-bond acceptor site -are clearly seen. The selection of the overlayed structure can be seen in part (green) as a stick model.

postulated (as represented by the entire ensemble used) five important sites for interactions are found: one possible H-bond donor sites, two H-bond acceptor sites and two hydrophobic interaction sites (Fig. 5). It was shown that of the matching oxygens only those which are postulated as important for bioactivity [one primary (or secondary) OH-group and one carbonyl oxygen] show clustering of non-bonded interaction in 3-D space. Thus these results back up the pharmacophore model postulated(Fig. 6).

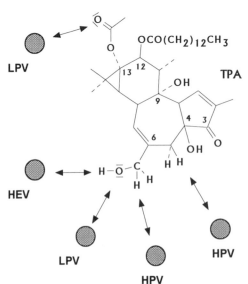

Figure 5. The possible interaction sites of the pharmacophore model illustrated for TPA. H-extension vector (HEV, H-bond donors) for H-bond acceptor sites in the protein, lone-pair vectors (LPV, H-bond acceptor) for H-bond donor sites in the protein and hydrophobicity vectors (HPVs) for hydrophobic interactions with the corresponding sites in the protein.

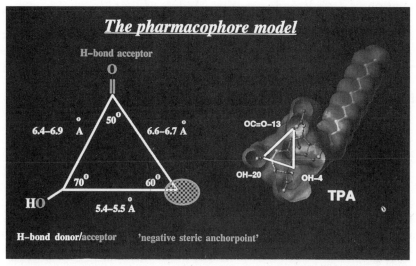

Figure 6. Schematic representation of the pharmacophore model defined by using the directionality of common sites of non-bonded interaction in the six prototypes of agonists; for more details see text.

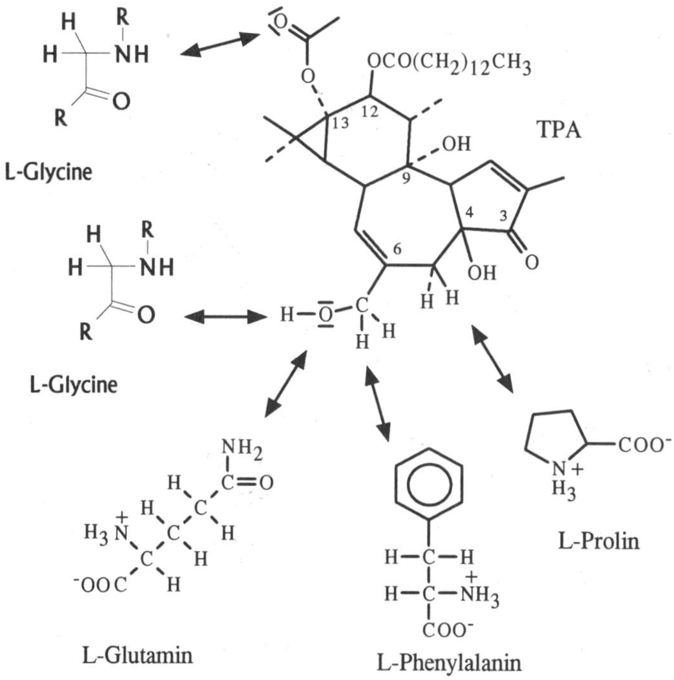

Figure 7. Pseudoreceptor model 1; Schematic 2D representation of the active site of PKC. For simplicity only the 2D structure of 12-O-Tetradecanoylphorbol-13-acetate is displayed as a ligand. With the CH_2OH group at position 20 (phorbol numbering) one glycine (H-bond acceptor), one glutamine (H-bond donator) and one phenylalanine (Hydrophobic interaction) interact. The carbonyl in the ester group at position 13 (phorbol numbering) interacts with the second glycine function (H-bond donor). The proline exhibits mainly hydrophobic interaction with the hydrophobic site of phorbol at position 5.

Construction of Two Pseudoreceptor Models of PKC

The important interaction sites identified on the side of the pharmacophore and the sequence alignment of the zinc finger like motifs on the side of the receptor may be used to construct 3-D-pseudoreceptor models of PKC. Two models are possible.

Model 1

The C1 region of PKC (comprising the zinc finger rich like motif) was shown to contain zinc[17] leaving open the question in which way the zinc ions are co-ordinated. Hubbard et al. report direct evidence by extended X-ray absorption fine structure (EXAFS) data that PKCβ1 binds four zinc ions (Zn^{2+}) per molecule (i.e. two zinc ions per zinc finger motif). They postulate an average Zn^{2+} co-ordination of one nitrogen and three sulphur atoms[27].

If these results are taken into account in the consensus sequence (13 amino acids) as determined by sequence alignment eight amino acids are required to co-ordinate two zinc ions. The remaining five amino acids (2 Glycines, 1 Glutamine, 1 Proline, 1 Phenylalanine) are available for the construction of the pseudoreceptor model. Therefore the templates of these amino acids are retrieved from a database and automatically docked, orientated and optimised in 3-D space around the ensemble of PKC agonists using as anchor points the

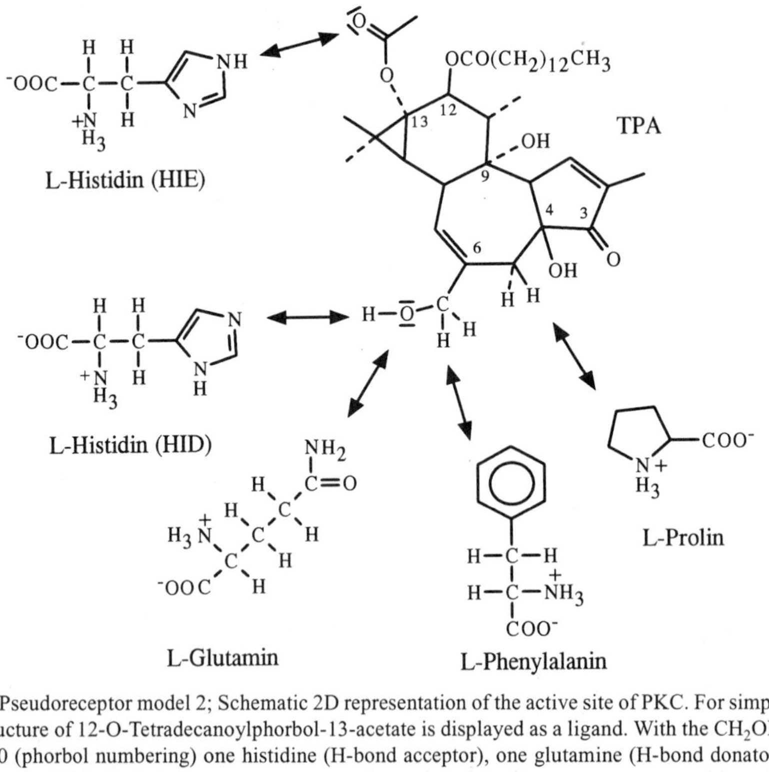

Figure 8. Pseudoreceptor model 2; Schematic 2D representation of the active site of PKC. For simplicity only the 2D structure of 12-O-Tetradecanoylphorbol-13-acetate is displayed as a ligand. With the CH_2OH group at position 20 (phorbol numbering) one histidine (H-bond acceptor), one glutamine (H-bond donator) and one phenylalanine (Hydrophobic interaction) interact. The carbonyl in the ester group at position 13 (phorbol numbering) interacts with the second histidine function (H-bond donor). The proline exhibits mainly hydrophobic interaction with the hydrophobic site of phorbol at position 5.

important interaction sites. In Fig.7 a schematic representation of pseudoreceptor model 1 is shown (using phorbol as one representative of the ensemble): one glycine (as H-bond acceptor), one glutamine (as H-bond donor) and one phenylalanine (as hydrophobic interaction site) interact with the CH_2OH group at position 20, whereas the second glycine interacts as a H-bond donor with the carbonyl oxygen of the ester at position 13 (phorbol numbering). The proline is engaged (preferentially) in hydrophobic interaction with the CH_2 group at position 5 (phorbol numbering).

Model 2

In this model an alternative possibility of co-ordination of zinc in the amino acid sequence is assumed referring to an interesting a similarity of the cystein-rich regions of PKC isoforms to the zinc finger motif of the oestrogen receptor. Thus also a zinc co-ordination of four sulphur atoms may be taken into account co-ordinating the zinc ions by cyteins only. In addition, contrasting model 1, either one or both glycines are postulated to be not essential as receptor binding site of the ligand. If, in analogy to model 1, five amino acids are used to model the agonist binding site this two assumptions lead to alternative pseudoreceptor models. In the extreme case, in model 2, the two glycines may be replaced by histidines whereas the other three residues remain at the same interaction sites as in model 1. The exchange of either one of the two glycines with histidine was also considered. In Fig. 8 a schematic 2-D representation of the pseudoreceptor model 2 is shown and in Fig. 9 a 3-D representation of this pseudoreceptor model 2 is depicted.

Assessment of the Models Proposed

The interaction energies for the different ligands are calculated for both models (Table II). In model 1 the ligands show a wide range of interaction energies (-20,1 kcal for12-O-Acetylphorbol-13-acetate to -10,7 kcal for .1,2-sn-Diacetylglycerole). In this model 1 there is a good agreement between 'important interaction sites' as determined by analysis of

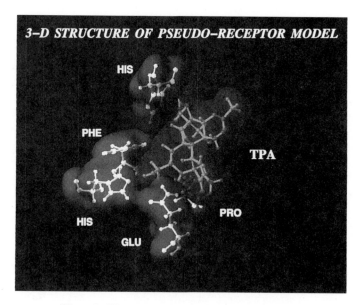

Figure 9. 3D structure of the pseudoreceptor model 2.

Table 2. Interaction energies of the ligand receptor complex and relative biological activities of corresponding typical agonist

Ligands used for Modeling	Interaction energies (ΔE kcal/mol)			Biological activities of typical agonists	
	Model 1	Model 2		[nM][1]	[nM][2]
12-O-Acetylphorbol-13-acetate	-20.1	-14.8	high	5	8
3-O-Acetylingenol	-15.4	-11.7		8	n.d.[3]
Apylsiatoxin	-11.0	-13.6		12	19
(-) Indolactam V	-12.4	-11.6		18	17
Lactone (from Teng)	-12.6	-3.8	low	n.d.	2500
1,2-sn-Diacetylgycerole	-10.7	-3.9		n.d.	1300

(1) Concentration of typical agonists (nM) giving 50% of maximum stimulation of protein kinase C
(2) 50% inhibition of PKC binding (labeled Phorbol-12,13-dibutyrat) measured for typical agonists corresponding to the ligand prototypes investigated : TPA, 3-Tetradecanoylingenol, Aplysiatoxin, Teleocidin B4, Pentolactone (Teng et al.) [33], 1-Tetradecanoyl-2-acetyl-glycerole
(3) not determined

directionality of non-bonded interactions and relevant amino acids (sequence alignment associated with specific co-ordination of zinc). However, in model 1 by interaction energies ligands corresponding to typical high and low bioactive agonists can not be discriminated. In model 2 one group of ligands exhibit energy gains between about 11 to 15 kcal/mol, the other show energy gains around 4 kcal/mol. Thus, model 2 reflects reasonably well in interaction energies of ligands the differences in the biological activities of corresponding typical agonists measured independently. They correlate reasonably well with tumor promoting activity. Thus according to the criteria of interaction energies model 2 may be considered the more appropriate template to develop new lead structures for PKC agonists and antagonists.

DISCUSSION AND CONCLUSION

In the absence of a 3-D protein structure of PKC the pseudoreceptor concept tries to merge the results of investigations at the ligand (agonist) - and at the protein - level. For the present trial, six prototype structures representing typical PKC agonists (mostly bioactive PKC activators) were selected. Their superposition and the analyses of the directionality of their non-bonded interactions allowed to rectify two (three) common sites of interaction with the receptor thus defining a (minimum) pharmacophore. This partial result together with a sequence of 13 amino acids constituting the agonist binding site in all PKC-isoenzymes and reasonable assumptions of Zn^{2+} co-ordination in the amino acid sequence lead to two pseudoreceptor models. They provide corroborated information of the amino acids important for binding of agonists by the receptor and on the other hand back up the pharmacophore model postulated.

The first model shows good agreement between the essential interaction sites of the ensemble and available relevant amino acids. Yet due to the criteria of interaction energies (Table II) it does not allow to discriminate between high and low affinity ligands similarly as in case of the bioactivities of corresponding typical agonists which they represent. In contrast the second model exhibits a satisfactory relationship between the interaction energies of the ligands and the bioactivities of corresponding typical agonists. This model appears to provide a more appropriate template of the pseudoreceptor for development of new and potent PKC agonists and antagonists.

It should be pointed out that — not in principle but in practice — the pseudoreceptor concept is limited to relative rigid ligand molecules, of which only one or a few bioactive conformations are possible. Thus at least one ligand of the entire ensemble should be rigid. This may be used as a template for the more flexible structures (e.g. DAG, bryostatin). As a ligand bryostatin was excluded because of its flexibility and numerous functional oxygen groups. Also the present trial excludes the more general possibilities of (i) inducible fits and (ii) multiple binding modes.

The similarity or identity of the pseudoreceptor 3-D structure derived here and its natural counterpart remains to be determined by crystallographic analysis of the PKC structure. However, independently, the pseudoreceptor models may be used already as a basis for further investigations, e.g. as template for de novo design of ligands, even if the the surrogate is only similar but not identical with their natural counterpart. At least the pseudoreceptor model represents a copy of the (inverse) recognition pattern of the ensemble of PKC agonists - the pharmacophore. For example the receptor surrogate may be used as an input for various automatic approaches of de novo ligand design, using various algorithms and fragment data bases to construct new lead structures in known receptor sites, such as LUDI, CLIX, HOOK, GenStar etc.[28-32].

Thus the pseudoreceptor model of PKC not only gives more detailed information on the agonist binding site in PKC-but it may also be a helpful tool for the de novo ligand design of new lead structures for PKC-agonists and antagonists, some of which may open up new classes of antineoplastic agents.

REFERENCES

1. Basu, A. (1993) *Pharmac. Ther.*, 59: 257-280
2. Tritton,T.R. and Hickman, J.A. (1990) *Cancer Cell*, 2 : 95-105
3. Powis, G. (1991) *Trends Pharmacol. Sci.*, 12: 188-193
4. Azzi, A., Boscoboinik, D. and Hensey, C. (1992) *Eur. J. Biochem.*, 208: 547-557
5. Hug, H. and Sarre, T.F. (1993) *Biochem J.,* 291: 329-343
6. Hecker, E. (1985) *Arzneim.-Forsch./Drug Res.*, 35: 1890-1903
7. Takai, Y., Kishimoto, A. , Inoue, M. and Nishizuka, Y. (1977) I, *J. biol. Chem.*, 252: 7603-7609. II, *J. Biol. Chem.*, 252: 7610-7616.
8. Nishizuka, Y. (1995) *FASEB J.*(in press).
9. Hecker, E. (1978) In: *Carcinogenesis (Vol. 2) , Mechanisms of tumor promotion and cocarcinogenesis*, (Slaga T.J., Sivak A. and Boutwell R.K. Hrsg.), pp.11, Raven Press, New York.
10. Castagna, M., Takai, Y., Kaibuchi, K., Sano, K., Kikkawa, U. and Nishizuka, Y. (1982) *J. Biol. Chem.*, 257: 7847-7851.
11. Berridge, M.J. (1987) *A. Rev. Biochem.*, 56: 159-193.
12. Stabel, S. and Parker, P.J. (1991) *Pharmac. Ther.*, 51: 71-95
13. Burns, D.J., Basta, P.V., Holmes, W.D., Ballas, L.M., Rankl, N.B., Barbee, J.L., Bell, R.M. and Loomis, C.R. (1992) In: *Adenine Nucleotides in Cellular Energy Transfer and Signal Transduction,* (Papa, S., Azzi, A. and Tager, J.M. ed.) pp. 207-217, Birkhäuser Verlag, Basel/Switzerland.
14. Blumberg, P.M. (1988) *Cancer Res.*, 48: 1-8.
15. Rando, R.R. and Kishi, Y. (1992) *Biochemistry,* 31: 2211-2218
16. Burns, D.J. and Bell, R.M. (1991) *J. Biol. Chem.*, 266: 18330-18338
17. Ahmed, S., Kozma, R., Lee, J., Monfries, C., Harden, N. and Lim, L. (1991) *Biochemical J.*, 55: 233 -241
18. Vedani, A., Zbinden, P. and Synder, P. (1993) *J. Receptor Res*, 13: 163-177
19. Taylor, R. and Kennord, O. (1984) *Acc. Chem. Res.* 17: 320-326
20. Murry-Rust, P. and Glusker, J.P. (1984) *J. Am. Chem. Soc.* 106: 1018-1025
21. Vedani, A. and Dunitz, J.D. (1985) *J. Am. Chem. Soc.* 107: 7653-7958
22. Baker, E.N. and Hubbard, R.E. (1984) *Prog. Biophys. Mol. Biol.* 44: 97-179
23. Tintelnot, M. and Andrew, P. (1992) *J. Comput. Aided Mol. Design.* 3: 67-84

24. v.d. Lieth, C.W., Krauter, G. and Hecker, E. (1995) In: *Novel Appraoches in Anticancer Drug Design, Molecular Modelling-New Treatment Strategies*. Contrib. Oncol., (Zeller W.J., D'Incali, M. and Newell, D.R. ed.) 49: 25-39,.Basel, Karger.

25. Krauter, G. (1993) Über Struktur und Wirkung von Agonisten der Protein Kinase C: Untersuchungen mittels klassischer quantitativer Strukturwirkungsbeziehungen (QSAR) und rechnerunterstützter molekularer Modellierungen; doctoral thesis, Naturwissenschaftlich-Mathematische Gesamtfakultät, University of Heidelberg

26. Rippmann, F. (1990) *Quant. Struct. -Act. Relat.* 9: 1-5.

27. Hubbard, S.R., Bishop, W.R., Kirschmeier, P., Goerge, S.J., Cramer, S.P. and Hendrickson, W.A. (1991) *Science*, 254: 1776 - 1779

28. Böhm, H.J. (1992) *J. Comput. Aided Mol. Design*, 6: 61-78

29. Böhm, H.J. (1992) *J. Comput. Aided Mol. Design*, 6: 593-606

30. Rotstein, S.H. and Murcko, M.A. (1993) *J. Comput. Aided Mol. Design*, 7: 23-43.

31. Eisen, M.B., Wiley, D.C., Karplus, M. and Hubbard, R.E. (1994) *Structure, Function, and Genetics*, 19: 199-221

32. Lawrence, M.C. and Davis, P.C. (1992) *Proteins: Structure, Function, and Genetics*, 12: 31-41.

33. Teng, K., Marquez, V.E., Milne, G.W.A., Barchi, J.J., Kazanietz, M.G., Lewin, N.E., Blumberg, P.M. and Abushanab, E. (1992) *J. Am. Chem. Soc.*, 114: 1059-1070

34. Ono, Y., Fujii, T., Ogita, K., Kikkawa, U., Igarishi, K. and Nishizuka, Y. (1987) *FEBS Lett.* 226: 125-128

35. Osada, S., Mizuno, K., Saido, T. C., Akita, Y., Suzuki, K., Kuroki, T. and Ohno, S. (1990) *J. Biol. Chem.*, 265: 22434-22440

36. Ohno, S., Kawasaki, H., Imajoh, S., Suzuki, K., Inagaki, M., Yokokura, H., Sakoh, T. and Hidaka, H. (1987) *Nature*, 325: 161-166

37. Bacher, N., Zisman, Y., Berent, E. and Livneh, E. (1991) *Molecular and Cellular Biology,* 11: 126-133

38. Coussen, L., Parker, P.J., Rhee, L., Yang-Feng, T.L., Chen, E., Waterfield, M.D., Franke, U. and Ullrich, A. (1986) *Science*, 233: 859-865

39. Tabuse, Y., Nishiwaki, K. and Miwa, J. (1989) *Science*, 243: 1713- 1716

40. Chen, K., Peng, Z., Lavu, S. and Kung, H. (1989) *Sec. Mess. Phosphoprot.*, 12: 251-260

41. Schaeffer, E., Smith, D., Mardon, G., Quinn, W. and Zuker, C. (1989) *Cell*, 57: 403-412

42. Maruyama, N.I. and Brenner, S. (1991) *Proc. Natl. Acad. Sci. USA*; 55: 5729 - 5733

43. Hall, C., Monfries, C., Smith, P., Lim, H.H., Kozma, R., Ahmed, S., Vanniasingham, V., Leung, T. and Lim, L. (1990) *J. Mol. Biol.*, 211: 11-16

13

THE USE OF SYNCHROTRON RADIATION IN PROTEIN CRYSTALLOGRAPHY

Peter F. Lindley

DRAL Daresbury Laboratory
Warrington WA4 4AD, United Kingdom

ABSTRACT

Synchrotron radiation provides a source of high intensity, wavelength tuneable, highly collimated radiation which can be used to investigate structure-function relationships in biological macromolecules in a number of ways. This article will deal mainly with the applications of the hard X-ray region (0.5 - 2.5 Å) of the synchrotron spectrum to three-dimensional structure analysis of single crystals of macromolecules. The main themes highlighted will be, (a) why synchrotron radiation is necessary for structural studies on biological macromolecules, (b) examples of experimental stations used for protein crystallography at the Synchrotron Radiation Source, DRAL Daresbury Laboratory, (c) how the wavelength tunability can be exploited to solve the "phase problem", (d) high resolution data using cryogenic methods and (e) studies on large structures.

INTRODUCTION. WHY SYNCHROTRON RADIATION IS NECESSARY FOR STRUCTURAL STUDIES ON BIOLOGICAL MACROMOLECULES

Three-dimensional molecular structure underpins many aspects of modern molecular biology. Prime examples are the work on viruses such as the common cold virus[1] and the foot-and-mouth disease virus[2], and the work of Steitz and colleagues[3] on the complex of *E.coli* glutaminyl-tRNA synthetase with tRNA[Gln] and ATP, [aminoacyl-transfer RNA synthetases are the enzymes responsible for translating the genetic code]. More recently structural studies on actin[4] and myosin[5] have provided templates to assist in the understanding of the mechanism of muscle contraction. Other work has profound medical implications, for example, structural studies on trypanothione reductase[6] may well lead to the design of inhibitors which incapacitate parasitic trypanosomes and leishmanias. Many of these key structural studies involve the use of synchrotron X-radiation and as the macromolecules become more complex, for example, the 50S ribosomal particles[7], then the use of synchrotron radiation will become increasingly important. For high quality, high resolution data, studies on large unit cells, samples particularly sensitive to radiation damage and/or weakly diffracting, multi-wavelength phase determination methods and time-resolved studies, the unique properties of synchrotron are almost mandatory. This article attempts to explain the importance of synchrotron radiation techniques in the crystallographic analysis of biological macromolecules.

The main reasons for the impact of synchrotron radiation on the structure analysis of biological macromolecules can be readily seen from the expression for the total energy, E_{hkl}, diffracted by an ideally mosaic crystal rotated at a constant velocity, ω, through the diffracting position;[8]

$$E_{hkl} = K \cdot (1/\omega) \cdot I_o \cdot \lambda^3 \cdot ALP \cdot \frac{(V_{crystal})}{(V_{unit\,cell}^2)} \cdot |F_{hkl}|^2 \tag{1}$$

where $K = [e^2/(4\pi e_o.mc^2)]^2$, the "radius" of the electron.

I_o is the intensity of an incident beam of wavelength λ,

A is an absorption factor, λ dependent, for the crystal specimen and its

L is the Lorentz factor, dependent on the time that the crystal spends in the diffracting position and inversely proportional to λ,

P is a polarisation factor dependent on the state of polarisation of the incident beam, which for synchrotron sources is predominantly in the plane of the synchrotron ring,

$V_{unit\,cell}$ is the unit cell volume in a crystal of volume $V_{crystal}$,

and $|F_{hkl}|^2$ is the diffracting power of the set of planes (hkl) in the crystal lattice.

For biological macromolecules the unit cell volume, $V_{unit\,cell}$, tends to be large, ranging from about 59,000 $\mathring{A}^3$ for a small enzyme such as ribonuclease A to 173×10^6 $\mathring{A}^3$ for simian virus 40[9], one of the largest structures so far attempted. A direct consequence of these structural units containing large numbers of atoms, is that the diffracting power, $|F_{hkl}|^2$, of each set of planes is generally weak. It is also often the case that for technical reasons it is very difficult to grow large crystals and $V_{crystal}$ therefore tends to be small. As can be seen from equation [1], the combination of these factors leads to a reduction in E_{hkl}. On the other hand increasing the value of I_o will increase E_{hkl} and since currently available synchrotron sources are typically 10^3 to 10^4 times more intense than conventional sources, their immediate advantages are apparent.

Equation [1] also shows that E_{hkl} is dependent on the wavelength λ, and the tunability of a synchrotron source enables wavelengths typically in the range 0.5 - 2.5 $\mathring{A}$ to be selected as required. A reduction in λ leads to a reduction in E_{hkl}. However, reducing the wavelength brings the great advantage of minimising systematic errors due to sample absorption, (the variation in transmission is typically reduced to some 5 - 10 % at 0.9 $\mathring{A}$ compared to 30 - 40 % at 1.5 $\mathring{A}$). A further consideration is radiation damage which is thought to occur mainly by the diffusion of free radicals (solvated electrons), through the crystal lattice causing disruption of the mosaic block structure and eventually loss of crystallinity. Its effect on the diffraction data is generally non-linear, but the high resolution data is lost initially and eventually the entire crystal ceases to diffract. High values of I_o will cause an increased number of free radicals, again a judicious choice of wavelength may be desirable, but because I_o is high the speed of data collection will be rapid and it is often possible to measure the high resolution data before radiation damage can have an appreciable effect, *i.e.* the rate of free radical diffusion appears more important than the overall number of radicals formed. The tunability of the source in the context of multi-wavelength anomalous dispersion techniques will be discussed in a later section.

Finally, the low beam divergence and high degree of collimation, that can be routinely achieved with a synchrotron source not only enable good spatial resolution for crystals with large unit cells (where the diffracted beams are very close together), but also permit the collection of meaningful and usable data from crystals too small in size for study by conventional sources. These characteristics lead to a significant improvement in the signal-to-noise ratio for the diffraction data and this is particularly important for the relatively weak, high resolution data. In general it has been found that use of synchrotron radiation leads to an increase in the amount of data that can be obtained per crystal specimen. This can be in terms of more diffraction patterns per crystal for a given resolution and often in an increase in the resolution of the measurable data.

EXPERIMENTAL PROTEIN CRYSTALLOGRAPHY AT THE SRS

Station 9.5

This station derives its synchrotron radiation from a 3-pole wiggler magnet operating at 5 tesla and was designed as a dual-purpose facility catering either for focussed white-beam experiments or for readily tuneable wavelength studies[10], the station can also be routinely used to collect data at an optimised wavelength. The first element is a platinum coated fused quartz toroidal mirror placed some 18 m from the wiggler magnet source. The mirror with an acceptance aperture of 1.2 mrad horizontal and 0.1 mrad vertical produces a white beam focal spot size of 1.3 x 0.4 mm with an intensity of some 10^{11} photons/sec/mm^2 at the sample roughly 32 m from the source. This is the normal arrangement for focussed Laue experiments and the station is equipped with an ultra-fast shutter capable of reaching opening times down to 50 msec. However, a water cooled channel cut Si(111) double crystal monochromator can be interposed at about 30 m from the source to give rapidly tuneable monochromatic X-rays, minimum band pass $\Delta\lambda/\lambda = 0.00015$, suitable for multi-wavelength anomalous dispersion studies. For time-resolved white-beam studies the station can be used with a modified Arndt-Wonacott oscillation camera[11,12]. For normal data collection purposes a 30 cm Mar-Research image plate detector is employed[13]. A recent development has been the installation of three-circle Nonius goniostat for specimen orientation and the detector is mounted on a fourth circle; the detector and crystal movement are operated under modified MADNES software.

Station 9.6

Station 9.6 also derives its synchrotron radiation from a wiggler magnet and the optical components are a mirror and a monochromator. The platinum coated fused cylindrically curved quartz mirror is some 12 m from the wiggler magnet and provides 1:1 vertical focusing. The bent triangular Si(111) monochromator is located about 21 m from the mirror and has a 3 mrad acceptance giving 8:1 horizontal defocussing at 0.895 Å with a wavelength band pass, $\Delta\lambda/\lambda = 0.0004$. The size of the focal spot at the specimen, 2 m from the monochromator, is 0.5 x 0.3 mm in the horizontal and vertical directions respectively. Typically for the SRS operating at 2 Gev and 200 mA the intensity at the sample position is 10^{12} photons/sec/mm^2. Although the monochromator can be tuned for different wavelengths, its dominant use is as a fixed-wavelength station. The use of the short wavelength reduces absorption errors (and radiation damage), enabling the collection of high quality, high resolution diffraction data. The station is fitted with a 30 cm diameter Mar-Research instrument.

WAVELENGTH TUNABILITY: ANOMALOUS DISPERSION TECHNIQUES

Two major problems confront the protein crystallographer in determining the three-dimensional structure of a biological macromolecule. Firstly, single crystals have to be obtained of sufficiently high quality to enable the collection of diffraction data to high resolution and, secondly, the "phase problem" has to be surmounted. The amplitudes of the diffracted beams can be derived directly from the measurement of the diffraction intensities, but in order to construct an image of the diffracting object, the relative phases of the diffracted beams are also required. Unlike the case of the optical microscope, this relative phase information cannot be obtained directly for the X-ray case. The classic method of solving this problem in protein crystallographic involves the preparation of heavy atom derivatives[14]. With only one heavy atom derivative and in the absence of significant anomalous disperion there is a phase ambiguity and this problem is often solved through additional heavy atom derivatives. The preparation of good heavy atom derivatives can be problematic due to lack of isomorphism, low levels of heavy atom substitution, multiple sites *etc.*, and each derivative requires more protein and the collection and processing of a further set of intensity data. As a consequence protein crystallographers have long sought alternative approaches and one such depends on the significant anomalous scattering characteristics of heavy atoms. It is in this context that synchrotron sources where the wavelength is readily tuneable offer unrivalled opportunities.

The atomic scattering factor of an atom, f, is a measure of the amplitude of the electromagnetic radiation scattered by that atom when radiation of a given amplitude impinges upon it, and is given by the expression:

$$f = f_o + f' + if''$$

where f' is a dispersion component and f'' is a component which has a phase advance of $\pi/2$.

When the frequency of the incident radiation is not close to an absorption edge of the scattering atom, the dispersion and absorption components are small and $f \approx f_o$ the "normal" scattering of the atom. However, at or near an absorption edge f' and f'' can be large and this leads to differences in intensities for certain classes of symmetry related reflections. Several important properties of the coefficients f' and f'' are pertinent to anomalous dispersion measurements. Firstly, the electrons predominating the anomalous scattering effects are the tightly bound core electrons and hence f' and f'' are far less dependent on $\sin q/l$ than the normal scattering component. Thus, the relative importance of the anomalous dispersion should increase with increasing resolution.

Secondly, f' and f'' depend on the wavelength of the incident radiation. They vary most rapidly, and reach their maximum values, in the immediate vicinity of an absorption edge. Values of f'' remote from the absorption edge are readily calculated using a relativistic wave function formulation[15], but calculations at or near the edge are too complex. However, f'' can be derived directly from measurements of the X-ray absorption, *i.e.* from an EXAFS spectrum, usually in fluorescence mode on the same single crystal that is being used for data collection, and these values scaled to the calculated off-edge values. Values of f' can then be obtained through the Kramers-Kronig dispersion relationship. A knowledge of the values of f' and f'' can be very important for predicting the magnitude of anomalous effects and therefore the accuracy required for data collection, but in phase determination procedures these values are normally refined as an integral part of the phase determination process[16]. Finally it should be noted that f' and f'' are polarisation dependent.

The use of the anomalous effect with carefully selected reflections is commonplace in conventional multiple isomorphous replacement phasing, but clearly, tuning the incident wavelength, readily feasible with a synchrotron source with little change in intensity, can be used to optimise the anomalous signal[17]. Single isomorphous replacement with anomalous scattering. SIRAS, can be a particularly powerful technique if used in conjunction with solvent flattening[18]. A typical example at the SRS was the determination of the structure of a heptanucleotide, d(GCATGCT); this structure was not amenable to direct methods and it was only possible to prepare a bromine derivative. Data collection at the bromine edge and the use of SIROAS (the O indicates optimised) gave phases with a figure of merit of 0.73 and led to a partially interpretable electron density map. Phase recombination gave an FOM of 0.80 and led to the complete structure. This structure contains a novel loop conformation with unusual base pairing (W.N.Hunter,1994, personal communication) and the crystal packing confirms the difficulties in preparing heavy atom derivatives. A more recent example is that of the plasma protein, ceruloplasmin.

However it is also possible to determine phase information by making multi-wavelength measurements.

Multiwavelength Anomalous Dispersion (MWAD)

A comprehensive account of this technique is given by Moffat[19] and Fourme & Hendrickson[16] and references therein. If several data sets are collected for a protein incorporating a suitable heavy atom, at different wavelengths chosen so that they yield optimal changes in both f' and f'' [20,21] the effect is similar to that of having several isomorphous derivatives. In the simplest form of the MWAD technique, diffraction data are collected for the Freidel related reflections (hkl) and (-h-k-l) from a protein containing a heavy atom. The wavelength, λ_1, is selected using fluorescence data, to coincide with an absorption edge of the heavy atom when the component f'' is a maximum; f' will also be large which will normally reduce the length of the real part of the scattering vector, F_H. These measurements give rise to the circles of radius $|F_1+|$ and $|F_1-|$ for a given reflection (hkl) and its Freidel pair

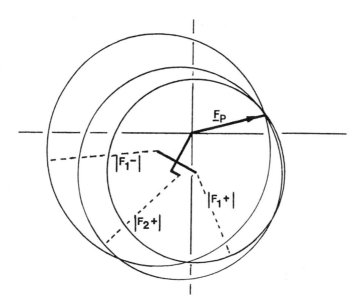

Figure 1. The simplest case for the multi-wavelength anomalous dispersion (MWAD) technique.

(-h-k-l) respectively in the Argand diagram. The wavelength is then tuned away from the edge to λ_2 so that f " (and f ') are small and measurements made for the reflections (hkl) only. These measurements give rise to the circle of radius $|F_2+|$ in the Argand diagram. The three circles cut at a single point which corresponds to the correct phase as indicated in the figure 1.

In practice the application of the MWAD method requires a high standard of experimental technique, up to four wavelengths are used, and great care has to taken to minimise systematic errors. Thus, all data for a given phase must be recorded from the same crystal to eliminate inter-crystal scaling and absorption errors. Data for symmetry related reflections (Freidel and Bijvoet pairs) should be collected close together in time for all the wavelengths to minimise the effects of radiation damage and slow variations in the experimental conditions. Indeed, it is highly desirable that all the data for all the wavelengths are recorded from the same crystal and this implies that the crystal must be relatively resistant to radiation damage and chemically stable; in this context cryogenic techniques may well play an important role, (see section below). Crystals should be mounted to exploit both the morphology and symmetry, to reduce absorption errors arising from different path lengths for components of symmetry related pairs, to enable such reflections to be recorded at the same time, and to minimise the overall amount of data to be collected. Suitable selection of wavelengths can not only optimise differences in f ' and f ", but also minimise differences in absorption between wavelengths; absorption differences at different wavelengths can be taken into account by experimental measurement of transmission factors. Finally it should be stressed that very great care must be taken in scaling between data collected at different wavelengths, since it is relatively easy to lose the anomalous information.

Several successful multiwavelength experiments have been carried at the SRS using station 9.5, including the structure determination of OppA[22]. OppA is a 58 kDa bacterial transport protein involved in the transport of peptides across the cytoplasmic membrane of Gram negative bacteria. It binds peptides from 2 - 5 residues in length, but with little specificity. OppA from *Salmonella typhimurium* has been cloned and expressed in *E. coli* and the protein co-crystallised with uranyl acetate, producing two distinct crystal forms. MWAD data were collected about the uranium L_{III} edge to a nominal resolution of 2.3 Å. Maximum likelihood phasing methods gave phase information from the MWAD data leading to an electron density map which was readily interpretable without density modification.

CRYOGENIC MEASUREMENTS AND HIGH RESOLUTION DATA

Relatively few macromolecular structures have been determined to a resolution of better than 2.0 Å, yet the acquisition of high quality, high resolution diffraction data can yield structural detail almost on a par with small molecule crystallography where atomic resolution is the norm. Some proteins can be crystallised in tightly packed arrays with a relatively low solvent content, 35-40%, and in these cases it is often possible to collect diffraction data to 1.5 Å or even higher. One of the major obstacles to obtaining high quality, high resolution diffraction data from single crystals of biological macromolecules is their susceptibility to radiation damage. Inevitably, one of the first effects of radiation damage is the deterioration of the high resolution regions of the pattern. Where the crystal morphology is suitable, translation of the crystal in the highly collimated X-ray beam can assist this problem, but an alternative method is to freeze the protein crystal so that the diffusion of free radicals is very slow and radiation damage considerably reduced. Since crystals of biological macromolecules contain substantial amounts of solvent, typically between 35 and 80% by volume, the technical problem is to force the solvent to cool in an amorphous state so that it does not physically destroy the crystallinity by expansion.

Three main methods of cooling have been used to date, flash freezing, fast freezing and slow freezing; which method is best for a particular macromolecular system is normally a matter of empirical determination. The flash freezing method has been employed by Ada Yonath and colleagues to obtain data from the physically fragile and very radiation sensitive crystals of 50S ribosomal particles[7]. At room temperature the crystal lifetime in the synchrotron beam was too short for useful data to be collected. The crystals were transferred to an inert hydrocarbon environment, or to solutions similar to the crystallisation medium but with higher viscosities, and flash frozen on a thin glass spatula by immersion in liquid propane; they were then transferred to a cold nitrogen gas stream for data measurement. A variation on the technique involved soaking cryosolvents into the crystals prior to freezing. Fast freezing has been successfully used to collect diffraction data from bovine γB-crystallin crystals. In this case a crystal was coated in oil by transferring it from its mother liquor to an oil droplet, adhered to a glass fibre by surface tension effects and introduced into a nitrogen gas stream at 150K. The oil coating prevents loss of solvent from the crystal during transfer. It was found that large crystals, 0.5 x 0.5 x 1.0 mm, often became opaque after freezing indicating gross damage to the crystallinity, or showed appreciable mosaic spread in the subsequent diffraction patterns rendering them useless for data collection. Smaller crystals, 0.2 x 0.2 x 0.8 mm, gave good diffraction patterns with an increase in the mosaic spread of only a factor of about two, compared to room temperature measurements, presumably because of a smaller angular and size distributions of the mosaic blocks. Data were collected photographically on station 9.6 at the SRS[23] to a resolution at the edge of the film of 1.2Å; the crystal showed little indication of any deterioration in the synchrotron beam. Processing and refinement of this "atomic resolution" data has led to a remarkably detailed structure of the γB-crystallin molecule. Local disorder in certain side-chains has been identified and 80% of the solvent molecules have been readily located enabling an accurate study of protein-water and water-water interactions.

Slow freezing using a liquid helium cryostat, over a period of hours, has been successfully used with crystals of the coenzyme of vitamin B_{12} to 15K[24]. Whether biological macromolecular crystals can be annealed to low temperatures with progressive sets of cooling, heating and cooling stages is not well researched.

APPLICATION OF SYNCHROTRON RADIATION TO LARGE STRUCTURES

Viruses

One of the most important areas of structural biology that has benefitted from the application of synchrotron radiation techniques is that of virus structure and this article would not be complete without a brief reference to such studies; the importance of synchrotron radiation to structural studies on large assemblies such as ribosomes has already been indicated in the previous section. Virus structures are usually associated with large unit cells, weak diffracting power, and small crystal size, the three main parameters on the right hand side of equation [1] which lead to weak diffraction intensities. In addition they suffer from severe radiation damage so that many crystals have to be used for a complete data set and the acquisition of high resolution data is technically difficult. In these cases the use of synchrotron radiation is mandatory.

Typical examples for which data were collected on station 9.6 at the SRS, Daresbury, are FMDV, the foot-and-mouth disease virus[2] and SV40, the simian virus 40[9]. FMDV belongs to the family of picornaviruses which are responsible for a wide variety of animal and human diseases including polio and the common cold. These viruses are spherical with diameters of some 300 Å and have a 40 Å thick protein shell which protects a genome of a single strand of RNA. The RNA genome can be translated directly by the host cell to give the viral proteins and hence initiate an infection. FMDV crystallises in a body-centred cubic cell with dimension, a = 345 Å. The crystals are small, 0.12 x 0.12 x 0.06 mm and extremely radiation sensitive so that only a single small angle oscillation photograph (0.4 -0.5°) could be recorded per crystal. The low beam divergence, high intensity and relatively short (0.9 Å) wavelength, characteristics of station 9.6 made data collection feasible. Many crystals were exposed in fairly random orientations to minimise the effects of radiation damage during crystal alignment, the so-called American method[25], and eventually some 106 useable film packs were obtained to yield data for a structure analysis at 2.9 Å. However, recent technical improvements to station 9.6 including a replacement cylindrically curved mirror and an image plate detector, have enabled diffraction patterns to be collected from a serotype of FMDV (D.Stuart, 1992, personal communication) showing excellent spatial resolution even at 2.3 Å, (the unit cell appears to be primitive orthorhombic with dimensions approximately 325 x 325 x 360 Å), and permitting some 15 -16 oscillation images, 0.4°, to be obtained per crystal; a significant increase in data yield per crystal. Whereas FMDV presents probably the smallest ratio of,

$$| F_{hkl} |^2 \cdot \frac{(V_{crystal})}{(V_{cell}^2)}$$

for data collected at the SRS, the SV40 virus one of the largest "objects" yet solved. SV40 is a member of the polyoma virus family, the simplest of the viruses with double-stranded genomes. The virus particles are some 500 Å in diameter and the unit cell is again body-centred cubic with a cell dimension of 558 Å. Data to a resolution of 3.8 Å were collected on station 9.6 in 0.2° steps with a wavelength of 0.88 Å; upto 4 photographs were recorded per crystal. The structure, solved by multiple isomorphous replacement methods, shows that the 72 pentamers of the viral protein VP1, which form the outer shell, have identical conformations except for the C-terminal arms. Five arms emerge from each pentamer and form links to adjacent pentamers. This linking of standard building blocks permits the required variability in packing geometry without sacrificing specificity. More recently, the Oxford group led by David Stuart have collected data on station 9.6 from crystals of "blue-tongue" virus core particles where the cell dimensions exceed 700 Å; structure elucidation is in progress.

Proteins and Enzymes

In addition to viruses the structures of several biologically important, but large, protein and enzyme molecules have also been determined using the SRS and brief details of typical examples now follow.

(a) F_1-ATPase from Bovine Heart Mitochondria[26]. Adenosine triphosphate synthetase is the central enzyme in energy conversion in mitochondria, chloroplasts and bacteria. It uses a proton-motive force, generated across the membrane by electron flow, to drive the synthesis of ATP from ADP and inorganic phosphate. The multisubunit assembly comprises a globular domain and an intrinsic membrane domain linked by a slender stalk about 45 Å long. Disruption of the stalk releases the globular domain, a complex of 5 different proteins which in bovine mitochondria are designated 3α (510 residues), 3β (482 residues), 1γ (272 residues), 1δ (146 residues) and 1ε (50 residues) with a total molecular weight of 371 kDa. Data collection on the native crystals and methylmercury derivatives has led to a structure in which the δ and ε subunits, and 145 residues of the γ subunit, are not visible, but which nevertheless enables a useful insight into the functional mechanism.

(b) *B800-850 Light Harvesting Membrane Complex from Photosynthetic Bacteria (Papiz, Lawless, Isaacs, Freer, Mcdermot, Prince & Cogdell, Personal Communication, 1995).* The B800-850 complex crystallises in space group R32 with a = 121.1 and c = 296.7 Å; the molecular weight of the asymmetric unit is approximately 125 kDa. The structure reveals an inner barrel of helices surrounded by a less tightly packed layer of helices interspersed with chlorophyll and carotenoid moieties. The chlorophyll moieties are oriented so that they can interact with light from all directions; structure-functions studies are now underway to determine the mechanism(s) of energy transfer.

(c) *Avian Lens Protein δ-Crystallin*[27]. The avian lens protein, δ-crystallin, is a tetramer with a molecular weight of some 200 kDa and non-crystallographic 222 symmetry, (space group $P2_12_12_1$ with a = 89.6, b = 147.0 and c = 153.1 Å). The subunit exhibits a new type of fold composed of three essentially α-helical domains; domain 2 comprises a bundle of 5 helices which contribute to a 20-helix bundle at the core of the molecule. δ-Crystallin has a 90% sequence homology with the enzyme argininosuccinate lyase, (which catalyses the reversible elimination of L-argininosuccinate with the formation of fumarate), indicating that it is an example of a "high-jacked" enzyme.

(d) *Ceruloplasmin (Zaitseva, Zaitsev, Bax, Card, Ralph & Lindley, Personal Communication, 1995).* Human ceruloplasmin is a copper-containing glycoprotein with a molecular weight of some 132 kDa. Four functions have been attributed to this plasma protein, (i) copper transport, (ii) ferroxidase activity, (iii) amine oxidase activity, and (iv) anti-oxidant activity, but its precise biological functions have not yet been elucidated. Low, or zero levels, of the protein in the plasma are associated with Wilson's disease, as a result of accumulation of copper in the body tissues including the brain. The protein is comprised of a single polypeptide chain of 1046 amino acid residues and has a carbohydrate content of between 7 and 8 %. Sequence analysis indicates an extraordinary homology with the A-type domains of blood clotting factors V and VIII. Recently, a new crystal form, (space group $P3_221$ with a = 213.5 and c = 86.0 Å), has been under study at the SRS. Data have been collected for the native crystals and a gold derivative to 3.1 Å and for a mercury derivative to 3.7 Å. MIR and density modification techniques have yielded an electron density map which reveals some 90% of the structure. The protein is comprised of 6 domains, each with a plastocyanin-type fold. Domains 2, 4 and 6 contain mononuclear copper sites and there is a trinuclear copper centre between domains 1 and 6.

SUMMARY

This article has not attempted to give an exhaustive account of the application of synchrotron radiation to protein crystallography, but has selected a few topics aimed to show how synchrotron radiation has become an integral and invaluable tool for the macromolecular crystallographer. Hopefully, the references will enable a fuller appraisal of this subject. Macromolecular crystallography is in a growth phase, due partly to the increased ability to genetically engineer molecules and partly to programmes such as the human genome project from which some 50,000 new biological macromolecules may be expected to emerge whose structure function relationships will be need to be investigated in detail. Structural interests are beginning to focus on problems of even larger size and complexity, viruses, ribosomes and receptor-protein complexes, and new techniques and instrumentation are being developed to handle the problems of obtaining suitable single crystals and acquiring diffraction data despite sensitivity to radiation damage or denaturation. These experiments should lead to an important and exciting new range of structure-function studies.

In third generation of synchrotrons, such as ESRF, routinely equipped with insertion devices such as undulators and multiple wigglers and give even brighter beams. Provided that detector technology can keep pace with these developments experimental areas such as time-resolved measurements to investigate reaction mechanisms and the measurement of high quality data from very small crystals (< 50 mm) should considerably strengthen the ability of the crystallographer to relate macromolecular structure to biological function.

ACKNOWLEDGEMENTS

The author gratefully acknowledges the scientific and technical staff at SERC Daresbury Laboratory, both past and present, without whom this article would not have been possible. In particular he would like to thank Colin Nave, Miroslav Papiz, Sean McSweeney, P.Rizkallah, Liz Duke, John Campbell, Steve Kinder, Steve Buffey and all the other DRAL staff associated with the Protein Crystallography Project Team.

REFERENCES

1. Rossmann, M.G., Arnold, E., Erickson, J.W., Frankenberger, E.A.,Griffith, J.P., Hecht, H.J., Johnson, J.E., Kamer, G., Luo, M., Mosser, A.G., Rueckert, R.R., Sherry, B. and Vriend, G. (1985) *Nature* (London) 317: 145.
2. Acharya, R., Fry, E., Stuart, D., Fox, G., Rowlands, D. and Brown, F. (1989) *Nature* 337: 709.
3. Rould, M.A., Perona, J.J., Söll, D. and Steitz, T.A. (1989) *Science* 246: 1135.
4. Kabsch, W., Mannherz, H.G., Suck, D., Pai, E.F. and Holmes, K.C. (1990) *Nature* (London) 347: 37.
5. Winklemann, D.A., Rayment, I., Holden, H.M. and Baker, T.S. (1992) Structure of Myosin Subfragment-1. *The 4th International Conference on Biophysics & Synchrotron Radiation*, BSR92 Tsukuba, Japan, Abstract F2-09.
6. Hunter, W.N., Bailey, S., Habash, J., Harrop, S.J., Helliwell, J.R., Aboagye-Kwartang, T., Smith K. and Fairlamb, A.H. (1992) *J. Mol. Biol.* 227: 322.
7. Hope, H., Frolow, F., van Böhlen, K., Makowski, I., Kratky, C., Halfon, Y., Danz, H., Webster, P., Bartels, K.S., Wittmann, H.G. and Yonath, A. (1989) *Acta Cryst.* B45: 190.
8. Woolfson, M.M. (1970) In: *An Introduction to X-Crystallography*, Chpt 6, Cambridge University Press.
9. Liddington, R.C., Yan, Y., Moulai, J., Sahli, R., Benjamin, T.L. and Harrison, S.C. (1991) *Nature* 354: 278.
10. Thompson, A.W., Habash, J., Harrop, S., Helliwell, J.R., Nave, C., Atkinson, P., Hasnain, S.S., Glover, I.D., Moore, P.R., Harris, N., Kinder, S. and Buffey, S. (1992) *Rev. Sci. Instrum.* 63: 1062.
11. Arndt, U.W., Champness, J.N. and Wonacott, A.J. (1973) *J. Appl. Cryst.* 6: 457.
12. Arndt, U.W. and Wonacott, A.J. (1977) *The Rotation Method in Crystallography*, Amsterdam: North-Holland.
13. Hendrix, J. (1992) *Rev. Sci. Instrum.* 63: 641.
14. Green, D.W., Ingram, V.M. and Perutz, M.F. (1954) *Proc. Roy. Soc.* (London) A225: 287.
15. Cromer, D.T. (1983) *J. Appl. Cryst.* 16: 437.
16. Fourme, R. and Hendrickson, W.A. (1990) In: *Synchrotron Radiation and Biophysics* (Hasnain, S.S. ed.) Chpt 7: pp. 156, Ellis Horwood Ltd., Chichester, UK.
17. Baker, P.J., Farrants, G.W., Stillman, T.J., Britton, K.L., Helliwell, J.R., and Rice, D.W. (1990) *Acta Cryst.* A46: 721.
18. Wang, B.C. (1985) *Methods Enzymol.* 115: 90.
19. Moffat, K. (1988) *Nature* 336: 422.
20. Templeton, L.K., Templeton, D.H., Phillips, J.C. and Hodgson, K.O. (1980) *Acta Cryst.* B36: 436.
21. Narayan, R. and Ramaseshan, S. (1981) *Acta Cryst.* A37: 636.

22. Glover, I.D., Denny, R.C., Nguti, N.D., Kinder, S., McSweeney, S.M., Thompson, A.W., Dodson, E.J., Wilkinson, A.J. and Tame, J.R.H. (1995) *Acta Cryst.*, D51 (in press)
23. Lindley, P.F., Bateman, O., Glover, I.D., Myles, D., Najmudin, S., and Slingsby, C. (1993) *J. Chem. Soc. Faraday Trans.* 89: 2677.
24. Bouquiere, J.P., Finney, J.L., Lehmann, M.S., Lindley, P.F. and Savage, H.F.J. (1993) *Acta Cryst.* B49: 79.
25. Rossmann, M.G. and Erickson, J.W. (1983) J. Appl. Crystallogr. 16(6): 629-636
26. Abrahams, J.P., Leslie, A.W.G., Lutter, R. and Walker, J.E. (1994) *Nature* 370: 621.
27. Simpson, A., Bateman, O., Driessen, H., Lindley, P., Moss, D., Mylvaganam, S., Narebor, E. and Slingsby, C. (1994) *Nature Struct. Biol.* 1: 724.

14

NMR APPROACHES TO THE STUDY OF STRUCTURE-FUNCTION RELATIONSHIPS IN IRON-SULFUR PROTEINS

Rubredoxin, [2Fe-2s] Ferredoxins, and a Rieske Protein

John L. Markley,[*1] Bin Xia,[1] Young Kee Chae,[1] Hong Cheng,[†1]
William M. Westler,[1] Jeremie D. Pikus,[2] and Brian G. Fox[2]

[1] Department of Biochemistry
[2] Institute for Enzyme Research
University of Wisconsin-Madison
Madison Wisconsin 53706

ABSTRACT

Newer NMR methods, particularly in conjunction with stable isotope labeling (with ^{2}H, ^{13}C, and ^{15}N), offer powerful approaches to the elucidation of structure-function relationships in paramagnetic proteins such as iron-sulfur proteins. The optimization of NMR pulse sequences for rapidly-relaxing spins and the utilization of multidimensional multinuclear NMR spectroscopy have made it possible to determine sequence-specific assignments for a large number of NMR signals in rubredoxins, ferredoxins, and high-potential iron proteins, including some from the cysteine residues that ligate the iron ions. Such assignments are the key to a wealth of information derived from NMR parameters, such as the temperature and pH dependence of chemical shifts and the relaxation properties of the resonances that report on interactions between nuclei of the protein and unpaired electron density from the metal center. This information can be used to test theoretical descriptions of electron distribution within these molecules and to model the structures and dynamic properties of the proteins in solution. Mutagenesis of these proteins, followed by biochemical evaluation of their functional properties and biophysical analysis of their structures and stabilities, is beginning to reveal which residues are important for cluster formation and which residues play a role in electron transfer to and from redox partner proteins. This review discusses recent NMR studies from our laboratory of six iron-sulfur proteins: *Clostridium pasteurianum* rubredoxin, the vegetative and heterocyst ferredoxins from *Anabaena* 7120, human ferredoxin, and the *Pseudomonas mendocina* KR1 *tmoC* Rieske protein.

---

[*] Represents person presenting Paper

[†] Present addresses: Mayo Research Institute, Rochester Minnesota

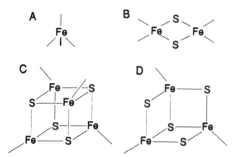

Figure 1. Four forms of ligation found by x-ray crystallography in iron-sulfur proteins. In all types of ligation, the ligands from the protein are cysteinate sulfur atoms. As indicated in the figure, all but the three-iron cluster have four cysteinyl ligands; the three-iron cluster is ligated by three cysteines. (A) One-iron complex found in rubredoxins. (B) [2Fe-2S] cluster found in plant-type, respiratory, and 2Fe bacterial ferredoxins. (C) [4Fe-4S] cluster found in bacterial 4Fe ferredoxins, in high potential iron proteins (HiPIPs), as one of the clusters in 7Fe ferredoxins, and as both of the two clusters in 8Fe ferredoxins. (D) [3Fe-4S] cluster found as one of the clusters in 7Fe ferredoxins, and in aconitase[9]. (Reproduced with permission from ref. 9)

INTRODUCTION

Iron-sulfur clusters constitute structural/functional motifs found in a wide range of proteins with diverse activities such as electron transport carriers, enzymes, transcription factors, and iron responsive elements. Various patterns of iron-sulfur ligation found in proteins, as identified by X-ray crystallography, are shown in Fig. 1. Rubredoxins contain the simplest cluster, a single iron ion coordinated tetrahedrally by four cysteines. Ferredoxins are classified by their redox potentials and cluster types, which consist of two, three, or four iron ions and either two or four inorganic sulfides. The [2Fe-2S] clusters have two oxidation states: oxidized, with both metals as Fe(III), and reduced, with one Fe(III) and one Fe(II); the [4Fe-4S] clusters can have three oxidation states: 3Fe(III)/1Fe(II), 2Fe(III)/2Fe(II), and 1Fe(III)/3Fe(II). In high potential iron proteins (HiPIPs), only the first two are accessible as the oxidized and reduced states, respectively; in ferredoxins only the last two are accessible as the oxidized and reduced states. The iron ions in clusters with more than one iron can be ferromagnetically or antiferromagnetically coupled[1-7]. In rubredoxins, ferredoxins, and HiPIPs, each iron or iron-sulfur cluster is covalently attached to the sulfur atoms of four Cys residues (Fig. 1). In other classes of iron-sulfur proteins, other amino acids serve as cluster ligands (e.g., Rieske proteins have two histidines and two cysteines)[8].

We have developed efficient methods for overproducing several iron-sulfur proteins in *Escherichia coli* and for labeling the recombinant proteins selectively and uniformly with stable isotopes for NMR investigations. The proteins are produced primarily as the apoproteins; we isolate these and then reconstitute their iron centers *in vitro* prior to final purification of the holoproteins. We use stable isotope labeling in conjunction with modern multinuclear, multidimensional NMR techniques as a way of analyzing the structures of the diamagnetic parts of these proteins in solution and of investigating the electronic properties of the clusters. Finally, we make use of site-directed mutagenesis to investigate the roles of particular residues in determining the structural and functional properties of these proteins. The iron-sulfur proteins we currently are investigating are *(i)* *Clostridium pasteurianum* rubredoxin (Rdx), as an example of a single iron protein, *(ii)* *Anabaena* 7120 vegetative ferredoxin (VFd),as a prototype for a photosynthetic (plant-type) ferredoxin, *(iii)* *Anabaena* 7120 heterocyst ferredoxin (HFd), as a prototype for a nitrogen fixing ferredoxin, *(iv)* human ferredoxin (HuFd) as an example of a mitochondrial ferredoxin, and *(v)* the *tmoC* protein

Anabaena 7120 vegetative ferredoxin (VFd)
```
1   ATFKVTLINE AEGTKHEIEV PDDEYILDAA EEQGYDL-PFS -CRAG--ACST CAGKLVSG--
56  ------TVDQSDQSFLDDDQIEAGYVLTCVAYPTSDVVIQTHKEEDLY
```

Anabaena 7120 heterocyst ferredoxin (HFd)
```
1   ASYQVRLINK KQDIDTTIEI DEETTILDGA EENGIEL-PFS -CHSGS--CSS CVGKVVEG--
56  ------EVDQSDQIFLDDEQMGKGFALLCVTYPRSNCYIKTHQEPYLA
```

Human ferredoxin (HuFd)
```
1   SSSEDKITVHFINRDGETLTTKGKVGD-SLLDVVVENNLDIDGFGACE-GTLACSTCHLIFED
62  HIYEKLDAITDEENDMLDLAYG-LTDRSRLGCQICLTKSMDNMTVRVPETVADARQSIDVGKTS
```

Clostridium pasteurianum rubredoxin (Rdx)
```
1   fMKKYTCTVCGYIYNPEDGDPDNGVNPGTDFKDIPDDWVCPLCGVGKDQFEEVEE
```

Pseudomonas mendocina KR1 tmoC Rieske protein
```
1   MSFEKICSLD DIWVGEMETF ETSDGTEVLI VNSEEHGVKA YQAMCPHQEI LLSEGSYEGG
61  VITCRAHLWT FNDGTGHGIN PDDCCLAEYP VEVKGDDIYV STKGILPNKA HS*
```

Figure 2. Comparison of sequences of VFd[10], HFd[11], HuFd[12], Rdx[13], and the *tmoC* Rieske protein[14]. The iron ligands are indicated in boldface.

from *Pseudomonas mendocina* KR1, which is part of the toluene-4-monooxygenase enzyme complex, as an example of a Rieske protein. The sequences of these proteins are shown in Fig. 2 and our typical yields of the recombinant proteins are given in Table I.

NMR METHODOLOGY FOR STUDIES OF IRON-SULFUR PROTEINS

Iron-sulfur proteins were among the first proteins studied by means of NMR spectroscopy[22-24] and have since represented fruitful subjects for NMR investigations[25-43]. Iron-sulfur proteins figure prominently in discussions of the fundamentals of NMR spectroscopy of paramagnetic proteins[44,45]. Unpaired electrons present in the paramagnetic center interact with nuclei either by a contact mechanism through chemical bonds or by a pseudo-contact mechanism through space. These interactions provide efficient nuclear relaxation mechanisms that broaden NMR signals and/or provide new chemical shift mechanisms that give rise to "hyperfine-shifted resonances" located well outside the normal diamagnetic chemical shift range for a given type of spin. The paramagnetism can hinder the determination of sequence-specific assignments, particularly to resonances from amino acid residues that ligate the iron sulfur cluster. These, of course, may be the most important assignments,

Table 1. Typical yields of recombinant proteins isolated from *e. coli* and reconstituted *in vitro*

	Yield of recombinant protein (mg/L culture)		
Protein	Rich medium	Minimal medium	Reference
Anabaena 7120 vegetative Fd	20	15	(15)
Anabaena 7120 heterocyst Fd	20	15	(16, 17)
Human Fd	20-30	20	(18,19)
Rubredoxin	40	40	(13,18)
Rieske protein *tmoC*	2	—	(20)

because they are needed to interpret the hyperfine-shifted signals. Basically, all of the NMR methods used in the studies of diamagnetic proteins[46,47] can be adapted to the study of paramagnetic proteins. These must be supplemented with specialized experiments that deal with nuclei that are relaxed rapidly by the paramagnetic center and those that experience large hyperfine shifts[48,9]. Cross-peaks between pairs of hyperfine-shifted resonances and between hyperfine-shifted and diamagnetic resonances have been observed in magnitude COSY and TOCSY spectra. Because hyperfine-shifted resonances relax too rapidly to permit coherence transfer between two proton nuclei, cross-peaks seen in such spectra arise from cross-correlation[49] between dipole-dipole relaxation and Curie spin relaxation[50,51]. These results suggest that additional experiments designed to optimize the detection of cross-correlation effects may be useful in studies of iron-sulfur proteins.

RUBREDOXIN

A clone of the clostridial rubredoxin (Rdx) gene obtained from J.-M. Moulis was inserted into the pET9a vector used for overexpression[20]. The following samples were prepared: Rdx at natural abundance, [^{2}H$^\alpha$]Cys Rdx, [^{2}H$^{\beta2,\beta3}$]Cys Rdx, [U-^{13}C]Rdx, [U-^{15}N]Rdx, [^{15}N]Cys Rdx, and [^{13}C$^\beta$]Cys Rdx. Identifications of the hyperfine signals by residue type have been made thus only for the cysteines. Much more work needs to be done to sort out and interpret the much larger number of hyperfine ^{1}H, ^{15}N, and ^{13}C signals that

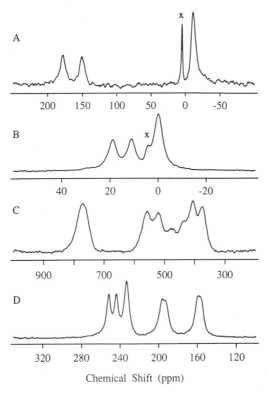

Figure 3. Hyperfine-shifted ^{2}H signals observed in spectra of Rdx samples labeled selectively with deuterium at the α- and β-positions of Cys. [^{2}H$^\alpha$]Cys: (A) oxidized, (B) reduced; [^{2}H^{b2,b3}]Cys: (C) oxidized, (D) reduced[20].

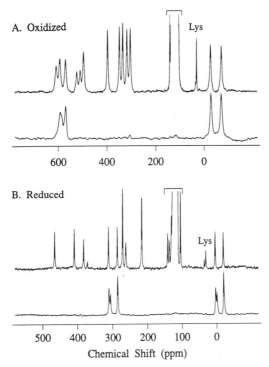

A. Oxidized   Lys

600   400   200   0

B. Reduced

Lys

500   400   300   200   100   0

Chemical Shift (ppm)

Figure 4. Hyperfine-shifted ^{15}N signals observed in the spectra of clostridial rubredoxin samples: (A) oxidized and (B) reduced. The top spectra are of samples labeled uniformly with ^{15}N; the bottom spectra are of samples labeled selectively by incorporation of [^{15}N]Cys^{20}. (Reproduced with permission from ref. 20)

have been resolved. Our ^{2}H NMR studies of the [^{2}H$^\alpha$]Cys Rdx and [^{2}H$^{\beta2,\beta3}$]Cys Rdx (Fig. 3) have provided the first complete and unambiguous assignments of signals from the cysteinate ligands to the iron. The results confirm signals observed by Werth et al.[52] for Fe(II) Rdx, which indicated the locations of the β-protons in the 150-260 ppm region and the α-protons in the 11-17 ppm region. However, our results show only two of the α-protons to be in the 11-17 ppm region (not all four as assumed by Werth et al.); the other two are located upfield of the diamagnetic region (0 to -8 ppm). The β-deuterons are better resolved than the corresponding ^{1}H signals reported earlier[50]. Ours is the first detection of the Fe(III) Rdx hyperfine signals from the eight β-protons (350-800 ppm region), and the first assignment of α-protons: two in the 150-190 ppm region and two in the -8 ppm region.

Selective labeling has been used to determine which hyperfine-shifted ^{15}N signals, which cover 750 ppm, arise from the cysteine residues (Fig. 4). Spectra of [U-^{13}C]Rdx (not shown) reveal signals from 30 very rapidly relaxing carbons, all of which presumably interact with the iron center.

TWO-IRON FERREDOXINS

Numerous mutants of [2Fe-2S] ferredoxins have been prepared and evaluated (Table II). Of the mutants designed to test whether other residues could be substituted for cysteine as ligands to the cluster, the most successful have been the Cys to Ser mutants. All four single mutants (C41S, C43S, C49S, and C79S) of VFd have been studied by optical spectroscopy,

Table 2. Catalog of ferredoxin mutations made and reconstituted proteins isolated

type of ferredoxin rationale	mutation carried out (mutant gene prepared)	stable protein isolated
Anabaena 7120 vegetative ferredoxin		
Reconstituted wild-type recombinant		wt wt
Mutations of the cysteines that ligate iron atoms to test range of different possibilities that can be accommodated	C79A,D,H,S,Y C41H+C46H	C41A,D,H,S,Y C41S,Y C46A,D,H,S,Y C46S,Y C49A,D,H,S,Y C49S C79S
Mutations to test role of peptide backbone H-bonds to the cluster		S40P, R42P, A43P G44P, A45P, S47P
Mutations to test role of conserved arginine residue		R42A,H,E R42A,H,E
Mutations to test role of other conserved residues	A50N,V, F65A,I,Y,W F65A+S64Y, L77T A43S+A45S+T48S+A50N	Q33S, S40A, A43S, T48S Q33S, S40A, A43S T48S, A50N,V F65A,I,Y,W F65A+S64Y, L77T
Mutations to test the plasticity of cluster attachment points (cysteine positions)	C79A+V87C S40C+C41A+C79S+F86V	S40A+C41A S40A+C41A C79A+D86C
Mutations to test the role of residue 78 assumed to be important for activity	T78P, T78S,T78L	T78R, T78N, T78G, T78R, T78N, T78G, T78C, T78E, T78A T78C, T78E, T78A T78P, T78S
Anabaena 7120 heterocyst ferredoxin		
Reconstituted wild-type recombinant		wt wt
Mutations of conserved Ser	S45A, S47A, S48A	S40A, S43A S40A, S43A S45A, S47A, S48A
Mutation of conserved His		H42S, H42R, H42K H42S, H42R, H42K
Mutation to change polarity around the cluster		V50S,L78T V50S
Mutation of the extra cysteine		C87A C87A
Hybrids between *Anabaena* 7120 vegetative and heterocyst ferredoxins		
Heterocyst to vegetative		VH, CVH VH, CVH
Vegetative to heterocyst		HV, CHV HV, CHV

HV = vegetative Fd background with R42H+A43S+A45S+T48S+A50V
CHV = vegetative Fd background with R42H+A43S+A45S+T48S+A50V+T78L
VH = heterocyst Fd background with H42R+S43A+S45A+S48T+V50A
CVH = heterocyst Fd background with H42R+S43A+S45A+S48T+V50A+L78T

Human ferredoxin		
Mutation of the extra cysteine		C95A C95A
Mutations of cysteines that ligate the iron atoms	C46S+C95A, C52S+C95A C55S+C95A, C92S+C95A C92A+C95S	C46A, C52A, C55A, C92A

by EPR (reduced, frozen), and by NMR (oxidized)[53]. It was not possible, however, to study the least stable mutant (C79S) in solution by NMR in the reduced state[53]. The crystal structure of one mutant (C49S) has been solved in Hazel Holden's laboratory (Enzyme Institute, Universiy of Wisconsin-Madison)[54]. All four mutants show evidence for oxygen ligation to the cluster. This is seen in the bond distance (O-Fe distance shorter than S-Fe)[54], in the magnitudes of the hyperfine shifts, and in the hyperfine shifts of the serine β-protons[53]. The three most stable mutants were found to support wild-type-like electron transfer to ferredoxin reductase[54]. Interesting correlations between the pattern of hyperfine shifts and the reduction potential have been found in our studies of the Cys to Ser mutations[53].

Our collaborative studies with Gordon Tollin and coworkers at the Univeristy of Arizona on mutants of *Anabaena* 7120 VFd have revealed the critical importance of two residues (Glu94 and Phe65) for electron transfer to FNR. Mutations at these two sites do not interfere with binding of the Fd to FNR (the first step in electron transfer), but decrease the rate of electron transfer by 10^4 [55]. Glu94 is involved in an important salt bridge between two loops adjacent to the cluster; activity is lost when it is mutated to Lys. Several residues in the vicinity of Glu94 can be mutated without loss of activity. Phe65, the other critical residue, can be replaced by another aromatic residue (Tyr or Trp) without loss of activity but not by an aliphatic residue (Ala, Ile)[56]. We have obtained definitive evidence, from ^{15}N NMR chemical shifts and T_2 values from a sample of VFd labeled selectively with ^{15}N in the only arginine residue, that the backbone nitrogen of Arg42 hydrogen bonds to the cluster in solution[15]. We had previously noted this interaction in the crystal structure of this protein[57].

In collaboration with Larry Vickery (University of California, Irvine), we have investigated His56 in mitochondrial Fds. Miura and Ichikawa[58] had reported that His56 of bovine Fd does not titrate between pH 5 and 8 in the oxidized form of the protein and attributed this effect to interaction of His56 with Ser88[59]. We confirmed an abnormal pK_a for His56 of human Fd and determined from ^{15}N chemical shifts that the residue is neutral in both oxidation states in the accessible pH range. These results rule out an earlier suggestion that His56, which is adjacent to the iron-sulfur cluster, is responsible for the pH dependence of the redox potential. Our studies made use of selective ^{15}N and ^{13}C labeling of the histidines of human Fd plus single-site His56 mutants[19].

A full solution structure of *Anabaena* 7120 VFd has been determined[17,60,61] and compared to the highly-refined X-ray structure[54]. Analysis of NMR structural data for *Anabaena* 7120 HFd revealed an interesting local difference between the structure in solution and that in the crystal[62] which we attributed to crystal packing[17]. Analysis of the T1 relaxation rates of hyperfine-shifted ^{15}N signals of *Anabaena* 7120 HFd have suggested a promising approach to accurate distance measurements between nuclei and iron atoms in solution[63].

Unexpected results have come from guanidinium chloride (GdnCl) denaturation studies of wild-type and mutant ferredoxins[64]. Low levels of the chemical denaturant unfold the protein without disrupting the cluster. Studies of the unfolding reaction have been used to determine the relative stabilities of various mutants of the *Anabaena* vegetative and heterocyst ferredoxins. Most interestingly, the single-site mutation that caused the greatest *decrease* in stability (3 kcal/mol) was the replacement of Arg42 with His; this substitution is seen in all nitrogen-fixing ferredoxins. The complementary mutation of His42 of the heterocyst ferredoxin to Arg led to a corresponding *increase* in stability (by 3 kcal/mol)[64].

RIESKE PROTEIN

Our NMR investigations of the *tmoC* protein from *Pseudomonas mendocina* KR1[4,65] are just getting underway. The protein is being overproduced has been from the pUC-18 derived expression vector pRS184 in the *E. coli* host strain DH5αF' after

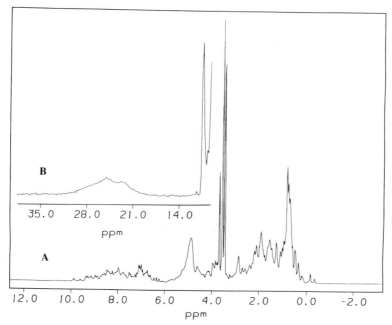

Figure 5. ^{1}H spectrum of the *tmoC* protein (Rieske protein) in oxidized state: (A) diamagnetic region; (B) expanded region showing hyperfine ^{1}H signals in the region of 25 ppm. (Y. K. Chae, B. Xia, J. D. Pikus, B. G. Fox, & J. L. Markley, unpublished data.)

induction with IPTG. Conventional ion-exchange and gel filtration purification methods provide ~50 mg of homogeneously pure protein per 100 g of bacterial cell paste, along with two other protein components of the toluene-4-monooxygenase enzyme complex[21]. The *tmoC* protein exhibits optical absorption maxima at 325 and 460 nm, typical of 2Fe ferredoxins. The presence of a Rieske-type center has been ascertained from Mössbauer and EPR measurements. The spectral features of these two redox states are nearly identical to those observed for the *Thermus thermophilus* Rieske protein[65] and the Rieske-type center found in the *Pseudomonas putida* benzoate dioxygenase[66]. The presence of the primary sequence motif $CXHX_{16/17}CX_2H$ (Fig. 2, which is found in all of the known Rieske-type proteins[67], is also consistent with the presence of a Rieske-type center in the *tmoC* protein). One-dimensional ^{1}H NMR spectra of the *tmoC* protein (Fig. 5) reveal the presence of resolved hyperfine signals both upfield and downfield of the diamagnetic region. The fingerprint region of the ^{1}H COSY spectrum of *tmoC* (Fig. 6) is well resolved, indicating that the protein will be amenable to detailed NMR analysis. This presents exciting prospects because no structure is known for any Rieske protein, and no extensive NMR studies of Rieske proteins have been published.

SUMMARY AND FUTURE PROSPECTS

Fundamental aspects of structure-function relationships in iron-sulfur proteins can be framed by the following questions. What are the critical aspects of the peptide sequence that code for iron-sulfur centers of different geometry and electronic properties? What aspects of the peptide sequence are responsible for tuning the redox potential to that of the physiologically active protein? (Another way of looking at the previous question is to ask

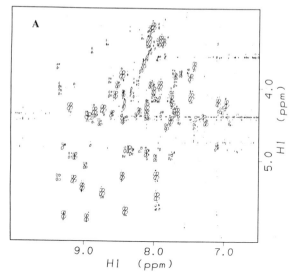

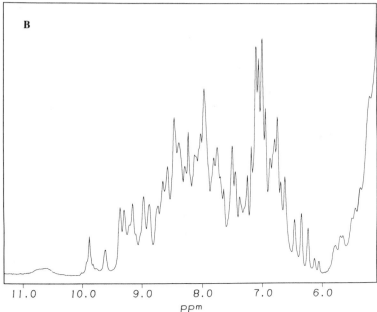

Figure 6. Spectra of the *tmoC*: (A) expansion of the H^N region of the one-dimensional ^1H NMR spectrum; (B) fingerprint region of COSY data. The sample used in both experiments was dissolved in H_2O. (Y. K. Chae, B. Xia, J. D. Pikus, B. G. Fox, & J. L. Markley, unpublished data.)

what are the determinants of the relative stabilities of the oxidized and reduced forms of the Fe-S protein?) What are the structures of the iron centers, and what role is played by the rest of the protein? What are the patterns of electron delocalization onto atoms of the protein, and what role do these play in physiological electron transport? What residues of the protein are responsible for the specificity of electron transport, and what are the structures of the relevant protein-protein complexes?

Results from our laboratories and others around the world are beginning to provide partial answers to these questions. Our general approach has been to refine methods for overproducing prototypes of different classes of iron so that large quantities can be produced for biophysical and functional studies. By producing the proteins in *E. coli*, we are able to label them uniformly or selectively with ^{2}H, ^{13}C, and ^{15}N for stable-isotope assisted magnetic resonance investigations. Moreover, this approach gives us the ability to change the protein sequences by mutagenesis in order to test ideas concerning the roles of certain residues in determining protein structure, stability, and function.

We have concentrated recently on developing methods for detecting and assigning hyperfine-shifted resonances. Considerable progress has been made[15,20], but more selective labeling methods (for example by incorporation of cysteine with chiral ^{1}H and ^{2}H labeling in the β-position and by residue selective labeling by peptide synthesis) will be required in order to complete the assignments. Once assignments have been made for hyperfine-shifted signals, the information from their NMR parameters (chemical shift, temperature and pH dependence of the shift, linewidth, and longitudinal relaxation rate) can be interpreted in molecular terms. Advances are being made in the theoretical understanding of these NMR parameters, but some gaps remain between experimental results and their explanation. The current theoretical model[68,69] yields a satisfactory interpretation of the temperature dependence of paramagnetic cysteinyl β-proton resonances in NMR spectra of many Fds and HiPIPs[29,33,34,70-80]. Despite its many successes, the current model[69] fails to explain the anti-Curie temperature dependence observed for the cysteine hyperfine-shifted resonances in oxidized [2Fe-2S] Fds[19]. This effect has been attributed to anti-Curie contributions from excited paramagnetic states, which the model ignores. Moreover, the model fails to predict the observed temperature dependences of the cysteinyl carbon and nitrogen resonances of reduced VFd[19], which are opposite those of the protons of the same. It is hoped that the availability of detailed assignments of hyperfine-shifted resonances will stimulate the development of more comprehensive theoretical models for iron-sulfur proteins.

Acknowledgments

This research was supported by grant MCB-9215142 from the U.S. National Science Foundation to J.L.M.. NMR data were collected at the National Magnetic Resonance Facility at Madison which has partial support from the Biomedical Research Technology Program of the National Institutes of Health under grant number RR02301.

REFERENCES

1. Beinert, H. (1990) *FASEB J.* 4: 2483-2491.
2. Bruschi, M., and Guerlesquin, F. (1988) *FEMS Microbiology Reviews* 54: 155-176.
3. Howard, J. B. and Rees, D. C. (1991) *Adv. Protein Chem.* 42: 199-208.
4. Lovenberg, W., ed. (1973) *Iron Sulfur Proteins*, Vol. III, Academic Press, New York.
5. Spiro, T. G., ed. (1982) *Iron Sulfur Proteins*, Vol. IV, John Wiley, New York.
6. Sweeney, W. V. and Rabinowitz, J. C. (1980) *Annu. Rev. Biochem.* 49: 139-161.
7. Thompson, A. J. (1985) In: *Metalloproteins* (Hanson, P., ed.) Part I, pp. 79-120, Verlag Chemie, Weinheim.
8. Gurbiel, R. J., Batie, C. J., Sivaraja, M., True, A. E., Fee, J. A., Hoffman, B. M. and Ballou, D. P. (1989) *Biochemistry* 28: 4861-4871.
9. Cheng, H. and Markley J.L. (1995) *Ann. Rev. Biophys. Biomol. Structure* 24: 209-237.
10. Alam, J., Whitaker, R. A., Krogmann, D. W. and Curtis, S. E. (1986) *J. Bacteriol.* 168: 1265-1271
11. Böhme, H. and Haselkorn, R. (1988) *Mol. Gen. Genet.* 214: 278-285.
12. Mittal, S., Zhu, Y.-Z. and Vickery, L. E. (1988) *Archives Biochem. Biophys.* 264: 383-391.
13. Mathieu, I., Meyer, J. and Moulis, J.-M. (1992) *Biochem. J.* 285: 255-262.

14. Yen, K.-M., Karl, M. R., Blatt, L. M., Simon, M. J., Winter, R. B., Fausset, P. R., Lu, H. S., Harcourt, A. A. and Chen, K. K. (1991) *J. Bacteriol.* 173: 5315-5327.
15. Cheng, H., Westler, W.M., Xia, B., Oh, B.-H. and Markley, J.L. (1995) *Arch. Biochem. Biophys.* 316: 619-634.
16. Böhme, H. and Haselkorn, R. (1989) *Plant Mol. Biol.* 12: 667-672.
17. Chae, Y.K., Abildgaard, F., Mooberry, E.S. and Markley, J.L. (1994) *Biochemistry* 33: 3287-3295.
18. Coghlan, V. M. and Vickery, L. E. (1989) *Proc. Natl. Acad. Sci. USA* 86: 835-839.
19. Xia, B., Cheng, H., Skjeldal, L., Coghlin, V.M., Vickery, L.E. and Markley, J.L. (1995) *Biochemistry* 34: 180-187.
20. Xia, B., Westler, W. M., Cheng, H, Meyer, J., Moulis, J.-M. and Markley, J. L. (1995) *J. Am. Chem. Soc.*, in press.
21. Fox, B. G. and Pikus, J. D. unpublished results.
22. Phillips, W. D. (1973) In: *NMR of Paramagnetic Molecules*, eds. (La Mar, G.N., Horrocks, W.D., Holm R.H., eds.) pp. 421-478. Academic Press, New York.
23. Phillips, W. D. and Poe, M. (1972) In: *Methods in Enzymology*, (San Pietro, A. ed.) vol. 24: pp. 304-317, Academic Press, New York.
24. Phillips, W. D. and Poe, M. (1973) In: *Iron-Sulfur Proteins*, (Lovenberg, W. ed) vol. 2: pp. 255-284, Academic Press, New York.
25. Bertini, I., and Luchinat, C. (1986) *NMR of Paramagnetic Molecules in Biological Systems*, Benjamin/Cummings, Menlo Park, California.
26. Bertini, I., Banci, L. and Luchinat, C. (1989) *Methods Enzymol.* 177: 246-264.
27. Blake, P. R., Park, J.-B., Adams, M. W. W. and Summers, M. F. (1992) *J. Am. Chem. Soc.* 114: 4931-4933.
28. Markley, J. L., Chan, T.-M, Krishnamoorthi, R., & Ulrich, E. L. (1986) In: *Iron-Sulfur Protein Research*, (Matsubara, H., Katsube, Y., Wada, K. eds.) pp. 167-181, Springer-Verlag, Tokyo/Berlin.
29. Nagayama, K., Ohmori, D., Imai, T., & Oshima, T. (1986) In: *Iron-Sulfur Protein Research*, (Matsubara, H., Katsube, Y., Wada, K. eds.) pp. 125-138, Springer-Verlag, Tokyo/Berlin.
30. Krishnamoorthi, R., Markley, J. L., Cusanovich, M. A., Przysiecki, C. T. and Meyer, T. E. (1986) *Biochemistry* 25: 60-67.
31. Dugad, L. B., Le Mar, G. N., Banci, L. and Bertini, I. (1990) *Biochemistry* 29: 2263-2271.
32. Bertini, I. (1994) The Tridimensional Structure of a Paramagnetic HiPIP in Solution from NMR, International Conference on the Use of Stable Isotopes in NMR Studies of Protein Structure, Dynamics and Function, Palais du Luxembourg, Paris, March 21-23, abstract.
33. Bertini, I., Briganti, F., Luchinat, C. and Scozzafava, A. (1990) ^{1}H NMR Studies of the Oxidized and Partially Reduced 2[Fe-4S] Ferredoxins from *Clostridium pasteurianum. Inorg. Chem.* 28: 1874-1880.
34. Bertini, I., Briganti, F., Luchinat, C., Messori, L., Monnanni, R., Scozzafava, A. and Vallini, G. (1991) *FEBS Lett.* 289: 253-256.
35. Bertini, I., Briganti, F., Luchinat, C., Messori, L., Monnanni, R., Scozzafava, A. and Vallini, G. (1992) *Eur. J. Biochem.* 28: 1874-1880.
36. Busse, S. C., La Mar, G. N. and Howard, J. B. (1991) *J. Biol. Chem.* 266: 23714-23723.
37. Gaillard, J., Moulis, J.-M. and Meyer, J. (1987) *Inorg. Chem.* 26: 320-324.
38. Pochapsky, T. C. and Ye, X. M. (1991) *Biochemistry* 30: 3850-3856.
39. Ye, X. M., Pochapsky, T. C. and Pochapsky, S. S. (1992) *Biochemistry* 31: 1961-1968.
40. Greenfield, J. J., Wu, X. and Jordan (1989) *Biochim. Biophys. Acta* 995: 246-254.
41. Miura, S. and Ichikawa, Y. (1991) *Eur. J. Biochem.* 197: 747-757.
42. Krishnamoorthi, R., Markley, J. L., Cusanovich, M. A. and Przysiecki, C. T. (1986) *Biochemistry* 25: 50-54.
43. Banci, L., Bertini, I. and Briganti, F. (1991) *New J. Chem.* 15: 467-477.
44. Satterlee, J. D. (1990) *Concepts in Magn. Reson.* 2: 69-79 and 119-129.
45. Bertini, I., Turano, P. and Vila, A. (1993) *Chem. Rev.* 93: 2833-2932.
46. Wüthrich, K. (1986) *NMR of Proteins and Nucleic Acids*, Wiley, New York.
47. Edison, A. S., Abildgaard, F., Westler, W. M., Mooberry, E. S. and Markley, J. L. (1994) *Methods. Enzymol.* 239: 3-79.
48. Lecomte, J. T. J., Unger, S. W. and La Mar, G. N. (1991) *J. Magn. Reson.* 94: 112-122.
49. Wimperis, S. and Bodenhausen, G. (1989) *Mol. Physics* 66: 897-919.
50. Qin, J., Delaglio, F., La Mar, G. and Bax, A. (1993) *J. Magn. Reson.* Ser. B 102: 332-336.
51. Bertini, I., Luchinat, C. and Tarchi, D. (1993) *Chem. Phys. Lett.* 203: 445-449.
52. Werth, M. T., Kurtz, D. M., Jr., Moura, I. and LeGall, J. (1987) *J. Am. Chem. Soc.* 109: 273-275.
53. Cheng, H., Xia, B., Reed, G. H. and Markley, J. L. (1994) *Biochemistry* 33: 3155-3164.

54. Holden, H. M., Jacobson, B. L., Hurley, J. K., Tollin, G., Oh, B.-H., Skjeldal, L., Chae, Y. K., Cheng, H., Xia, B. and Markley, J. L. (1994) *J. Bioeng. Biomemb.* 26: 67-88.

55. Hurley, J. K., Salamon, Z., Meyer, T. E., Fitch, J. C., Cusanovich, M. A., Markley, J. L., Cheng, H., Xia, B., Chae, Y. K., Medina, M., Gomez-Moreno, C. and Tollin, G. (1993) *Biochemistry* 32: 9346-9354.

56. Hurley, J. K., Cheng, H., Xia, B., Markley, J. L., Medina, M., Gomez-Moreno, C. and Tollin, G. (1993) *J. Am. Chem. Soc.* 115 11698-11701.

57. Rypniewski, W. R. Breiter, D. R., Benning, M. M., Wesenberg, G., Oh, B.-H., Markley, J. L., Rayment, I. and Holden, H. M. (1991) *Biochemistry* 30: 4126-4131.

58. Miura, S., Tomita, S. and Ichikawa, Y. (1991) *J. Biol. Chem.* 266: 19212-19216.

59. Miura, S. and Ichikawa, Y. (1991) *J. Biol. Chem.* 266: 6252-6258.

60. Chae, Y. K., Xia, B., Cheng, H., Oh, B.-H., Skjeldal, L., Westler, W. M. and Markley, J. L. (1995) In: *Nuclear Magnetic Resonance of Paramagnetic Macromolecules* (La Mar, G.N. ed.), NATO ASI Series C, Vol. 457: pp. 297-317, Kluwer, Dordrecht.

61. Chae, Y. K. (1994) Multinuclear Multidimensional NMR Studies of *Anabaena* 7120 vegetative and Heterocyst Ferredoxins, Ph.D. Thesis, University of Wisconsin-Madison.

62. Breiter, D. R., Meyer, T. E., Rayment, I. and Holden, H. M. (1991) *J. Biol. Chem.* 266: 18660-18667.

63. Chae, Y. K. and Markley, J. L. (1995) *Biochemistry* 34: 188-193.

64. Hurley, J. K., Caffrey, M. S., Markley, J. L., Cheng, H., Xia, B., Chae, Y. K., Holden, H. M. and Tollin, G. *Protein Sci.*, in press.

65. Fee, J. A., Findling, K. L., Yoshida, T., Hille, R., Tarr, G. E., Hearshen, D. O., Dunham, W. R., Day, E. P., Kent, T. A. and Münck, E. (1984) *J. Biol. Chem.*, 259: 124-133.

66. Altier, D. J., Fox, B. G., Münck, E. and Lipscomb, J. D. (1993) *J. Inorg. Biochem.* 43: 300.

67. Gurbiel, R. J., Ohnishi, T., Robertson, D. E., Daldal, F. and Hoffman, B. M. (1991) *Biochemistry* 30: 11579-11584.

68. Dunham, W. R., Palmer, G., Sands, R. H. and Bearden, A. J. (1971) *Biochim. Biophys. Acta* 253: 373-384.

69. Banci, L., Bertini, I. and Luchinat, C. (1990) *Struct. Bonding* 72: 113-136.

70. Skjeldal, L., Westler, W. M. and Markley, J. L. (1990) *Arch. Biochem. Biophys.* 278: 482-485.

71. Skjeldal, L., Markley, J. L., Coghlan, V. M. and Vickery, L. (1991) *Biochemistry* 30: 9078-9083.

72. Skjeldal, L., Westler, W. M., Oh, B.-H., Krezel, A. M., Holden, H. M., Jacobson, B. L., Rayment, I. and Markley, J. L. (1991) *Biochemistry* 30: 7363-7368.

73. Banci, L., Bertini, I., Briganti, F., Luchinat, C. and Scozzafava, A., et al. (1991) *Inorg. Chem.* 30: 4517-4524.

74. Banci, L., Bertini, I., Briganti, F., Scozzafava, A. and Oliver, M. V. (1991) *Inorg. Chim. Acta* 180: 171-175.

75. Banci, L., Bertini, I., Capozzi, F., Carloni, P. and Ciurli, S. et al. (1992) *J. Am. Chem. Soc.* 115: 3431-3440.

76. Banci, L., Bertini, I., Ciurli, S., Ferretti, S. and Luchinat, C. et al. (1993) *Biochemistry* 32: 9387-9397.

77. Bertini, I., Capozzi, F., Ciurli, S., Luchinat, C., Messori, L. and Piccioli, M. (1992) *J. Am. Chem. Soc.* 114: 3332-3340.

78. Bertini, I., Capozzi, F., Luchinat, C., Piccioli, M. and Oliver, V. (1992) *Inorg. Chim. Acta* 198-200, 483-491.

79. Bertini, I., Capozzi, F., Luchinat, C. and Piccioli, M. (1993) *Eur. J. Biochem.* 212: 69-78.

80. Busse, S. C., La Mar, G. N., Yu, L. P., Howard, J. B. and Smith, E. T., et al. (1992) *Biochemistry* 31: 11952-11962.

15

THERMODYNAMIC VIEWS OF INHIBITION, ACTIVATION AND DENATURATION OF ADENOSINE DEAMINASE BY RING OPENED NUCLEOSIDES AND DENATURANTS

Ali A. Moosavi-Movahedi

Institute of Biochemistry and Biophysics
University of Tehran, Tehran, Iran

ABSTRACT

Adenosine deaminase (ADA) is one of the important enzymes acting in the metabolism of nucleic acid components. Lack of ADA has been shown to be associated with inherited severe combined immunodeficiency (SCID) and acquired immunodeficiency syndrome (AIDS). A marked increase in ADA has also been implicated in a number of other clinical conditions including, hereditary haemolytic anaemia and leukemias. Therefore, the great importance of research on inhibition, activation of ADA is required.

Here, the inhibition and activation of ADA by derivatives of acyclic adenine nucleoside (Compounds I, II, C - IX) substituted at the ninth adenine position are studied and denaturation of ADA by sodium n - dodecyl sulphate (SDS) and dodecyl trimethylammonium bromide have been also considered kinetically and thermodynamically at phosphate buffer, pH 7.5 in various temperatures.

The type of inhibition was analysed thermodynamically, using melting point (T_m) as a sensitive point for conformational change. The activation and deactivation of ADA were also investigated by pK method to obtain the amino acids which are interacted with inhibitors and activator. The glutamic, aspartic acids and a little histidine was involved to C - IX for activation. The deactivation was occured by change on histidine only.

The denaturation of ADA was also studied to obtain more information on the structure of the enzyme. The interaction of SDS caused the folding, whereas DTAB caused the unfolding which is corresponded to activation and deactivation for ADA respectively. The folding of ADA by SDS induced the minimum solubility conformed at lower temperature which means the greater apolar interactions in the interior phase result in a lower value for T_H (temperature of minimum solubility), whereas it is inconsistent to unfolding state which is induced by ADA - DTAB complexes that is occured in a higher value for T_H.

INTRODUCTION

Adenosine deaminase (ADA) (E.C. 3.5.4.4) is one of the major enzyme in purine metabolism, catabolizing the reversible hydrolysis of adenosine or deoxyadenosine to the respective inosine product and ammonia[1]:

$$\text{Adenosine} + H_2O \rightarrow \text{Inosine} + NH_3$$

The enzyme is found in nearly all mammalian cells and plays a central role in maintaining competency of the immune system, amongst several other functions[2,3].

Many therapeutic agents of the adenosine type are known to be inactivated or activated by deamination with this enzyme. A marked increase in ADA has been associated with acute lymphoblastic leukaemia and with a hereditary form of haemolytic anaemia[4].

Lack of ADA activity in red cells is associated with a hereditary and ultimately fatal immunodeficiency disease in man[5]. The disease is called severe combined immunodeficiency (SCID) because there is dysfunction of both T and B lymphocytes with impairment of cellular and humoral immunity[6]. The derangement of the ADA gene debilitate the immune system and responsible for about 25 percent of all cases of SCID[7]. It is important to note that the depressed activities of purine enzymes in lymphocytes of patients infected with human immunodeficiency virus (HIV) has been reported. It should be mentioned that the growing interest in immunodeficiency (a feature of acquired immunodeficiency syndrome, AIDS) has led to the finding that decreased ADA activity is an enzyme marker for SCID[8,9]. These findings have stimulated an intense interest to produce analogues of adenosine that have the ability to bind to ADA as inhibitors or activators.

Since over the last few decades there are an enormously diverse range of the inhibitory effect of various compounds as well as purine nucleosides on this enzyme but there is no report on activation of ADA. In view of the fact that for better understanding of the function of the enzyme, need to recognize structure of the macromolecule. Therefore, the present paper studies the kinetic and thermodynamic views of ADA which is interacted with ring - opened analogues of adenine nucleoside as activators or inhibitors as well as anionic and cationic surfactants e.g. Sodium n - dodecyl sulphate (SDS) and dodecyl trimethylammonium bromide (DTAB) as a denaturants for denaturation study as a key understanding for structural analysis.

EXPERIMENTAL

Materials

Compounds I , IX were synthesized as reported by Hakimelahi[10,11] and compound II was prepared according to the method of Schaeffer[12]. Adenosine deaminase from calf intestinal mucosa, SDS and DTAB were obtained from Sigma. All the salts used in the preparation of the buffers were analytical grades and they were made up in doubly distilled water. Sodium phosphate buffers, 55mM at pH 7.5 for inhibition and activation by nucleosides and 2.5 mM at pH 6.4 for denaturation by surfactants were used. For pKa measurement was used the buffers ranging in pH from 3 to 12 with 0.1 as previously described[13]. Bromophenacyl bromide and diethyl pyrocarbonate as modifiers were obtained from Sigma.

Methods

The activity measurement was based on the Kaplan method[14] in which the change of absorbance at 265 nm was measured between 0.03 to 0.18 in 55mM phosphate buffer, pH

Table 1. Kinetic parameters of adenine nucleoside derivatives as substrate, inhibitors for calf intestinal adenosine deaminase compounds I , II. C - IX is activator and SDS, DTAB are denaturants

Compound	$K_i(\mu M)$	$K_m(\mu M)$
(I) Esterified Derivative Schaeffer Product	54.9	—
(II) Schaeffer Product	151.3	—
(III) Adenosine	—	41
Inosine	470	—
Guanosine	218	—

C - IX	Activator (nucleoside)
SDS	Denaturant (activator)
DTAB	Denaturant (inhibitor)

7.5 at various temperatures. The cuvettes had a 1 - cm path length and contained 3.0 ml of 15 - 200 µm adenosine to which was added 15µl of the ADA solution with and without the presence of nucleosides. A unit of enzyme activity is the amount of enzyme which convert 1 µ mol of adenosine to inosine per min. of pH 7.5 .The ultraviolet spectra and kinetic data were obtained with a Shimadzu 160 spectrophotometer. The concentration of ADA for activity measurement was 1.76 10^{-9} M.

The experiment for pKa measurement was made as previously described[13]. The use of bromophenacyl bromide, a modifier for aspartate and glutamate residues in proteins, diminished the peak at 3.9 and diethyl pyrocarbonate a modifier for histidine residues, reduced the peaks at pH 5.5 - 6.5.

For denaturation experiment using a Gilford model 2400 - 2 multiple sample absorbance spectrophotometer equipped with recorder for melting at constant heating rate of 2/3 °C per min . The hyperchromicity and hypochromicity refered to A_{280} was recorded at 5 degree intervals. All measurement reported to SDS and DTAB concentrations below the critical micelle concentration (cmc). The concentration of ADA for denaturation study was 0.02% (w/w).

In all calculations the molecular weight of 34,500 for ADA was taken[15].

RESULTS AND DISCUSSION

Table 1 shows kinetic parameters for effect of various compounds on ADA. K_m, V_{max} was obtained from Hofstee plot[16] based on equation:

$$V = V_{max} - K_m \frac{V}{S}$$

(1)

where V and V_{max} are velocity and maximum velocity respectively. K_m and V_{max} can be obtained from Hoftsee plot. Figure 1 indicates the effect of various concentrations of adenosine with ADA at different temperatures which is described previously[17]. ki was obtained from secondary plot based on equation:

$$\frac{K'_m}{V_{max}} = \frac{K_m}{V_{max}K_i} [i] + \frac{K_m}{V_{max}}$$

(2)

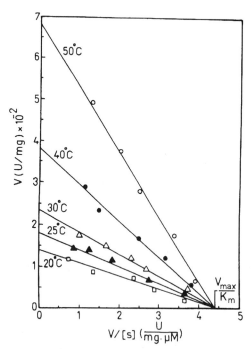

Figure 1. Hofstee plots of the effect of various concentrations of adenosine with adenosine deaminase (ADA) at different temperatures. This figure is directly taken from Ref 17.

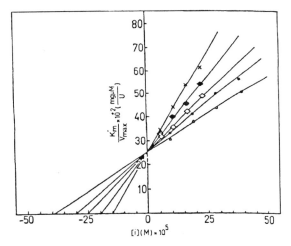

Figure 2. Secondary plots versus various concentration of guanosine at different temperatures. This plot is one example for guanosine which was used for obtaining Ki for nucleoside compounds in Table I. X , 20°C; u, 25°C; , 30°C; ●, 35°C; O, 40°C.

where K'_m is the apparent K_m and [i] is the concentration of inhibitor. k_i can be obtained by the intercept on the negative horizonal axis which is shown in Figure 2 for example.

The nucleosides in the Table I are competitive inhibitors except compound IX which is activator. The competitive inhibitory was checked by Lineweaver Burk plot.

It can confirm the competitivity or non - competitivity by thermodynamic view which is outlined here for first time.

The effect of temperatures on the specific activity of ADA in the absence and the presence of compound I and II is shown in Figure 3. The optimum activity occurred at a temperature of 333K; then it was deactivated. It was previously reported that optimum temperature (T_m) for ADA is 333 K[18]. T_m is a transition point for conformational change in protein. Among the conformational change, the T_m will be altered. There is no change in

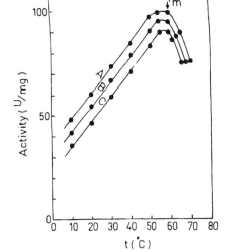

Figure 3. The effect of temperature on the specific activity of adenosine deaminase. (A) adenosine; (B) in the presence of compound II; (C) in the presence of compound I. The concentrations of adenosine 24.1 μM and inhibitors I and II, 23.7 μM. The arrow indicates the optimum temperature of ADA (T_m). This figure is directly taken from Ref 17.

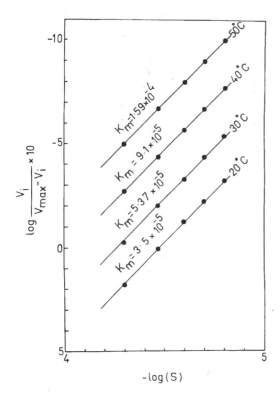

Figure 4. Hill plots for adenosine deaminase and adenosine complex at various temperature. It is important to note if substituting the V_i inhibitors, no change occurred for slope ($n_H = 1$) but the K_m value decreased. This figure is directly taken from Ref 17.

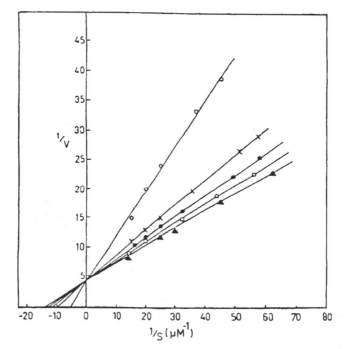

Figure 5. Lineweaver Burk plot for compound C-IX at various concentration. O, 20μM; X, 26.66 μM; ● 33.33; □ 40 μM; s, 46.73 μM.

Figure 3 for interaction of ADA with compound I and II, which means, the conformational change has not occured. Whereas, if denaturation happens, T_m will be changed which will be discussed later. Another record for this purpose is the Hill plot which is shown in Figure 4. This figure is for ADA - adenosine complex, but it is important to note if substituting the V_i inhibitors, no change occurred for slope ($n_H = 1$) but the K_m value decreased[18]. When the Hill coefficient[19] of $n_H = 1$, which means, no conformational change occured among the interaction of compounds I and II with ADA.

It is important to note that deactivation is probably due to thermal denaturation of ADA. Compounds I and II were not changed by melting point indicating their ineffectiveness on the conformational state of ADA, and confirming their binding to the same site (active site) to which adenosine was already bound.

It is important to note, there is no report on the activation of ADA in literature until now. The activation of ADA was occured by compound C - IX which is shown in Figures 5 and 6. The compound C - IX has also competitive action to adenosine as a substrate.

ADA is a glycoprotein consisting of a single polypeptide chain of 311 amino acids[20] and it was sequenced in 1984[21].

It is reported that Asp(295), His(238) and Glu(217) are located in ADA active site[3]. In either mechanism, Asp(295) acts as a general base and the zinc serves as a powerful electrophile to activate the water molecule. His(238) further assists in orienting the water and stabilizing the charge of the attacking incipient hydroxide. The protonated Glu(217) or the water hydrogen bonded to it could donate or share a proton with N - 1 of the substrate[22].

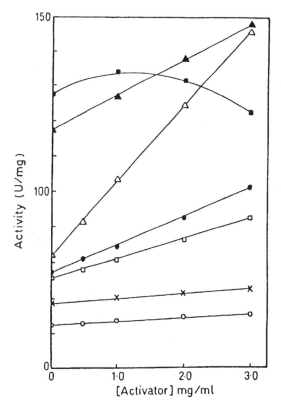

Figure 6. Specific activity versus activator C - IX concentration at different concentrations of substrate (adenosine). O , 20 μM; X, 26.66 μM; □ 33.33 μM; ● 40 μM; Δ, 46.73 μM; s , 53.33 μM ; ■ , 66.66 μM.

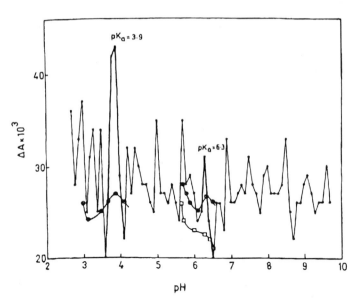

Figure 7. Two-dimensional representation of difference absorption studies of ADA versus pH's. ●, for the aspartate and glutamate modifier of 4 - bromophenacyl bromide and for the histidine modifier of diethyl pyrocarbonate at the presence of C -IX for activation of ADA. □, for the histidine modifier of diethyl pyrocarbonate at the presence of compounds I and II. No change for aspartate and glutamate at the presence of compounds I and II.

Figure 7 represents the plot of difference absorbance vs pH which indicating pKa's of each titrable groups in protein[13].

The pKa's at 3.9 and 6.3 belong to aspartate, glutamate and histidine respectively which are in ADA active site.

The use of bromophenacyl bromide as a modifier for aspartate, glutamate and diethyl pyrocarbonate as a modifier for histidine residues in ADA at the presence compound of C - IX diminished the peaks at 3.9 and a little at 6.3 respectively which means the interaction of C - IX with aspartate, glutamate largely and histidine slightly were occured during the activation of ADA. Although during deactivation by compounds I and II there is no change in peak belonging to aspartate and glutamate (pKa = 3.9) whereas there is alteration for histidine peaks in the area of pH's between 5.5 to 6.5.

For better understuding of ADA structure, here it is discussed the denaturation of ADA in the presence of denaturants of sodium n - dodecyl sulphate (SDS) and dodecyl trimethylammonium bromide (DTAB) thermodynamically.

Figure 8 shows the activation and deactivation of ADA in the presence of SDS and DTAB respectively. For better interpretation of these functions we choosed thermodynamic study of ADA denaturation.

Thermal denaturation of ADA in the presence of SDS and DTAB was extensively studied by us previously[23,18]. Table II represents the thermodynamic parameters for interaction of ADA with SDS and DTAB as denaturants. The data shows that SDS is induced folding for ADA whereas unfolding take happen at the presence of DTAB.

The reducing of ΔCp, T_s, T_H, $\Delta H_{vH}/\Delta H_{298}$ indicate the folding for ADA - SDS complexes and increasing the ΔCp, T_s, T_H induce the unfolding for ADA - DTAB interaction.

ADA is activated by SDS which is probably corresponded to folding and deactivated by DTAB which has agreement to unfolding.

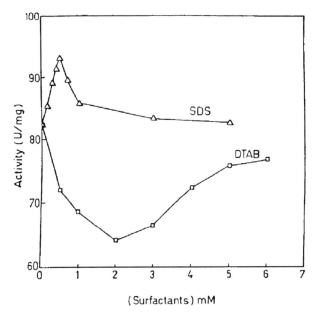

(Surfactants) mM

Figure 8. Specific activity versus SDS and DTAB concentrations. $\triangle$, SDS; $\square$, DTAB.

Table I represents the alteration for T_m during ADA denaturation by DTAB specially and SDS at higher concentration, whereas the T_m was unchanged in Figure 3 when compound I and II was interacted with ADA as competitive inhibitors.

The interaction of SDS and DTAB caused the folding and unfolding of ADA resulting in a decreasing and increasing of T_H (temperature of minimum solubility) respectively. The

Table 2. Thermodynamic parameters for interaction of ADA with SDS and DTAB.
This data was taken from Refs (18,23)

[SDS] mM	T_m (K)	ΔS_m $Jmol^{-1}K^{-1}$	ΔH_m $kJmol^{-1}$	ΔH_{vH} $kJmol^{-1}$	ΔH_{298} $kJmol^{-1}$	$\Delta H_{vH}/\Delta H_{298}$	ΔC_p $kJmol^{-1}K^{-1}$	T_s (K)	T_H (K)
0	333.5	468	152	129	30.4	4.22	3.43	292	289
0.24	333.5	300	100	109	52.1	2.08	1.35	267	259
1	333.5	320	107	106	66.0	1.60	1.15	252	240
2	333.5	330	110	109	76.6	1.42	0.944	235	216
4	333.5	340	113	109	100.5	1.08	0.739	211	180
5	342	290	102	109	76.7	1.40	0.582	199	166
[DTAB] mM									
1	321.25	570.21	183.18	137.28	–	–	5.59	291.51	288.47
2	313.5	819.5	256.91	195.13	–	–	9.33	288.19	285.97
3	312.5	1086.67	339.58	273.48	–	–	16.53	293.23	291.96
4	312.75	1192.3	372.89	351.66	–	–	26.75	299.40	298.81

folding of ADA by SDS, induced minimum solubility at lower temperatures indicating enhanced apolar interactions in the interior phase resulting in a lower value for T_H. In contrast with the interaction of ADA with DTAB led to the unfolding of the enzyme and a higher value of T_H. This interpretation is probably another feature for reason of activation and deactivation of ADA in the presence of SDS and DTAB respectively.

ACKNOWLEDGMENTS

This study has been financially supported by a grant from the Research Council of the University of Tehran.

REFERENCES

1. Worthington, C. (1988) *Worthington Manual, enzymes, related biochemicals*, pp.11, Worthington Bio-chemical Corporation Freehold, New Jersey.
2. Martin, D.W. and Gelfand, E. W. (1981) *Ann. Rev. Biochem.* 50: 845 - 877.
3. Sharff, A.J., Wilson, D.K., Chang, Z. and Quiocho, F.A. (1992) *J. Mol. Biol.* 226: 917 - 921.
4. Daddona, P.E. and Kelley, W.N. (1978) *J. Biol. Chem.* 253: 4617.
5. Giblett, E.R., Anderson, J.E., Cohen, F., Pollara, B. and Meuwissen, H. J. (1972) *Lancet* 2: 1067 - 1069.
6. Coleman, M.S., Donofrio, J., Hutton, J.J. and Hahn, L. (1978) *J. Biol. Chem.* 253: 1619 - 1626.
7. Verma, I.M. (1990) *Scientific American* 263, 34 - 41.
8. Cowan, M.J., Brady, R.O. and widder, K.J. (1986) *Proc. Natl. Acad. Sci.* U.S.A., 83: 1089 - 1091.
9. Can, T.E., Daddona, P.E. and Mitchel, B.S. (1987) *Blood* 69: 1376 - 1380.
10. Hakimelahi, G.H., Khalafi - Nezhad, A. and Mohanazadeh, F. (1990) *Int. J. Chemistry* 1: 9.
11. Hakimelahi, G.H., Zarrinehzad, M., Jarrahpour, A.A. and Sharghi, H., (1987) *Helvetica Chimica Acta* 70: 219.
12. Schaeffer, H.J. (1983) In: *Nucleoside, nucleotides and their biological application* (Rideout, J.L., Henry, D.W. and Beacham L.M. eds.) Academic press, New York.
13. Farzami, B., Moosavi - Movahedi, A.A. and Naderi, G.A. (1994) *Int. J. Biol. Macromol.* 16: 181 - 186.
14. Kaplan, N. O. (1955) *Methods Enzmol.* 2: 473.
15. Brady, T.G. and O'Sullivan, W. (1967) *Biochim. Biophys. Acta* 132: 127.
16. Dixon, M. and Webb, E.C. (1979) *Enzymes* 3rd Edn. pp. 61, Academic press, New York,.
17. Moosavi - Movahedi, A.A., Rahmani, Y. and Hakimelahi, G.H. (1993) *Int. J. Biol. Macromol.* 15: 125 - 129.
18. Moosavi-Movahedi, A.A., Samiee, B. and Hakimelahi, G.H. (1993) *J. Colloid Interface Sci.* 161: 53 - 59.
19. Hill, A.V. (1910) *J. Physiol.* 40: 4p.
20. Shimada, N., Hasegawa, S., Saito, S., Nishikiori, T., Fujii, A. and Takita, J. (1987) *J. Antibiotics* 40: 1788.
21. Daddona, P.E., Schewach, D.S., William, N.K., Argos, P., Markham, A.F. and Orkin, S.H. (1984) *J. Biol. Chem.* 259: 1210.
22. Wilson, D.K., Rudolph, F.B. and Quiocho, F.A. (1991) *Science* 252: 1278 - 1281.
23. Moosavi-Movahedi, A.A., Moghaddamnia, H. and Hakimelahi, G.H. (1994) *Iran. J. Chem. & Chem. Eng.* 13: 39 - 44.

16

DOMAIN MOTIONS IN PROTEIN CRYSTALS

David Moss,* Andrej Šali,† Christine Slingsby, Alan Simpson, and
Andreas Wostrack

Laboratory of Molecular Biology
Department of Crystallography
Birkbeck College, Malet Street, London, WC1E 7HX, England

INTRODUCTION

Protein domains and elements of secondary structure such as α-helices form pseudo-rigid bodies which can move as a result of environmental forces. The binding of inhibitor molecules to enzyme molecules can result in relative movement of domains and crystal packing forces may also cause conformational changes in proteins which may result in loss of activity in the case of enzymes. The first study reported here analyses the conformational changes which take place in a range of closely related crystal structures and compares them to the rigid body domain motion in one of the structures determined form X-ray analysis of rigid body thermal parameters. The second study reports the structure of an eye lens protein which is homologous to the enzyme argininosuccinate lyase which crystallises in low and high pH forms. The high pH form exhibits a structure consisting of hexagonal close packing of supramolecular helices. The domain motions in these helices give rise to a remarkable diffuse X-ray diffraction pattern with 14-fold symmetry indicating that these supramolecular helices may be stable entities outside the crystal.

ENDOTHIAPEPSIN COMPLEXES

Endothiapepsin is an aspartic proteinase derived from the fungus *Endothia parasitica* which infects chestnut trees. Its structure consists of two domains (Fig. 1) between which there is an active site cleft containing two aspartic residues which play a central role in the catalysing the hydrolysis of oligopeptide substrates which conform to the enzyme's specificity pockets. Comparison of the three-dimensional structures of native endothiapepsin (EC 3.4.23.6) and 15 endothiapepsin oligopeptide inhibitor complexes defined at high resolution by X-ray crystallography[1] shows that endothiapepsin exists in two forms differing in the relative orientation of a domain comprising residues 190-302. There are relatively few interactions between the two domains of the enzyme; consequently they can move as separate

---

* Represents person presenting Paper

† Present address: The Rockefeller University, 1230 York Avenue, New York, NY 10021-6399

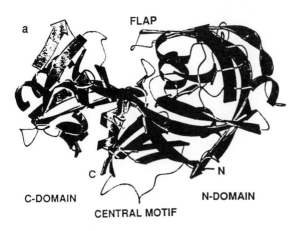

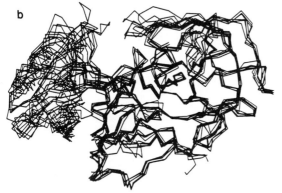

Figure 1. Structural organisation of aspartic proteinases. (a) Secondary structure ribbon diagram for endothiapepsin, (b) 15 aspartic proteinase structures superimposed on their central motifs and N-domains; for clarity, the amino-terminal part of pepsinogen up to the residue 12 (pepsin numbering) is not shown.

rigid bodies. A translational, librational, and screw analysis of the thermal parameters of endothiapepsin also supports a model in which the two parts can move relative to each other.

DEFINITION OF RIGID BODIES IN ENDOTHIAPEPSIN

A difference distance plot showed the conformational changes in endothiapepsin when the oligopeptide inhibitor CP-69,799 binds into its active site cleft. If one neglects local distortions, it was concluded that there are only two parts of the structure that retain their internal conformation but change their spatial relationship when the inhibitor binds. Thus the two rigid bodies in endothiapepsin with respect to the inhibitor binding are defined. The first rigid body comprises residues -2 to 189 and 303 to 326 and the second rigid body consists of residues 190 to 302. This division is in complete accord with the tripartite description of the aspartic proteinases fold: the first rigid body comprises both the central motif and the N-domain and the second rigid body corresponds to the C-domain.

DEFINITION OF THE RIGID BODY MOVEMENT IN ENDOTHIAPEPSIN

The description of the rigid body movement in terms of the screw motion (Fig. 2) shows that the conformational change involved a 4° rotation around and a negligible

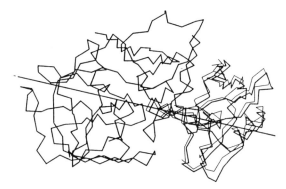

Figure 2. Geometry of the rigid body movement in endothiapepsin-inhibitor complexes. The movement of the second rigid body (C-domain) relative to the first one (central motif and N-domain) is described in terms of the screw motion. Thick line, native endothiapepsin; thin line, the second rigid body of the complex between endothiapepsin and H261. The orientation of this inhibitor complex is obtained from the superposition of its first rigid body with the first rigid body of the native endothiapepsin. The H261 complex is shown as a representative of the predominantly non-isomorphous group of inhibitor complexes that includes CP-69,799 inhibitor.

translation along the screw axis that passes approximately through the active site centre. The rms value for the two C-domains is 1.47 Å when only the central motifs and N-domains are used in the least-squares superposition. This is decreased to 0.46 Å (31% of the original value) when the two C-domains are separately superposed. This large reduction justifies description of the conformational transition as a rigid body movement as opposed to distortion within the C-domain.

The movement corresponds to changes in the relative position of the C_α atoms of up to 2.5 Å. Changes in the active site cleft become appreciable (≈ 1.5 Å) at the S_3 pocket formed in part by the α-helix 108-114; the cleft opens when the inhibitor binds. After the "flap" β-hairpin in the cleft, this helix undergoes the next largest decrease in thermal mobility of all regular secondary structure elements when the inhibitor binds. The average isotropic temperature factor of the helix falls from 37 to 16 Å^2. There is a further local movement of this helix of ≈ 0.8 Å in roughly the same direction as the global rigid body movement. This indicates that this could be the contact point on the enzyme used to trigger the gross conformational change by the ligand in the active site cleft. However, the reasons for the decrease in the average isotropic temperature factor and a positional change of this helix cannot be attributed unequivocally to the ligand binding because some of the helix residues are in contact with a neighbouring molecule in the unit cell of the inhibitor complex but not in that of the native enzyme.

Fig. 3 shows the magnitude of rigid body changes for all 15 endothiapepsin-inhibitor complexes. In all cases, the orientation of the screw axis is the same as for the example described above. It is apparent that two forms of endothiapepsin exist. The first one with a very small rotation and translation is very close to the native endothiapepsin; the second one with $\approx 4°$ rotation and a very small translation of ≈ 0.3 Å is the form discussed above. Several conclusions can be easily made. First, there is no correlation between the chemical type of the P_1-P_1' bond of the inhibitor and the relative orientation of the rigid bodies, except that the statine inhibitors tend to associate with lower displacement from native conformation within each group of the complexes. Second, there is a clear but not an exclusive association of the rigid body movement with the non-isomorphous crystallographic unit cell. Third, the P_3 and/or P_2 residues of the ligand seem to be a necessary but not a sufficient condition for

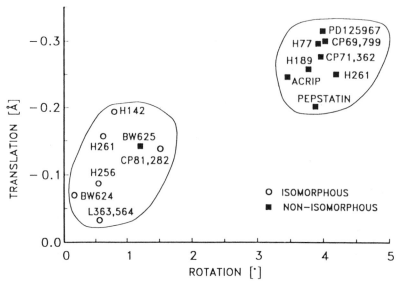

Figure 3. Magnitude of the rigid body movement in 15 endothiapepsin-inhibitor complexes. A rotation and a translation component are specified for all 15 inhibitor complexes. In addition, the type of crystallographic unit cell is indicated by an empty circle for an isomorphous and a filled square for a non-isomorphous unit cell.

the rigid body movement to occur, since the absence of the P_3 and P_2 residues is the only outstanding feature of the BW625 inhibitor that crystallises in the non-isomorphous unit cell and exhibits very small rigid body movement.

RIGID BODY THERMAL PARAMETERS FROM X-RAY ANALYSIS OF ENDOTHIAPEPSIN COMPLEXES

The next step in this analysis was to determine the rigid body thermal parameters from X-ray analysis of individual endothiapepsin complexes and to compare these motions with the displacements in the above complexes. The mean-square displacements of rigid bodies are described by T, L and S tensors which respectively describe the translational,

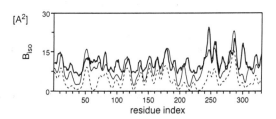

Figure 4. TLS analysis of the rigid body motion of the endothiapepsin-CP-69,799 complex. Three different types of main chain isotropic temperature factors smoothed by the running window of five residues are shown. The thick line represents the directly refined isotropic temperature factors. The thin line shows the isotropic temperature factors derived from the TLS tensors refined by RESTRAIN. The dashed line shows the isotropic temperature factors calculated from the single screw motion fitted to the refined TLS tensors.

librational and screw-rotational mean square displacements of rigid bodies. These tensors were determined by RESTRAIN[2,3] by refinement against the X-ray diffraction data for the endothiapepsin-CP-69,799 complex. This rigid body approximation to the main-chain displacements gave rise to calculated isotropic temperature factors which were in good agreement with temperature factors from a conventional refinement (Fig. 4). Furthermore the calculated TLS motion could be approximated by a single screw motion. The screw axis of the rigid body displacement of the first rigid body is only 1.2 Å away from the screw axis relating the static structures and the angle between the screw axes is only 16°. The corresponding numbers for the second rigid body are 0.9 Å and 45°. Thus the screw motions indicated by the X-ray temperature factors approximate to those relating the static structures.

DOMAIN MOTIONS IN AVIAN EYE LENS PROTEINS

A second example of domain motion which derived from X-ray analysis, is of an eye lens protein, δ-crystallin. In the endothiapepsin complexes, the domain motion could be derived from the Bragg reflections. In the case of the eye lens protein, the domain motion also gives rise to diffuse X-ray scattering which enables the rigid bodies in the crystal to be identified. First, we will consider the origin of this eye lens protein and its interesting relationship to a superfamily of enzymes.

The vertebrate lens is a transparent tissue containing high levels of soluble protein, packed with short range order in precisely aligned, elongated cells. In birds the main lens proteins are the α-, β- and δ-crystallins, which are differentially expressed during development, with δ-crystallin being predominant in the early stages, resulting in its concentration in the centre of the adult lens.

δ-crystallin initially evolved in a common ancestor of modern reptiles and birds, by the overexpression of argininosuccinate lyase in the lens, around which time a gene duplication took place. Since then the lens gene has accumulated mutations in the coding sequence that have made it enzymatically inactive and the control sequences have diverged allowing independent control. The crystallin (δ-1) and enzyme (δ-2) genes encode proteins that share 91% and 96% identical sequences between chick and duck respectively.

THE δ-CRYSTALLIN SUPERFAMILY

δ-crystallin and argininosuccinate lyase belong to a superfamily of metabolic enzymes which are all active as homotetramers of approximate M 200,000. Members also include fumarase, aspartase, adenylosuccinase and 3-carboxy-*cis,cis*-muconate lactonising enzyme (CMLE), with the separate enzyme families having highly conserves sequences. The different enzymes are distantly related, with human fumarase and argininosuccinate lyase having 12.6% sequence identities, but there are three stretches of close sequence similarity. Enzymes within the superfamily carry out homologous reactions; the substrates contain carboxyl groups at the C_α and C_β positions, which are joined to a leaving group by a labile C_α-N or C_α-O bond.

The three-dimensional X-ray structure has recently been determined for turkey δ-1 crystallin showing that it consists of a tetramer of 50 kDa subunits, with 222 symmetry[4]. Each subunit is composed of three domains, including one that is a bundle of five α-helices of around 25 residues each. These bundles associate about the 222 symmetry axes to produce a 20-helix bundle at the core of the structure surrounded by the other eight domains (Fig. 5) each of which is predominantly α-helical. This protein structure is extremely water soluble, as demonstrated by the fact that crystallisation was carried out close to its isoelectric point (~ pH 5) and required

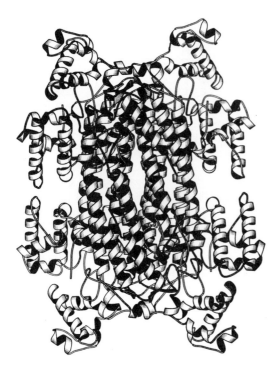

Figure 5. Ribbon diagram showing the quaternary structure of δ-crystallin. The molecule is a tetramer of 222 symmetry and the view is down one of the dyads with the other dyads horizontal and vertical respectively. Two of the five-helical bundles are visible and the four active site clefts can be seen near the corners of the tetramer.

high protein concentration (7-8% w/v) and moderate amounts of precipitating agent (~ 10% w/v PEG 6000). Obviously liquid-like lenses need soluble proteins. The question arises as to what types of intermolecular interactions occur at the high protein concentration in the lens where the overall protein concentration is around 20% (w/v). One experimental route is to study intermolecular interactions of δ-crystallin tetramers in a crystal lattice at a more physiological pH, with as many molecules as possible in the asymmetric unit to minimise the packing constraints of the crystallographic axes. Hence crystals were grown at a higher pH.

CRYSTAL STRUCTURE OF δ–CRYSTALLIN AT pH 8 AT 4.5 Å RESOLUTION

A different crystal phase was obtained when crystals were grown at pH 8 in 0.05 M Tris-HCl buffer. The crystals grew as elongated needles in space group C2 with unit cell dimensions $a = 263.5$ Å, $b = 152.2$ Å, $c = 204.5$ Å, $b = 100.36°$. The positions of the tetramers in the unit cell have been determined by molecular replacement. They form helices with $7_3 2$ symmetry in which adjacent tetramers are related by a screw axis with a clockwise rotation of 2/7 of a turn and a translation of 76 Å. These helices extend through the entire crystal, with 3.5 tetramers in the crystallographic asymmetric unit and crystallographic twofold axes relating adjacent asymmetric units so as to form a continuous helix (Fig. 6). Each helix approximates in shape to a cylinder with a width of one tetramer and these cylinders pack in the crystal with approximate hexagonal close packing.

The internal 222 symmetry of the tetramers results in them being related not only by the 7_3 screw axis, but also by an inter-tetramer twofold which lies perpendicular to the screw axis. One of the internal twofolds of the tetramer also lies perpendicular to the helix axis, so that the helix contains two sets of non-equivalent crystallographic twofolds. The internal

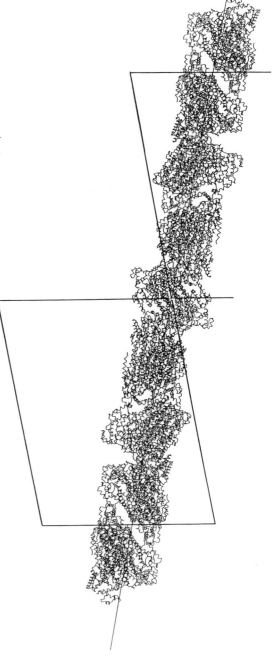

Figure 6. The helical arrangement of δ-crystallin molecules in the high-pH crystal phase. The view is down the b axis and the outline of the a and c cell edges is shown. The border group symmetry of the helix is $7_3 2$ and dyads can be seen at the corners of the cell and at the midpoints of the c cell edges. The complete unit cell of the helix consists of seven tetramers.

dyad which contributes to the helix symmetry relates the subunits within the tightly bound dimer. Thus, the helix can be described by the border group $7_3 2$, with a repeat distance of 530 Å and with two subunits in the asymmetric unit which are related by a twofold which is local to a single tetramer. The local unit cell of the helix with the above dimensions contains seven tetramers in two crystallographic asymmetric units and the translation relating adjacent helix cells is equal to the lattice vector $a + 2c$ (Fig. 6).

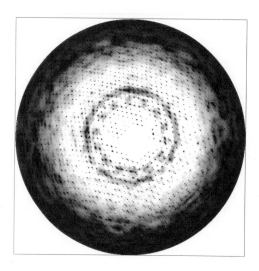

Figure 7. One-degree oscillation X-ray image of high pH crystals of the avian eye lens protein δ-crystallin recorded with synchrotron radiation at Daresbury Laboratory using a Mar Research imaging plate. The resolution at the edge of the image is 3.5 Å. The 14-fold pseudo-symmetry of the innermost diffuse ring arises from the 7_3 pseudo-screw axes relating the seven molecules in the unit cell. These axes are approximately parallel to the X-ray beam.

SUPRAMOLECULAR HELICES IN THE CRYSTAL GIVE RISE TO DIFFUSE SCATTERING WITH 14-FOLD SYMMETRY

Diffuse X-ray scattering occurs between the sharper Bragg peaks and is indicative of disorder and thermal motion in the crystal structure which breaks down the lattice symmetry. All protein crystals give rise to some diffuse X-ray scattering and the character of this scattering indicates the how motions are correlated in the crystal. A diffraction pattern obtained from a crystal of the high-pH phase of δ-crystallin oriented with the X-ray beam parallel to the helical axis shows 14-fold symmetry in the diffuse scattering at low resolution (Fig. 7). This pattern is consistent with the Fourier transform of the projection of the helix with 7-fold symmetry. Such diffuse scattering is indicative of the supramolecular helices undergoing a rigid body translational motion and suggests that the helices have some stability. Aggregates such as these might form in the avian eye lens without adverse effects providing the helices do not become long enough to scatter light.

CONCLUSION

The analysis of X-ray thermal parameters derived from the Bragg reflections of protein crystals can give information on the domain movements that a protein molecule can undergo when in different environments. Such analyses require reasonable assumptions to be made about the domains that will behave approximately as rigid bodies. Because it contains information about correlated motion, which is absent in the Bragg scattering, diffuse X-ray scattering can directly indicate which units are undergoing rigid body motion and hence indicate those aggregates that might be stable outside the crystal environment.

REFERENCES

1. Šali, A., Veerapandian, B., Cooper, J.B., Moss, D.S., Hofmann, T. and Blundell, T.L. (1992) *Proteins: Structure, Function, and Genetics* 12: 158-170.
2. Haneef, I., Moss, D.S., Stanford, M.J. and Borkakoti, N. (1985) *Acta Crystallographica* A41: 426-433.
3. Howlin, B., Moss, D.S. and Harris, G.W. (1989) *Acta Crystallographica* A45: 851-861.
4. Simpson, A., Bateman, O., Driessen, H., Lindley, P., Moss, D., Mylvaganam, S., Narebor, E. and Slingsby, C. (1994) *Nature Structural Biology* 1: 724-734.

17

INTERACTION OF BOVINE SERUM ALBUMIN AND HEXADECYL PYRIDINIUM BROMIDE STUDY BY SURFACTANT SELECTIVE ELECTRODE AND SPECTROPHOTOMETRY

Hossain Naderimanesh,[*1] Nader Alizadeh,[1] and Mojtabah Shamsipoor[2]

[1] Faculty of Science
   Tarbiat Modarres University, Iran
[2] Department of Chemistry
   Kermanshah University, Kermanshah, Iran

ABSTRACT

Surfactant selective electrode and spectrophotometric were used to study the effect of hexadecyl pyridinium bromide (HPB) on the conformation changes of bovine serum albumin (BSA) in aqueous solution. Our previous finding suggested multi-state serum albumin BSA. EMF plots - versus logarithm of HPB concentration revealed also four breaks. The saturated quantities of the HPB binding were 33,93,155,255 mol/mol. The Reynolds and Hill equations were applied to obtain co-operative HPB bindings to BSA. Gibbs free energy was calculated from binding data. Spectrophotometry study suggests hydrophobic interaction of HPB with BSA.

INTRODUCTION

The interaction between proteins and surfactants have been subject of many investigations since 1950[1-3]. Two sets of experimental techniques, in general, have been utilized for these studies; 1-methods which basically define macroscopic quantities, that are affected by all other components present in the solution such as viscosity, difference scanning and micro calorimetry, dialysis, conductivity, and e.m.f. measurements; 2-Methods which define changes in the molecular properties of the interacting species such as, spectroscopic changes; NMR, ORD, CD, IR and UV. Naturally, each one of these techniques has its own advantages

---

[*] Corresponding author.

and shortcomings. The result of these measurements are then used for constructing a binding isotherm, i.e. a curve representing the amount of bound surfactant or any other ligand per molecule of protein as a function of the concentration or activity of the free surfactant in equilibrium.

In this work we used Bovine Serum Albumin (BSA) as a model for protein. BSA has been one of the most extensively studied and applied protein due to its availability, low cost, stability, and unusual lingand binding. Unfortunately, the widespread interest in BSA has not been balanced by an understanding of its molecular structure. This becomes more important when one realizes that BSA has been recognized as a principle component of blood as early as 1939[4]. BSA is the most abundant protein in the circulatory system with 5g/100 ml concentration in blood. It has chief responsibility of blood pH maintenance[5], a half-life of 19 days in circulation[6], and outstanding ability to bind reversibly an incredible variety of ligands. Moreover, BSA is the main carrier of fatty acids which are insoluble in circulating plasma, and many other important functions. Nowadays, it is widely believed in the pharmaceutical industry that the overall distribution, metabolism, and efficacy of many drugs can be altered based on their affinity to BSA. In addition, many promising new drugs are rendered ineffective because of their unusually high affinity for this abundant protein. This will place BSA-ligand interaction in second step in rational drug designing.

Numerous mentioned techniques, even in principle, would not give an unambiguous estimate of surfactant ion activity, instead concentration of neutral surfactant. In the cases like dialysis, mean ionic activities are provided. Surfactant selective membrane electrode, on the other hand, offers the advantage of fast *in situ* determinations of non-bound surfactant ion in solution. It functions well both in monomer and micellar solution. Its high sensivity, ease of work, and speed suggest that BSA could be used as biomembrane sensor and therapeutic usage[7]. In this report we present part of our investigation to study BSA-Hexadecyl pyridinium bromide interaction by membrane electrode and spectrophotometry to further study BSA domain unfolding and the type of interaction.

MATERIAL

The four time recrystallized hexadecyl pyridinium bromide (HPB, BDH chemical Ltd.), sodium bromide (Merck chemical, Ltd.), and bovine serum albumin crystal (Sigma, batch # A3350) were used. Shimadzu 2100 and Unicam uv II spectrophotometer, Geneway 3030 pH meter, and standard sodium electrode (Orion 8411 Ross).

METHODS

EMF Measurement

The HPB selective electrode constructed base on earlier studies (8-10). The monomer surfactant concentration obtained by emf measurements from the following cell (I):

| HPB electrode | Test solution Containing a constant (0.0001 M) amount of sodium bromide | Electrode reverible to sodium Ion |

EMF(HPB/Na) in 0.004%BSA +0.0001M NaBr

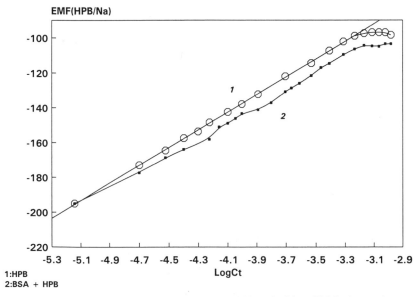

1:HPB
2:BSA + HPB

Figure 1. Electrode potential, EMF, for HPB with*, and without ❏ BSA interaction.

These measurements were performed over 0.00001-0.001 M HPB concentration range, in 0.004% BSA solution, relative to a standard sodium electrode. The pH variation was measured by Geneway pH meter in 0.004% BSA titration with 0.02 M HPB. In all the experiments the temperature was controlled to within 25+/-0.1C by circulating thermostated water through the jacketed glass cell, and the sample solution was continuously stirred with a magnetic stirrer. Given the ratio of a [sur]/a[Na] and activity coefficient of the surfactant monomer and its co-ions are approximately equal. The cell (I) data will measure [sur]/[Na]. Since [Na] is constant the emf of this cell is given by

$$E = E^0 + 2.303 \frac{RT}{nF} \log [HPB] \tag{1}$$

RESULTS AND DISCUSSION

Figure 1 shows the electrode potential against the logarithm of HPB concentration in 0.004% BSA solution. Binding isotherm was calculated from electrode potentials of HPB in pre-micellar concentration in presence of BSA (Fig. 2).

In the case of equivalent and non-interacting sites, the binding of surfactants to protein will follow the Reynolds equation[11-13].

$$\frac{1}{v} = \frac{1}{nk} \cdot \frac{1}{[HPB]} + \frac{1}{n} \tag{2}$$

When v is the number of surfactant moles bound per mole of protein, k is the intrinsic association constant of each site, and n is number of binding sites in a set. Both n and k can be calculated from Reynold equation (Fig. 3) called klotx plot, and deviation from linearity

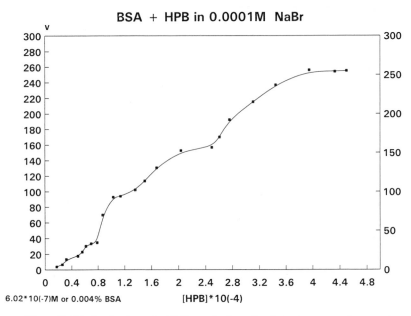

Figure 2. Binding isotherm for HPB as a funtion of surfactant concentration.

at higher v values can be explained either in terms of second set of sites or conformational changes in protein molecules. Fig. 2 reveals at least four deflection points which could be related to protein conformational changes due to different surfactant interaction sites or changes of binding equilibrium constant. The n for each site is calculated from Fig. 2 and are: $n_1 = 33$, $n_2 = 93$, $n_3 = 155$, and n_4 255. The number of surfactants per mole of protein

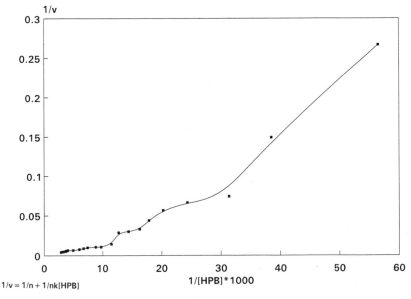

Figure 3. Klotz plot for HPB on interaction with BSA, 25°C.

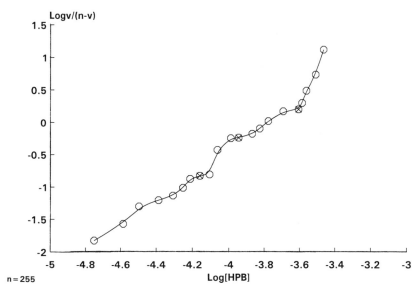

Figure 4. Tetraphasic hill plot for HPB on interaction with BSA.

for each site or conformation is $n_1 = 33$, $n^*_2 = n_2 - n_1 = 60$, $n^*_3 = n_3 - n_2 = 62$, $n^*_4 = n_4 - n_3 = 100$. Hill equation has been used to calculate Hill constant n_H and data is presented in Fig. 4[14,15]. We observe again four different n_H which changes from 1.4 to 6.3 and suggests a possible four separate positive co-operative binding sites. We can use different n for each site from klotz plot and replace the Hill equation for each site in Fig. 5.

$$Log\left(\frac{\overline{v}}{n - \overline{v}}\right) = n_H \cdot Logk + n_H \log [HPB]$$

(3)

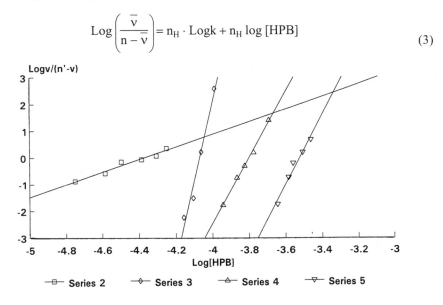

n1 = 33, n2 = 93, n3 = 155, n4 = 255
n'1 = 33, n'2 = 60, n'3 = 62, n'4 = 100

Figure 5. Hill plot for HPB on interaction with each separate BSA site.

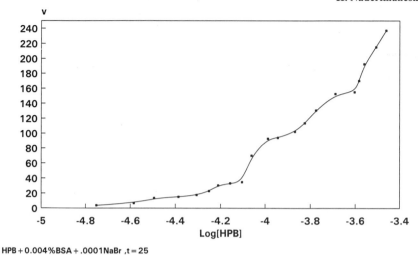

HPB + 0.004%BSA + .0001NaBr ,t = 25

Figure 6. Wyman plot for HPB on interaction with BSA.

It is also possible to calculate the Wyman binding potential π[16] from the area under the binding isotherm (Fig. 6) according to:

$$\pi = RT \int_{\bar{\nu}_i = 0}^{\bar{\nu}_i} \bar{\nu}_i \, d\ln [HPB]$$

(4)

This can be related to an apparent binding constant (K_{app})[17]

$$\pi = RT \ln(1 + K_{app} [HPB]^{\bar{\nu}_i})$$

(5)

From equation 4 and 5 Kapp has been calculated and used to determine the changes in binding free energy of HPB to BSA

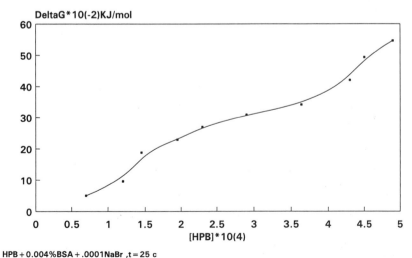

HPB + 0.004%BSA + .0001NaBr ,t = 25 c

Figure 7. Free Energy of binding for HPB on interaction with BSA.

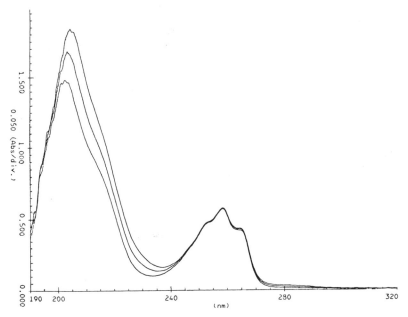

Figure 8a. Uv absorbance of 1.135×10^{-6}M BSA titrated with, 4.12×10^{-5}M, 8.22×10^{-5}M, 1.234×10^{-4}M of surfactant.

$$\Delta G_b = -RT \ Ln \ K_{app} \qquad (6)$$

Results are presented in and Fig. 7 and again a tetraphasic plot suggest four different binding sites. Uv experiments were performed under the same experimental conditions as a function of surfactant concentration. Since HPB has a pyridine ring it should have absorbance at 240-280 nm uv region due to $n \rightarrow \pi^*$ transfer. If surfactant attack the protein from polar head group the uv absorbance peak of BSA and HPB complex should be altered. In Fig. 8a and 8b, the uv absorbacne of BSA titrated with surfactant and surfactant titrated with BSA

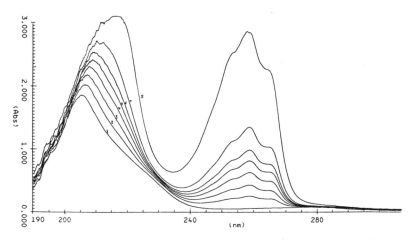

Figure 8b. Uv absorbance of 1.25×10M surfactant titrated with, 3.24×10^{-7}M, 4.85×10^{-7}M, 6.44×10^{-7}M of BSA.

are respectively presented. Absorbances are not showing any significant alteration due to titration which suggest a hydrophobic interaction between HPB and BSA. In conclusion, our results suggest: 1 - BSA interacts with surfactant in stepwise manner with possible change in conformations. Since this is a multidomain protein and as suggested by other groups[14-15], it is possible that different parts or domains interact in unequivalent time scale and initial interaction has positive co-operative effect on the latter stages. 2 - Our spectrophotometry results approve the results of other groups who believe this is a hydrophobic interaction[18].

REFERENCES

1. Anson, M.I. (1939) *Science* 90: 256.
2. Pallansch, M.J. and Briggs, D.R. (1954) *J. Amer. Chem. Soc.* 76: 1396.
3. Wasylewski, Z. (1979) *Acta Biochemica Polonica* Vol. 26: No. 3: 195-203.
4. Ancell, H. (1939) *Lancet* 1: 222-231.
5. Figge, J., Rossing, T.H. and Fencl, V. (1991) *J. Lab. Clin. Med.* 117: 453-467.
6. Carter, D.C. and Ho, J.X. (1994) *Adv.Pro. Chem.* 45, 153-203.
7. Birch, B.J. and Clark, D.E. (1973) *Anal. Chem. Acta.* 69: 387.
8. Davidson, C.J. Ph.D. thesis (University of Aberdeen, 1983).
9. Painter, D.M., Bloor, D.M., Takisawa, N., Hall, D.G. and Wynjones, E. (1988) *J. Chem. Soc. Far. Trans, 1*, 84: 2087.
10. Painter, D.M. Ph.D. Thesis (University of Salford, 1988).
11. Steinhardt, J., and Reynolds, J.A. (1968) *Multiple Equilibiria in Protein*, pp. 239-302, Academic press, New York.
12. Reynolds, J.A. Herbert, S., Polet, H. and Steinhardt, J. (1967) *Biochemistry* 6: 937.
13. Reynolds, J.A., Gallagher, J.P. and Steinbardt, J. (1970) *Biochemisrty* 9: 1232.
14. Takeda, K., Miura, M. and Takagi, T. (1981) *J. Colloid Interface Sci.* 82: 38.
15. Wada, A. and Takeda, K. (1990) *J. Colloid Interface Sci.* 138(1): 277-279.
16. Wyman, J. (1965) *J. Mol. Biol.* 11: 631-644.
17. Hill, A.V. (1910) *J. Phsio.*
18. Strop, (1987) *Chem. Commun.* 52(5): 1362-1374.

CILIATE TELOMERE PROTEINS

Carolyn M. Price

Department of Chemistry
University of Nebraska
Lincoln, Nebraska 68588

ABSTRACT

Telomeres are the specialized DNA-protein complexes at the termini of linear eukaryote chromosomes. The proteins that bind to the extreme end of the telomeric DNA have been isolated from two species of ciliated protozoa, *Euplotes* and *Oxytricha*. The proteins from both organisms recognize the sequence and structure of the telomeric DNA. As the proteins bind telomeric DNA very tightly, they appear to form a protective cap over the end of the chromosome. The DNA-binding specificity of the two proteins has been studied using gel-shift assays and nitrocellulose-filter binding, while their structural organization has been delineated using partial proteolysis and deletion mutagenesis. The two proteins share a conserved DNA-binding domain which is unlike any previously characterized DNA-binding motif.

INTRODUCTION

Telomeres, the natural ends of chromosomes, are vital for proper cell growth and development because they maintain chromosome integrity. They perform this essential function by protecting chromosome ends from degradation or end-to-end fusion, and by preventing loss of terminal sequences during DNA replication[1,2]. End-to-end fusion of chromosomes leads to severe genetic defects because fused chromosomes tend to fragment, causing rearrangement or deletion of genome segments. Telomeres prevent the fusion of chromosome ends by virtue of their structure. They exist as unusual DNA-protein complexes which are composed of the most terminal DNA sequences on a chromosome together with specialized telomere-binding proteins. These telomere-binding proteins protect the telomeric DNA from degradation and recombination.

Incomplete replication of the 5' end of a linear DNA molecule is a direct consequence of a standard replication reaction. Because DNA synthesis is normally primed by synthesis of an RNA primer, subsequent removal of this primer from the 5' end of each newly synthesized strand leads to a gap at one end of each daughter molecule (see Fig. 1). Thus, multiple rounds of replication are predicted to cause progressive shortening of the chromosome, eventual loss of coding sequence and disruption of the corresponding gene products.

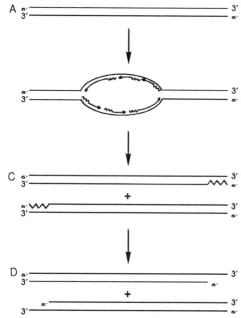

Figure 1. Primer initiated DNA replication results in loss of sequence from the 5' end of each newly replicated strand. A. Parent DNA molecule. B. Initiation of leading and lagging strand DNA replication by synthesis of RNA primers (represents the RNA primers). C. Completely replicated daughter molecules with RNA primers at the 5' termini of the newly synthesized strands. D. Daughter molecules that have had the RNA primers removed. Each newly synthesized strand is shorter than the parent molecule.

Telomeres solve this replication problem in several ways. First, they are composed of noncoding sequence DNA that serves as a buffer between the coding sequence and the terminus of a chromosome. Second, telomeric DNA is synthesized by a unique enzyme called telomerase which is not a part of the standard chromosome replication machinery.

TELOMERIC DNA

All telomeres are composed of repeated sequence DNA[2]. In most organisms the repeat unit is very short and simple consisting, of a 6-8 bp sequence that contains clusters of G-residues on the 3' strand. Although the length of the telomeric DNA present on any one chromosome is somewhat variable, there are well defined upper and lower limits that are specific for each species. For example, mouse telomeres consist of ~150 kb of the sequence T_2AG_3, human telomeres have 5-15 kb of T_2AG_3 repeats while yeast and *Tetrahymena* telomeres have <1000 bp of the sequences $G_{1-3}T$ and T_2G_4 respectively. Due to technical difficulties, the organization of the DNA at the extreme end of the telomere has only been examined in a few lower eukaryotes. However, in all of the species examined, the 3' strand was found to protrude 12-16 bases beyond the end of the 5' strand[3,4]. This 3' overhang is probably generated during telomere replication and it is likely to be a general feature of telomere structure. Proteins that bind specifically to the 3' overhang on the telomeric DNA have been detected in a number of organisms[5-9]. However, only the proteins from the ciliates *Euplotes* and *Oxytricha* are well characterized.

SYNTHESIS OF TELOMERIC DNA

The 3' strand of the telomere is synthesized by an RNA-containing enzyme called telomerase[10]. This enzyme is responsible both for maintaining telomere length during DNA

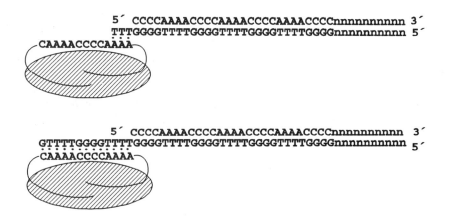

Figure 2. Extension of telomeric DNA by telomerase. **A.** The template region of telomerase RNA base pairs with the most terminal nucleotides of the telomeric DNA. **B.** Telomerase extends the telomere by catalyzing polymerization of dGTP and dTTP using the RNA as a template. The protein component of telomerase is represented by the shaded oval. The RNA component is represented by the curved line.

replication and for healing the ends of broken chromosomes by *de novo* telomere addition. The RNA component of telomerase contains a short template region that is complimentary to ~1½ repeats from the 3' strand. During addition of telomeric DNA, the 5' portion of the complementary region pairs with the terminus of the 3' strand. The remainder of the complementary region then templates repeat addition (Fig. 2). If single base changes are introduced into the template region of the telomerase RNA, the corresponding base change is incorporated into the telomeric DNA. In *Tetrahymena* cells, the resulting single base changes in the telomeric DNA often result in an unusual nuclear morphology and a senescent phenotype. This may be because telomere-binding proteins are unable to bind the mutant telomeric DNA.

At present it is not known whether the sole function of telomerase RNA is to act as a template molecule or whether it also catalytically active and participates in the polymerization reaction. Although the RNA component of telomerase has now been isolated from a variety of organisms including ciliates, yeast and humans, the protein component has only been identified from the *Tetrahymena* enzyme. *Tetrahymena* telomerase appears to contain two protein subunits, one of 80 kDa and one of 95 kDa[11]. The function of the two subunits remains to be determined.

EUPLOTES AND *OXYTRICHA* TELOMERE STRUCTURE

Euplotes and *Oxytricha* are excellent organisms for studying telomere biochemistry because of their unusual genomic organization. Like other ciliates, they have two functionally and structurally distinct nuclei: the germline micronucleus and the vegetative macronucleus. Although the macronucleus only contains a few large chromosomes, the macronucleus contains millions of linear gene-sized DNA molecules which have telomeres on each end[12]. This abundance of telomeres has greatly facilitated the characterization of telomere structure and the isolation of telomere proteins.

Euplotes and *Oxytricha* macronuclear telomeres are extremely short and their length is very tightly regulated[3,12]. As illustrated in Fig. 3, *Euplotes* telomeric DNA consists of exactly 28 bp of $C_4A_4 \times T_4G_4$ sequence. The G-rich strand is actually 42 nucleotides in length

5'CCCCAAAACCCCAAAACCCCAAAACCCCnnnnnnnnnn 3'
3' GGTTTTGGGGTTTTGGGGTTTTGGGGTTTTGGGGnnnnnnnnnn 5'

Figure 3. Organization of *Euplotes* telomeric DNA. nnnnnnnn represents unique sequence DNA.

and extends beyond the end of the C-rich strand to give a 14 nucleotide 3' overhang. In *Oxytricha*, the double-stranded region of the telomere consists of 20 bp of $C_4A_4 \times T_4G_4$ repeats and the 3' overhang is 16 nucleotides in length.

The existence of a nucleoprotein complex at *Euplotes* and *Oxytricha* telomeres was first demonstrated by nuclease and chemical footprinting[13,14]. Digestion of macronuclei with micrococcal nuclease revealed that the most terminal 100-130 bp of the macronuclear DNA were packaged into a nuclease resistant complex (Fig 4). Further footprinting experiments demonstrated that this telomeric complex had two separate structural domains. In the most terminal domain the 3' overhang of the telomeric DNA is bound by a protein that gives rise to a characteristic protection pattern with dimethylsulfate (DMS) footprinting. The DNA-protein interactions within this portion of the telomeric complex are not only sequence specific but also very salt stable as the footprint is not disrupted by high salt (2 M NaCl or 6 M CsCl). The DNA-protein interactions within the more internal domain of the telomeric complex are both weaker and non-sequence specific. Within this portion of the complex, the DNA appears to wrapped around the surface of a protein. This protein could be the protein that is bound to the 3' overhang as is depicted in Fig 4, or it could be a separate protein.

ISOLATION OF THE *EUPLOTES* AND *OXYTRICHA* TELOMERE-BINDING PROTEINS

Isolation of the *Euplotes* and *Oxytricha* proteins was greatly facilitated by their ability to remain bound to macronuclear DNA in the presence of high salt[5-7]. Essentially pure telomere-binding protein can be obtained by extracting macronuclei with 2 M NaCl or 6 M CsCl and isolating the resulting macronuclear DNA-telomere protein complexes by gel filtration or density gradient centrifugation. When purified telomere protein-DNA complexes are treated with DMS the 3' overhang of the telomeric DNA displays essentially the same pattern of methylation protection as is observed with isolated nuclei or whole cells[6,14].

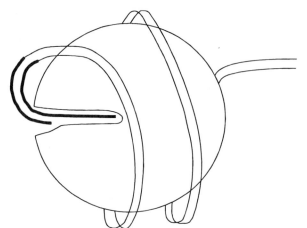

Figure 4. Model of the *Euplotes* telomeric complex. The G-strand overhang of telomeric DNA is tightly bound by the telomere-binding protein while the non-telomeric DNA within the complex is would around the surface of the protein.

Since both the *Euplotes* and *Oxytricha* proteins bind the telomeric DNA very tightly and protect it from Bal 31 digestion, the two proteins appear to form a protective cap over the end of the telomere.

Although the *Oxytricha* and *Euplotes* telomere proteins have similar DNA-binding specificities and similar functions, their structure is surprisingly different. The *Oxytricha* protein is a heterodimer with a 56 kDa α subunit and a 41 kDa β subunit[15], while the *Euplotes* protein is a monomer of 51 kDa[7]. The *Euplotes* protein shares 35% amino acid identity with the α subunit of the *Oxytricha* protein[16]. Given the similarity in DNA-binding specificity between the *Euplotes* and *Oxytricha* proteins, it came as a surprise when the *Euplotes* protein was isolated as a single polypeptide. Despite extensive searches at both the protein and gene level, no *Euplotes* "β" subunit has been found. More recent studies of the *Euplotes* and *Oxytricha* proteins have shed some light on this apparent conundrum. In *Oxytricha* the α subunit is responsible for sequence-specific recognition of telomeric DNA. However, both the α and β subunits are required to form a stable telomere protein-DNA complex[17]. In contrast, α β subunit does not appear to play a role in *Euplotes* telomere protein binding as the single 51 kDa polypeptide is sufficient to form a tight-binding DNA protein complex[7,18]. Thus, although the 51 kDa *Euplotes* protein has sequence homology with the α subunit of the *Oxytricha* protein, it seems to have the combined DNA-binding properties of both the α and β subunits.

DNA-BINDING SPECIFICITY AND STRUCTURAL ORGANIZATION OF THE *EUPLOTES* PROTEIN

The DNA-binding specificity of the *Euplotes* telomere-binding protein has been studied extensively using oligonucleotides that mimic natural and mutant versions of *Euplotes* telomeres[7]. Following release of the purified protein from macronuclear DNA by nuclease digestion, the affinity for the various oligonucleotide substrates was determined either by gel shift assay or nitrocellulose filter-binding. The protein binds tightly to oligonucleotides that have the same sequence as the 3' strand of the telomere (i.e. $(T_4G_4)_nT_4G_2$). Even slight modifications to this sequence reduce binding dramatically, for example, binding is completely abolished by changing the Gs to Cs. The protein recognizes the 3' terminus of the DNA and requires both that the DNA terminate with a 3' G_2 (as opposed to G_4 or T_4) and that the terminal 14 nucleotides be single-stranded. Overall more than 22 nucleotides of T_4G_4 sequence are required for efficient binding. Interestingly, the structure of the DNA internal to the 3' overhang has little effect on binding, as the protein binds equally well to a single-stranded $(T_4G_4)_{4.5}T_4G_2$ oligonucleotide as to an oligonucleotide construct that mimics the structure of wild type telomeres (i.e. 28 bp of $C_4A_4 \times G_4T_4$ with a 14 base $T_4G_4T_4G_2$).

When partial proteolysis is used to separate the *Euplotes* protein into structural domains, both trypsin and chymotrypsin generate a 35 kDa protease resistant fragment and a series of small peptides. The 35 kDa fragment retains most of the DNA-binding specificity characteristic of the native protein; ie it binds telomeric DNA in a salt-stable manner and specifically recognizes single-stranded T_4G_4-containing oligonucleotides. N-terminal sequence analysis demonstrated that the fragment came from the N-terminus of the native protein. Although the 35 kDa fragment clearly comprises the main DNA-binding domain, the protease sensitive C-terminal region does contribute to the specificity of the protein for the DNA terminus. Removal of the C-terminal 16 kDa leaves the protein able to bind oligonucleotides that terminate in $3'T_4$ or G_4 as well as to the natural $3'G_2$. Removal of the C-terminus of the protein also eliminates the DMS footprint observed with the native protein in a methylation interference assay. The different contributions of the N- and C-terminal

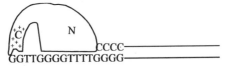

Figure 5. Model depicting how the N- and C-terminal domains of the *Euplotes* telomere-binding protein interact with telomeric DNA. The N-terminal domain binds the T_4G_4 repeats while the C-terminal domain recognizes the DNA terminus.

domains of the *Euplotes* protein to DNA-binding are illustrated in Fig. 5. As depicted, salt-stable sequence-specific binding to the T_4G_4 repeats is achieved by the N-terminal domain while specificity for the 3' terminus of the DNA is achieved by the very basic protease sensitive C-terminal domain.

DNA-BINDING SPECIFICITY AND STRUCTURAL ORGANIZATION OF THE *OXYTRICHA* PROTEIN

Like the *Euplotes* protein, the *Oxytricha* protein binds to single-stranded DNA that has the same sequence as the natural 3' G-strand overhang[18,19]. However, the *Oxytricha* protein has less stringent requirements in terms of the length of the telomeric DNA and the identity of the 3' terminal nucleotide necessary for binding. For example the *Oxytricha* protein will bind oligonucleotides that are as short as 12 bases in length or that terminate in a 3' T_4, G_4 or G_2.

The two subunits of the native *Oxytricha* telomere-binding protein are very tightly associated and cannot be separated unless they are first denatured[15]. For this reason, the relative contributions of the α and β subunits to DNA binding were determined using recombinant protein made in *E. coli* [20,21]. Mobility shift assays demonstrated that the α subunit is the main DNA-binding moiety as this polypeptide binds specifically to T_4G_4-containing oligonucleotides while the β subunit does not. However, in the absence of the β subunit, α binds telomeric DNA relatively weakly and the dimethylsulfate footprint differs significantly from the footprint obtained with both subunits together. Chemical and UV cross-linking experiments have shown that both the α and β subunits contact the telomeric DNA directly. Interestingly, formation of the recombinant $\alpha\beta$ heterodimer is DNA dependant and the two subunit remain as monomers if they are incubated together in the absence of telomeric DNA[22].

To learn more about which regions of the α and β subunits are involved both in DNA-binding and dimerization, a series of N- and C-terminal deletion mutants were generated and tested for their ability to bind DNA and to dimerize[17,22]. Removal of up to 178 amino acids (out of a total of 495) from the C-terminus of α had very little effect on DNA-binding but had a major effect on heterodimer formation. Removal of as few as 19 C-terminal amino acids completely prevented dimerization. These results indicate that α has two distinct structural domains: an N-terminal DNA-binding domain and a C-terminal dimerization domain. Deletion of up to 153 amino acids from the C-terminus of β has no effect on the ability of the subunit to form a tight-binding subunit with α and telomeric DNA. However, N-terminal deletions completely abolish ternary complex formation. Thus, the N-terminal 232 amino acids of β must contain both the dimerization and DNA-binding motifs. At present the function of the C-terminal domain is unknown. Figure 6 shows a model that depicts how the N- and C-terminal domains of α and β may interact with each other and with the telomeric DNA.

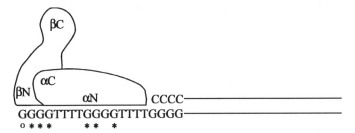

Figure 6. Model depicting how the α and β subunits of the *Oxytricha* telomere-binding protein interact with each other and with the telomeric DNA. α_n, β_n, α_c, β_n, represent the N- and C-termini of the α and β subunits respectively. Bases that are protected from DMS methylation are marked with a star (*). Information is not available for the base marked o.

THE DNA-BINDING MOTIF OF THE *EUPLOTES* AND *OXYTRICHA* PROTEINS

Like many other DNA-binding proteins the *Euplotes* and *Oxytricha* proteins are very basic. However, no known DNA-binding motifs are apparent within the sequence of either protein. This is not particularly surprising as the unique DNA-binding specificity of the two proteins suggests that they have novel DNA-binding motifs. The most extensive regions of sequence homology between the *Euplotes* protein and the α subunit of the *Oxytricha* protein lie within the DNA-binding domain of each protein[16]. This suggests that at least some of the conserved regions comprise portions of the DNA-binding site. This conclusion is strongly supported by UV cross-linking experiments performed with the *Oxytricha* protein[23]. These experiments identified three amino acids that allow the α subunit of the *Oxytricha* protein to be cross-linked to telomeric DNA. Two of these amino acids are present at identical positions within the highly conserved regions of the *Euplotes* DNA-binding domain.

It is noteworthy that a disproportionate number of the amino acids that are conserved between the *Euplotes* and *Oxytricha* proteins are hydrophobic (33%) and/or aromatic (18%)[16]. The ability of both proteins to remain bound to telomeric DNA in the presence of high salt suggests that hydrophobic interactions are important for DNA-binding. The UV cross-linking experiments with the *Oxytricha* protein provide support for this hypothesis as two of the three amino acids that can be cross-linked to telomeric DNA are in hydrophobic regions of the primary sequence[23].

REFERENCES

1. Zakian, V. A. (1989) *Ann. Rev. Genet.* **23**: 579-604.
2. Biessmann, H. and Mason, J.M. (1992) *Adv. Genet.* **30**: 185-249.
3. Klobutcher, L.A., Swanton, M. T., Donini, P. and Prescott, D.M. (1981) *Proc. Natl. Acad. Sci. USA* **78**: 3015-3019.
4. Henderson, E.R. and Blackburn, E.H. (1989) *Mol. Cell Biol.* **9**: 345-348.
5. Gottschling, D.E. and Zakian, V.A. (1986) *Cell* **47**: 195-205.
6. Price, C.M. and Cech, T.R. (1987) *Genes Dev.* **1**: 783-793.
7. Price, C., Skopp, R., Krueger, J. and Williams, D. (1992) *Biochemistry* **31**: 10835-10843.
8. Cardenas, M.E., Bianchi, A. and deLange, T. (1993) *Genes Dev.* **7**: 883-898.
9. Petracek, M.E., Konkel, L.M.C., Kable, M.L. and Berman, J. (1994) *EMBO J.* **13**: 3648-3658.
10. Blackburn, E. (1993) *Ann. Rev. Biochem.* **61**: 113-129.
11. Greider, C. Personal communication.

12. Klobutcher, L.A. and Prescott, D.M. (1986) In: *Molecular biology of ciliated protozoa.* (Gall, J.G. ed.) Academic Press, New York.
13. Gottschling, D.E. and Cech, T.R. (1984) *Cell* **38**: 501-510.
14. Raghuraman, M.K., Dun, C.J., Hicke, B.J. and Cech, T.R. (1989) *Nucl. Acids Res.* **17:** 4235-4253.
15. Price, C.M. and Cech, T.R. (1989) *Biochemistry* **28**: 769-774.
16. Wang, W.-L., Skopp, R., Scofield, M. and Price, C. (1992) *Nucl. Acids Res.* **20**: 6621-6629.
17. Fang, G., Gray, J.T. and Cech, T.R. (1993a) *Genes Dev.* **7**: 870-882.
18. Price, C.M., Unpublished results.
19. Price, C.M. (1990) *Mol. Cell Biol.* **10:** 3421-3431.
20. Hicke, B., Celander, D., Macdonald, G., Price, C.M. and Cech, T. (1990) *Proc. Natl. Acad. Sci. USA* **87**: 1481-1485.
21. Gray, J.T., Celander, D.W., Price, C.M. and Cech, T.R. (1991) *Cell* **67**: 807-814.
22. Fang, G. and Cech, T.R. (1993b) *Proc. Natl. Acad. Sci. USA* **90**: 6057-6060.
23. Hicke, B.J., Willis, M.C., Koch, T.H. and Cech, T.R. (1994) *Biochemistry* **33**: 3364-3373.

19

DIVERSE ACTIONS OF HUMAN LIPOPROTEINS ON ARACHIDONIC ACID METABOLISM

Sheikh A. Saeed,[*] Riaz A. Memon, and Anwar H. Gilani

Department of Pharmacology
Faculty of Health Sciences
The Aga Khan University, Karachi-74800, Pakistan

ABSTRACT

Products of arachidonic acid (AA) metabolism i.e. prostaglandins (PGs), prostacyclin, thromboxane-A_2 (TXA_2) and lipoxygenase metabolites play an important role in platelet aggregation leading to ischemic heart disease and thrombosis. Since lipoproteins are intimately involved in thromboembolic phenomena, we investigated the effects of human lipoproteins (HDL, LDL, VLDL) on metabolism. Lipoproteins were separated by density gradient zonal ultracentrifugation. The effects of lipoproteins on the synthesis of various PGs by isolated PG synthase were measured using radioimmunoassays. The effects of lipoproteins on production of AA metabolites by human platelets i.e. TXA_2 and hydroxyeicosa-tetraenoic acids (HETEs) was examined using radiometric thin layer chromatography coupled with automated data integrator system. All three lipoproteins produced concentration-related inhibition of PG biosynthesis. Relative to HDL and LDL; VLDL was five times more potent in inhibiting PGE_2 synthesis. All three lipoproteins also exerted an inhibitory influence on the production of PGF_{2a} although not in a concentration related manner. In human platelets, HDL inhibited 12-HETE and TXA_2 in concentration-related manner. LDL had a strong inhibitory effect on TXA_2 production and had a weak inhibitory effect against 12-HETE production. VLDL had no effect on platelet AA metabolism. These findings point to a new facet of lipoproteins action which may have a role in the pathophysiology of thromboembolic disorders.

INTRODUCTION

The products of arachidonic acid (AA) metabolism i.e., prostaglandins (PGs), prostacyclin, thromboxane A_2 (TXA_2) and hydroxyeicosatetranoic acids (HETEs) are well

---

[*] Represents person presenting Paper

characterized mediators of inflammatory and immunological reactions in the pathophysiology of several diseases where they may act in an autocrine or paracaine fashion[1,2]. Prostacyclin (PGI$_2$) is a vasodilator and inhibits platelet aggregation[2]. PGE$_2$ is a vasodilator and a potent inhibitor of lymphocyte and macrophage activation[2]. It is also a mediator of fever[1,2]. TXA$_2$ is a vasoconstrictor and stimulates platelet aggregation whereas 12-HETE is a chemotactic factor[1,2].

The primary function of lipoproteins is the transport of various lipids from intestine and liver to most of the tissue for oxidation and to the adipose tissue for storage. Recent studies have shown that lipoproteins may also have other functions. It has been reported that high density lipoproteins (HDL) stimulate whereas low density lipoproteins (LDL) inhibit the PGI$_2$ synthetase activity of pig aortic microsomes[3,4]. On the other hand, Shakhov et al.[5] have recently demonstrated that both LDL and HDL stimulate the formation of 6-keto-PGF$_{1\alpha}$ in cultured human and rabbit aortic smooth muscle cells. The effects of lipoproteins on other products of AA metabolism have not been characterized. Since lipoprotein levels are elevated in several diseases, it is likely that they may have a modulatory effect on products of AA metabolism.

In this study we have investigated the effects of human lipoproteins on PG synthesis by isolated PG synthetase using radioimmunoassays. We have also examined the effects of lipoproteins on AA metabolism by human platelets.

MATERIALS AND METHODS

Materials

Arachidonic acid (grade 1: 99% pure) and reduced glutathione were purchased from Sigma Chemical Company (St. Louis, MO, USA). ^{3}H-PGE$_2$ (SA 50-60 mCi/mmol), ^{3}H-PGF$_{2\alpha}$ (SA 160-180 Ci/mmol), ^{3}H-6-keto-PGF$_{1\alpha}$ (SA 120-180 Ci/mmol), [1-^{14}C] AA (SA 58 mci/mmol), [^{3}H] thromboxane B$_2$ (SA 120 Ci/mmol) and 12-hydroxy-[^{3}H]-eicosatetraenoic acid (12-HETE) (SA 100 Ci/mmol) were obtained from Amersham International plc, U.K. All other chemical used were of highest purity grade.

Preparation of Lipoproteins

Blood obtained from 12 hours fasted healthy volunteers (age range 20-30 years) was collected in EDTA (1 mg/ml). The plasma lipoproteins were separated by sequential density gradient ultracentrifugation[6]. After isolation each lipoprotein fraction was sterilized by filtration through a Millipore (0.45 μm) filter (Millipore Corp; Bedford, MA) and stored at 4°C. The protein content of each lipoprotein fraction was determined by using human serum albumin as standard and adjusted to initial protein concentration for each batch[7]. The homogeneity of each lipoprotein fraction was confirmed by agarose gel electrophoresis and the purified lipoproteins were used within one month of preparation.

Assay of Prostaglandin Biosynthesis

The effects of lipoproteins on the biosynthesis of various PGs from arachidonic acid were measured against PG synthetase activity of BSV micorsomes (Miles Laboratories Inc., USA). The standard assay mixture contained in a final volume of 1.0 ml., 50 mM phosphate buffer, pH 7.4, 1.3 mM reduced glutathione, an appropriate amount of test lipoprotein, 4 mg of BSV microsomes and 61 μM arachidonic acid (as the sodium salt). After an incubation period of 20 minutes with gentle shaking at 37°C, the reaction was terminated by addition of 0.5 ml of 0.4 M citric acid and 7 ml of ethyl acetate. The reaction mixture was vortexed and centrifuged for 5 minutes at 600 x g. The ethyl acetate layer was removed, evaporated to dryness under nitrogen and the residue dissolved in 1 ml of phosphate buffer, (pH 7.4). PGE_2, $PGF_{2\alpha}$ and 6-keto-$PGF_{1\alpha}$ were measured by specific radioimmunoassays as described earlier[8]. All experiments were performed with appropriate controls and each lipoprotein was tested at three different protein concentrations to determine the IC_{50} values (concentration required to inhibit PG synthesis by 50%) so as to get an estimate of relative potency of different lipoproteins.

Arachidonic Acid Metabolism by Human Platelets

Human blood platelets from donors were routinely obtained in plastic bags containing 30-40 ml of concentrated platelet-rich plasma (PRP) from The Aga Khan University Hospital Laboratory. The PRP was centrifuged at 100 g for 5 min and the supernatant discarded. The remaining PRP was then centrifuged at 1,200 g for 20 min and the sedimented platelets were washed twice with an ice-cold phosphate buffer (50mM, pH 7.4) containing sodium chloride (0.15 M) and EDTA (0.2 mM).

After centrifugation platelets were resuspended in the same buffer without EDTA at the initial PRP cell concentration. The PRP suspension was homogenized at 4°C using a polytron homogenizer for 15 s and the homogenate centrifuged at 1,200 g for 20 min. 300 μl of the supernatant (containing 0.4 mg of protein) was incubated with 10 μg unlabelled AA and 0.1 μCi [1-^{14}C] AA in the absence and presence of different concentrations of human lipoproteins. After 15 min with gentle shaking at 37°C the reaction was stopped by adding 0.4 ml of citric acid (0.4M) and ethyl acetate (7 ml). After mixing and centrifuging at 600 g for 5 min at 4°C, the organic layer was separated and evaporated to dryness under nitrogen.

Residues were dissolved in 50μl ethanol and 20 μl were applied to silica gel G thin layer chromatography (TLC) plates (Analtech, Delaware, USA). The AA, TXB_2 (a stable degradation product of TXA_2) and 12-HETE standards were spotted separately. The plates were developed in ether/petroleum ether/acetic acid (50:50:1 by volume) to a distance of 17 cm. By use of this solvent system the various lipoxygenase products (HETEs) were separated with TXB_2 and PGs remaining at the origin[9].

The solvent system used for the separation of various PGs and TXB_2 in dried organic extracts of platelet incubates, as above, was ethyl acetate: isoctane :water:acetic acid (11:5:10:2, v/v, upper phase). Radioactive zones were located and quantified by use of a Berthold TLC linear analyzer and chromatography data system (Model LB 511, Berthold, Germany).

Statistics

The statistical significance was assessed by analysis of variance. Comparison of control values with those obtained with lipoproteins at different protein concentrations (which is different for every lipoprotein) was made by the Newman-Keuls method of multiple range comparisons. A probability of a = 0.05 was chosen as the level of statistical significance.

RESULTS

The effect of different concentrations of various lipoproteins on the biosynthesis of PGE_2, 6-keto-$PGF_{1\alpha}$ and $PGF_{2\alpha}$ is shown in Figure 1. HDL decreased PGE_2 synthesis by 25 ±4%, 51 ± 7% and 83 ±4% at the given concentration of 0.11, 1.1 and 3.2 (mg protein/ml

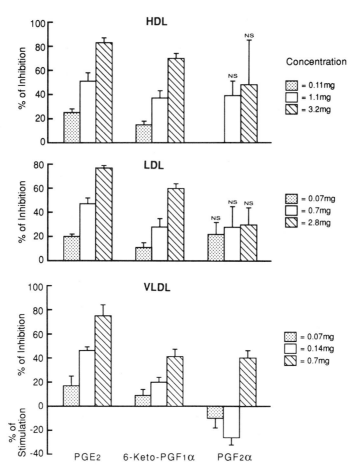

Figure 1. Effect of various concentrations of HDL, LDL and VLDL on the biosynthesis of PGE_2, 6-keto-$PGF_{1\alpha}$ and $PGF_{2\alpha}$. Plasma lipoprotein fractions were isolated and their effect on the biosynthesis of PGs from arachidonic acid was measured against PG synthase activity of BSV microsomes as described in materials and methods.

assay volume) respectively, whereas at the same protein concentrations HDL produced 15 ± 3%, 37 ± 6% and 70 ± 4% decrease in 6-keto-$PGF_{1\alpha}$ synthesis (Fig. 1). The IC_{50} values for HDL- induced inhibition of PGE_2 and 6-keto-$PGF_{1\alpha}$ were 1.05 + 0.09 and 1.95 ± 0.14 (mg protein/ml assay volume) respectively. HDL also inhibited $PGF_{2\alpha}$ synthesis by 48 ± 19% at the highest protein concentration.

LDL produced 20 ± 2%, 47 ± 5% and 77 ± 2% decrease in PGE_2 synthesis at the given concentrations of 0.07, 0.7 and 2.8 (mg protein/ml assay volume) respectively, whereas it produced 11 ± 4%, 28 ± 7% and 60 ± 4% decrease in 6-keto-$PGF_{1\alpha}$ synthesis at the same protein concentrations (Fig. 1). The IC_{50} values for LDL-induced inhibition of PGE_2 and 6-keto-$PGF_{1\alpha}$ were 0.92 ± 0.12 and 2.15 ± 0.11 (mg protein/ml assay volume) respectively. The effect of LDL on $PGF_{2\alpha}$ synthesis, though inhibitory in nature, was not concentration-dependent.

VLDL decreased PGE_2 synthesis by 17 ± 7%, 46 ± 3% and 75 ± 9% at the given concentration of 0.07, 0.14 and 0.7 (mg protein/ml assay volume) respectively (Fig. 1). The IC_{50} for VLDL induced inhibition of PGE_2 synthesis was 0.21 ± 0.01 (mg protein/ml assay

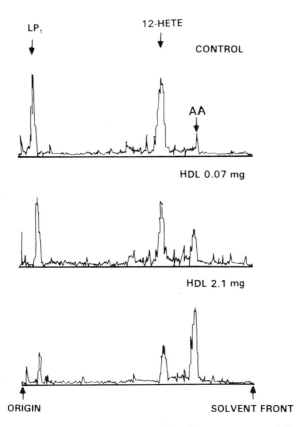

Figure 2. Radiochromatographic scan showing the profile of lipoxygenase metabolites produced by homogenates of human platelets incubated with [1-^{14}C] AA in the absence and presence of different concentrations of HDL.

volume). VLDL inhibited 6-keto-PGF$_{1\alpha}$ by $41 \pm 6\%$ at the maximum protein concentration obtainable through the isolation procedure. At lower protein concentrations, VLDL stimulated PGF$_{2\alpha}$ production, whereas at higher protein concentration it inhibited PGF$_{2\alpha}$ synthesis by $40 \pm 6\%$.

In human platelets, [1-^{14}C] AA is primarily metabolized by lipooxygenase and cyclooxygenase pathways into 12-HETE and TXB$_2$ respectively. [1-^{14}C] AA is also metabolized to a more polar lipoxygenase product designated as LP$_1$. It is possible that LP$_1$ is tri-hydroxy-eicosatrienoic acid (THETE) as described by Bryant and Bailey[10].

The effects of different concentrations of lipoproteins on [1-^{14}C] AA metabolism in human platelets is presented in the form of representative radiochromatographic scans obtained after thin layer chromatography of the products of incubation. The data presented show that HDL inhibited 12-HETE formation in a concentration-dependent manner (Figure 2). HDL decreased 12-HETE formation by 12 ± 1 %, 58 ± 4 % and 81 ± 3 at the given concentration of 0.07, 0.7 and 2.1 (mg protein/ml assay volume) respectively. HDL also decreased the formation of TXB$_2$ in human platelets in a concentration-dependent manner (Figure 3). HDL produced a 23 ± 3 %, 37 ± 6 % and 72 ± 3 % decrease in TXB$_2$ formation at the same protein concentrations.

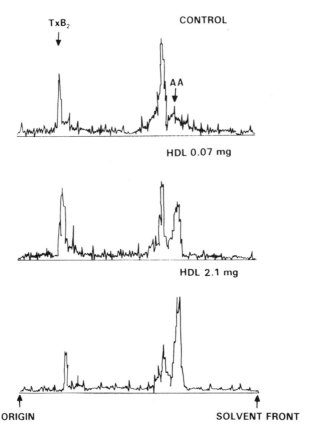

Figure 3. Radiochromatographic scan showing the profile of cyclooxygenase metabolites produced by homogenates of human platelets incubated with [1-^{14}C] AA in the absence and presence of different concentrations of HDL.

In similar experiments, LDL significantly decreased the formation of TXB$_2$ (Figure 4) whereas it had a weak inhibitory effect on the formation of 12-HETE (Figure 5). LDL decreased TXB$_2$ formation by 15 ± 2 %, 33 ± 5 and 65 ± 3% at the given concentrations of 0.07, 0.7 and 2.1 (mg protein/ml assay volume). There was no effect of VLDL on [1-^{14}C] AA metabolism in human platelets.

DISCUSSION

The result of the present study demonstrate that human lipoproteins exert an inhibitory effect on PGE$_2$ and 6-keto-PGF$_{1\alpha}$ synthesis. The magnitude of inhibition depends upon the lipoprotein concentration indicating that it is a concentration-dependent effect. On the basis of protein concentration, VLDL was 4-5 times more potent in inhibiting PGE$_2$ synthesis when compared to HDL and LDL. In addition to their inhibitory effects on PGE$_2$ and 6-keto-PGF$_{1\alpha}$, HDL and LDL also inhibited PGF$_{2\alpha}$ synthesis, though it was not concentration-dependent. VLDL stimulated PGF$_{2\alpha}$ production at lower protein concentrations whereas it inhibited PGF$_{2\alpha}$ synthesis at higher protein concentration. The reason for this paradoxical effect of VLDL on PGF$_{2\alpha}$ synthesis is not clear. VLDL consists of mainly triglycerides as

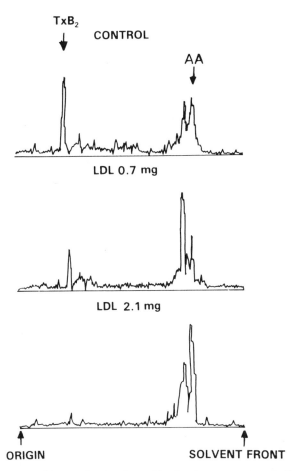

Figure 4. Radiochromatographic scan showing the profile of cyclooxygenase metabolites produced by homogenates of human platelets incubated with [1-^{14}C] AA in the absence and presence of different concentrations of LDL.

compared to LDL and HDL which contain cholesterol as their primary lipid component. It is likely that different lipid constituents of lipoproteins may exert differential effects on PGF$_{2\alpha}$ synthesis.

Our results are in agreement with those of Habenicht et al.[11] who showed that in 3T3 fibroblasts, LDL inhibits the activity of PGH synthetase, the rate limiting enzyme in PG synthesis. However, our results are in marked contrast to those of Shakov et al.[5] who reported that both LDL and HDL stimulate 6-keto-PGF$_{1\alpha}$ formation in rabbit and human aortic smooth muscle cells. A stimulatory effect of HDL on PGI$_2$ synthesis in endothelial cells has also been reported[12]. Others have shown that HDL stimulates and LDL inhibits PGI$_2$ synthetase activity in pig aortic microsomal fraction[3,4]. The reasons for these diverse stimulatory or inhibitory effects of lipoproteins in a variety of tissues are not clear, however, it is possible that PG synthetase complex from different tissues or different species may have different sensitivities to various lipoproteins.

The inhibitory effects of lipoproteins are not confined to 3T3 fibroblasts or BSV microsomes. In this study we demonstrate that HDL inhibits 12-HETE and TXA$_2$ production in human platelets in a concentration-dependent manner. On the other hand, LDL is a potent

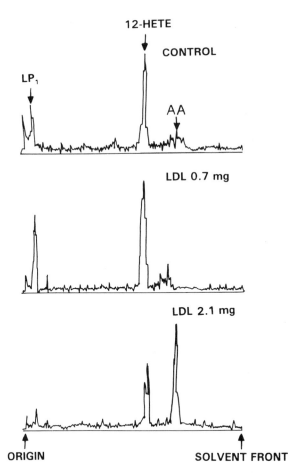

Figure 5. Radiochromatographic scan showing the profile of lipoxygenase metabolites produced by homo-genates of human platelets incubated with [1-^{14}C] AA in the absence and presence of different concentrations of LDL.

inhibitor of TXA$_2$ production in human platelets whereas it has a weak inhibitory effect on 12-HETE formation. Unlike HDL and LDL, VLDL has no effect on AA metabolism in human platelets. Our results are complementary to the recent observations of Fujimoto *et al.*[13] who demonstrated that HDL inhibits 12-lipoxygenase activity in rabbit platelets and decreases 12-HETE formation. The potent inhibitory effect of HDL on 12-HETE and TXA$_2$ formation may be beneficial as both of these AA metabolites play an essential role in platelet aggregation and thrombus formation.

In addition to their role in lipid transport, lipoproteins can also exert other beneficial effects. Lipoproteins can bind to endotoxin and prevent its lethal effects in rodents[14,15]. Lipoproteins can also decrease endotoxin-induced cytokine secretion by macrophages[16] and thereby decrease their toxic effects[17]. The ability of lipoproteins to inhibit the synthesis of AA metabolites (mediators of several inflammatory responses) may be an additional non-specific mechanism operative in host defense in several disease states and inflammatory conditions.

In summary, our results indicate that lipoproteins have significant inhibitory effects on various PGs, TXA_2 and 12-HETE suggesting that lipoproteins may have a physiological role in the regulation of AA metabolism. Since lipoproteins levels are elevated in several diseases such as coronary heart disease and diabetes mellitus[18,19], it is likely that they may modulate a wide range of biological effects of different products of AA metabolism in the pathophysiology of these diseases.

REFERENCES

1. Smith, W.L. (1989) *Biochem. J.* 259: 315-324.
2. Marcus, A.J. and Hajjar, D.P. (1993) *J. Lipid Res.* 34: 2017-2031.
3. Beitz, J. and Forster, W. (1980) *Biochem. Biophys. Acta.* 620: 352-355.
4. Beitz, J. and Forster, W. (1981) *Prostaglandins Med.* 6: 515-518.
5. Shakhov, Y., Larrue, J., Perova, N., Dorian, B., Darett, D., Shcherbakova, I., Bricaud, H. and Oganov, R. (1989) *J. Molec. Cell. Cardiol.* 21: 461-468.
6. Havel, R.J., Eder, H.A. and Bragdon, J.H. (1955) *J. Clin. Invest.* 34: 1345-1353.
7. Lowry, O.H., Rosebrough, N.J., Farr, A.L. and Randall, R.J. (1951) *J. Biol. Chem.* 93: 265-274.
8. Saeed, S.A., Strickland, D.M., Young, D.C., Dang, A. and Mitchell, M.D. (1982) *J. Clin. Endocrinol. Metab.* 55: 801-803.
9. Saeed, .S.A., Simjee, R.U., Mahmood, F., and Rahman, N.N. (1993) *J. Pharm. Pharmacol.* 45: 715-719.
10. Bryant, R. W., and Bailey, J.M. (1979) *Prostaglandins.* 17: 9-18.
11. Habenicht, A.J.R., Salbach, P., Goerig, M., Zeh, W., Janssen-Timmen, U. and Glomset, J.A. (1990) *Nature* 345: 634-636.
12. Fleischer, L.N., Tall, A.R., Witte, L.D., Miller, R.W. and Cannon, P.J. (1982) *J. Biol. Chem.* 257: 6653-6655.
13. Fujomoto, Y., Tsunomori, M., Muta, E., Yamamoto, T., Nishida, H., Sakuma, S., and Fujita, T. (1994) *Res. Commun. Molec. Pathol. Pharmacol.* 85: 355-358.
14. Harris, H.W., Grunfeld, C., Feingold, K.R. and Rapp, J.H. (1990) *J. Clin. Invest.* 86: 696-702.
15. Harris, H.W., Grunfeld, C., Feingold, K.R., Read, T.E., Kane, J.P., Jones, A.L., Eichbaum, E.B., Bland, G.F. and Rapp, J.H. (1993) *J. Clin. Invest.* 91: 1028-1034.
16. Flegel, W.A., Wolpl, A., Mannel, D.N. and Northof, H. (1989) *Infect. Immun.* 57: 2237-2245.
17. Memon, R.A., Feingold, K.R., and Grunfeld, C. (1995) In: *Human cytokines. Their role in disease and therapy* (Aggarwall, B.B. and Puri, R.K. eds.) pp 239-251, Blackwell Science, Cambridge.
18. Grundy, S.M. (1991) *Arterioscler. Thromb.* 11: 1619-1635.
19. Memon, R.A., Grunfeld, C., Moser, A.H., Feingold, K.R. (1994) *Horm. Metab. Res.* 26: 85-87.

20

MECHANISM OF GONADOTROPHIN RELEASING HORMONE RECEPTOR MEDIATED REGULATION OF G-PROTEINS IN CLONAL PITUITARY GONADOTROPHS

Bukhtiar H. Shah,[*][1] David J. Macewan,[2] and
Graeme Milligan[2]

[1] Department of Physiology
The Aga Khan University Medical College
Karachi-74800, Pakistan
[2] Molecular Pharmacology Group
Division of Biochemistry and Molecular Biology
Institute Biomedical and Life Sciences
University of Glasgow, Glasgow, Scotland, United Kingdom

ABSTRACT

Guanine nucleotide binding proteins (G-proteins) function to transduce hormonal and sensory signals from plasma membrane receptors to effectors to regulate cellular functions. Binding of gonadotrophin releasing hormone (GnRH) receptor agonist to its receptor causes production of the second messengers inositiol-1,4,5-triphosphate (IP_3) and sn-1,2-diacyglycerol (DAG) through the hydrolysis of phosphatidylinositol-4,5-bisphosphate because this receptor is coupled to G_q and G_{11} G-proteins. Prolonged exposure of αT3-1 pituitary gonadotrophs to a GnRH agonist results in marked downregulation of α-subunits of $G_{q\alpha}/_{11}$. The agonist-mediated downregulation is concentration and time-dependent and is not altered by treatment of cells with protein kinase C (PKC) activators and inhibitors. However the turnover of $G_{q\alpha}/G_{11\alpha}$ is substantially accelerated in the presence of agonist as revealed through ^{35}S-labelled pulse chase experiments. By contrast the rate of degradation of the G-protein $G_{i2\alpha}$ is unaffected by agonist treatment. Analysis of $G_{q\alpha}/G_{11\alpha}$ mRNA levels by reverse transcription/polymerase chain reaction (RT/PCR) demonstrated no differences between control and agonist-treated cells. These studies indicate that GnRH receptor agonist-mediated down regulation of $G_{q\alpha}/G_{11\alpha}$ is independent of PKC and is a reflection of enhanced proteolysis of the activated G-proteins.

---

[*] Represents person presenting Paper

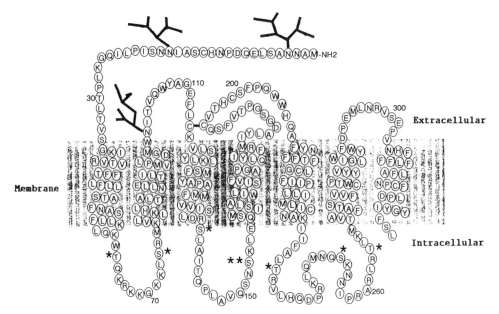

Figure 1. Schematic representation of the mouse GnRH receptor in pituitary gonadotrophs (2). Phosphorylation (*) and glycosylation (**) sites are shown.

INTRODUCTION

Guanine nucleotide binding proteins (G-proteins) function to transduce hormonal and sensory signals from plasma membrane receptors to effectors to regulate cellular function[1]. The gonadotrophin releasing hormone receptor (GnRH) is a member of the G-protein coupled, seven-transmembrane domain receptor superfamily[1,2]. The cloning of GnRH receptor by Tsutsumi et al.[2] revealed that the receptor has the predicted structure characteristic of G-protein-coupled receptors (GPCRs), consisting of a single polypeptide chain containing seven hydrophobic transmembrane domains connected by hydrophilic extracellular and intracellular loops (Fig. 1). A number of potential phosphorylation sites which might mediate heterologous desensitization are present on the GnRH receptor[2]. GnRH plays a pivotal role in regulation of biosynthesis and secretion of gonadotropins, FSH and LH[3]. GnRH analogs have wide therapeutic applications which include precocious puberty, endometriosis, uterine leiomyomata, hirsutism, infertility and sex hormone dependent malignant tumors such as prostate and breast carcinoma[4]. Binding of GnRH to its receptor causes production of the second messengers, inositol-1,4,5-triphosphate (IP$_3$) and sn-1,2-diacylglycerol (DAG) through the hydrolysis of phosphatidylinositol-4,5-bisphosphate because this receptor is coupled to G$_q$ and G$_{11}$ G-proteins[5]. G$_q$ and G$_{11}$ proteins appear to be essentially universally co-expressed. Such co-expression raises questions as to the individual functions of these two G-proteins.

Previous studies have shown that the maintained presence of an agonist at a G-protein-linked receptor frequently results in a reduction in cellular levels of the receptor. This process is known as downregulation and contributes to the battery of adaptive changes, collectively called desensitization, which cells and tissues can utilize to prevent chronic full scale response to a stimulus. While the focus of studies of downregulation have been at the level of the receptor, due partially to the availability of radiolabelled ligands able to identify and quantitate these polypeptides,[6-8] it has recently become clear that cellular levels of the guanine nucleotide binding proteins which transduce information from receptors to effector systems can also be regulated in response to the sustained presence of an agonist[7,8]. These effects are noted primarily as a selective downregulation of the G-protein or G-proteins with which the receptor for the agonist interacts but in certain cases may also be manifested by an upregulation of a G-protein which opposes the function of the receptor[9].

Study of the second messenger pathways involved in GnRH action has been greatly facilitated by the availability of immortalized anterior pituitary gonadotrope cell line, αT3-1 cells[2,3,10]. In this study we have examined the mechanism of agonist regulation of $G_{q\alpha}$ and $G_{11\alpha}$ in the murine gonadotroph αT3-1 cells.

MATERIALS AND METHODS

The GnRH analogue des-Gly[10],[D-Ala[6]]-luteinizing hormone releasing hormone ethylamide was from Sigma. All materials for tissue culture were from Gibco/BRL.

Cell Culture and Preparation of Membranes

Cells were grown in Dulbecco's modified Eagle's medium (DMEM) containing pyruvate supplemented with 10% foetal-calf serum, penicillin (100 units/ml) and strepto-mycin (100 μg/ml) in 5% CO_2 at 37 °C. Cells were grown in 75 cm^2 tissue-culture flasks and were harvested just before confluency. Membranes were prepared from the cells by homog-enization with a Teflon-on-glass homogenizer and differential centrifugation as de-scribed[11,12]. The membrane pellet was collected by centrifugation and protein concentration was measured according to the method of Lowry et al.[13].

Inositol Phosphate Assays

Cells were seeded in 24-well plates and labelled to isotopic equilibrium with [^{3}H]inositol (1 μCi/ml) in inositol-free Dulbecco's medium containing 1% dialysed foetal-calf serum for 48 h. The total inositol phosphates generated in response to GnRH agonist were measured by batch chromatography on Dowex-1 formate as previously described[11].

Western Blotting

The generation and specificities of the various antisera used in this study for immunoblotting G-proteins are shown in Table 1. Each antiserum was produced in a New Zealand White rabbit using a conjugate of a synthetic peptide with keyhole-limpet haemo-cyanin (Calbiochem) as antigen. Details of this process have been described previously[14].

Table 1. Specificities and generation of anti-G protein antisera. Each antiserum was generated in a New Zealand White rabbit

Antiserum	Peptide used	G protein sequence	Antiserum identifie
CQ2	QLNLKEYNLV	$G_{q\alpha}$ 351-360 $G_{11\alpha}$ 350-359	$G_{q\alpha}$, $G_{11\alpha}$
SG1	KENLKDCGLF	Transducin α 341-350	Transducin, $G_{i1\alpha}$, $G_{i2\alpha}$
CS1	RMHLRQYELL	$G_{s\alpha}$ 385-394	$G_{s\alpha}$
BN3	MSELDQLRQE	β 1-10	β_1, β_2
IQB	EKVSAFENPYDAIKSGq	α 119-134	$G_{q\alpha}$

Membrane samples were resolved by SDS/PAGE (in 10% (w/v) acrylamide gels overnight at 60 V and immunoblotting was done as previously.[11,12]

^{35}S-Pulse-Chase Labelling Studies

The turnover of G-proteins in gonadotroph cells in response to GnRH analogue was assessed as described previously[12]. Briefly cells were seeded in 6 well plates and were labelled with 50 µCi/ml Trans[^{35}S]-label (final concentration in well) for 20-48 h. The radioactive medium was removed and cells were washed once with 1 ml of normal DME culture medium. They were subsequently maintained in the presence or absence of the GnRH receptor agonist, des-Gly^{10}-[D-Ala^6] luteinizing hormone releasing hormone (1 µM). At appropriate times, the medium was removed and cells were dissolved in 1% SDS (200 µl/well). The proteins immunoprecipitated with specific G-protein antiserum were resolved by electrophoresis by SDS/PAGE (10% acrylamide). The gels were dried and exposed to phosphor storage screen autoradiography according to manufacturer's instructions using a Fujix Bio-imaging Analyser linked to an Apple macintosh Quadra 650 personal computer.

Reverse Transcriptase/Polymerase Chain Reaction (RT/PCR)

The RT/PCR procedure was carried out as described previously.[9,13] Total RNA was extracted from cells and 1-5 µg RNA (20 µl) were denatured by incubation at 65° C for 10 min followed by chilling on ice and reverse transcribed in 33 µl of reaction mixture using first strand cDNA synthesis kit (Pharmacia LKB Biotechnology). PCR reactions were carried out using the following 24-mer primers (20-40 pmol) in 100 µl reaction mixture:

HPRT-sense	5' CCTGCTGGATTACATTAAAGCACT 3'
HPRT-antisense	5' CCTGAAGTACTCATTATAGTCAAG 3'
$G_{q\alpha}$–sense	5' ATGACTTGGACCGTGTAGCCGACC 3'
$G_{11\alpha}$-sense	5' ACGTGGACCGCATCGCCACAGTAG 3'
$G_{q\alpha}/_{11\alpha}$-antisense	5' CCATGCGGTTCTCATTGTCTGACT 3'

PCR cycles were as follows: 95 °C/5 min, 60 °C/30 s, 72 °C/1 min (1 cycle); 95 °C/30 s, 60°C/30 s, 72 °C/1 min (25 cycles) 95 °C/30 s, 60 °C/30 s, 72 °C/5 min (1 cycle). Amplification of β-actin was carried out using similar conditions except the annealing temperature was 52 °C. Reaction products were then separated by 1.5-1.75% agarose gel electrophoresis.

RESULTS

αT3-1 cells express high levels of the GnRH receptor which exhibit binding characteristics similar to those found in normal mouse and rat pituitary[2,3]. Immunoblotting membranes of αT3-1 cells resolved by 10% (w/v) acrylamide SDS-PAGE with non-selective antiserum CQ2 identified an apparently single 42 kDa polypeptide. Treatment of αT3-1 cells with the GnRH receptor agonist, des-Gly^{10}-[D-Ala^6]-luteinizing hormone releasing hormone ethylamide (1 µM, 16h) decreased immunodetectable levels of this 42 kDa polypeptide whether detection was performed with antiserum IQ1 or CQ2 (Figure 2). Time-course of the GnRH analogue-mediated reduction of levels of a subunit of both G_q and G_{11} indicated that half-maximal loss of these polypeptides was produced by 5 h and the maximal reduction was achieved by 8 h (Fig. 3). Incubation with the ligand for longer time periods did not cause further loss of $G_{q\alpha}/G_{11\alpha}$. In contrast to levels of $G_{q\alpha}/G_{11\alpha}$, immunologically detected levels

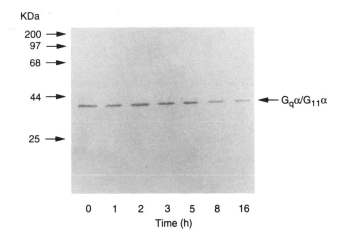

Figure 2. a) Treatment of αT3-1 cells with GnRH analogue (1 μM) for various time periods markedly downregulates the $G_{q/11a}$ proteins[11].

of the α-subunits of both the adenylyl cyclase stimulatory G-protein $G_{s\alpha}$, which in these cells is expressed predominantly as a 45 kDa long form of the polypeptide, and of the pertussis toxin-sensitive G-protein $G_{i2\alpha}$ were not modulated by equivalent agonist treatment[11]. Equally such agonist treatment had no effect on levels of the G-protein β subunit (data not shown).

The binding of GnRH and analogues to its receptor in αT3-1 cells leads to activation of phospholipase Cb1 resulting in elevated intracellular levels of inositol phosphates and 1,2-diacylglycero[2,11]. GnRH agonist markedly stimulated inositol phosphate production in LiCl-treated αT3-1 cells which had been labelled for 48 h with myo-[³H]inositol. This occurred in a dose-dependent fashion with half-maximal stimulation produced by 1nM

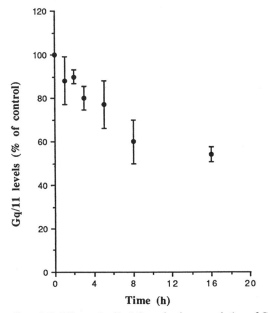

Figure 3. Time course effect of GnRH agonist (1 μM) on the downregulation of $G_{q\alpha/11\alpha}$ proteins. (data from 4 immunoblots).

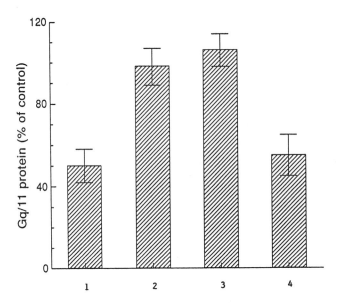

Figure 4. Effect of 1) GnRH analogue (1 μM), 2) phorbol ester (PMA; 100nM), 3) PKC inhibitor (chelerythrine; 10 μM) and 4) GnRH agonist plus chelerythrine on the $G_{q\alpha}/\Gamma_{11\alpha}$ proteins in αT3-1 cells. The data is presented as effect in percentage compared to control untreated cells (100%).

agonist. Incubation with agonist for 5 min period generated some 7150 dpm above basal levels of inositol phosphates/100 000 dpm of inositol-containing phospholipids[11].

Since protein kinase mediated phosphorylation is well known to regulate the expression of receptors, parallel to that we investigated the role of PKC in the downregulation of G-proteins. Sustained treatment of αT3-1 cells with the PKC activator phorbol ester, PMA (100 nM, 16h), was unable to modulate cellular levels of $G_{q\alpha}/G_{11\alpha}$ (Figure 4) indicating that agonist-induced down-regulation of $G_{q\alpha}/G_{11\alpha}$ did not occur subsequent to activation of PKC. Treatment with the selective PKC inhibitor, chelerythrine (10 μM) for 16 h was also unable to regulate cellular $G_{q\alpha}/G_{11\alpha}$ levels, and coincubation of αT3-1 cells with both GnRH analogue (1 μM) and chelerythrine (10 μM) produced the same levels of $G_{q\alpha}/G_{11\alpha}$ as did analogue alone.

To determine whether alteration in $G_{q\alpha}/G_{11\alpha}$ mRNA levels is responsible for agonist-mediated regulation of cellular levels of these G-proteins, we performed reverse transcriptase/PCR using RNA isolated from both untreated cells and those exposed to the GnRH analogue (1 μM) for varying times. RT/PCR results demonstrated that the agonist induced loss of the two G-protein a subunits was not accompanied by a corresponding decline in mRNA levels encoding these polypeptides (data not shown) indicating that regulation of the G-proteins is not due to alterations in mRNA levels.

To explore the mechanism of $G_{q\alpha}/G_{11\alpha}$ downregulation, αT3-1 cells were incubated with Trans[^{35}S]-label for 24 h and the decay of radiolabel with time in immunoprecipitated $G_{\alpha}/G_{11\alpha}$ was measured in control cells and those maintained in the presence of GnRH agonist (1 μM). Analysis of rate of decay of [^{35}S]-labelled $G_{\alpha}/G_{11\alpha}$ clearly demonstrated a more rapid rate of decay in the cells maintained in the presence of the analogue[12] as shown in (Fig. 5). Quantitative analysis indicated that in untreated cells, this process was described adequately by a mono-exponential decay with a half-time of 26.5 ± 3.2 h (n=4). However the decay of radiolabel from $G_{q\alpha}/G_{11\alpha}$ in the cells treated with agonist showed a profile where the data

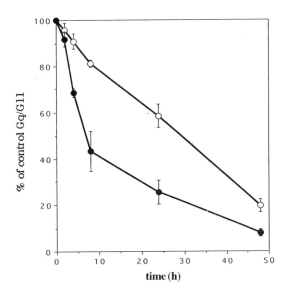

Figure 5. ^{35}S-pulse chase labelling to monitor the turnover of Gq/G$_{11\alpha}$ proteins in untreated and GnRH agonist (1 μM) treated cells as detailed in the Methods. The data show that G$_{q/11\alpha}$ turnover is accelerated in agonist treated cells12.

could be best fitted in a two compartment model. The initial rapid phase had a half-life of G$_{q\alpha}$/G$_{11\alpha}$ around 6.7 h whereas the slower secondary phase of decay corresponded to a half-life for G$_q$/G$_{11\alpha}$ of 17.2 h (n=4). In contrast to the effects on G$_q$/G$_{11\alpha}$, GnRH analogue treatment did not alter the rate of turnover (p=0.78) of G$_{i2\alpha}$-protein (data not shown).

DISCUSSION

The immortalized GnRH responsive murine gonadotroph pituitary cell line αT3-1, produced by targetted oncogenesis in transgenic mice[2,3], has been widely used to study cell signalling from the GnRH receptor[5,10,11]. Our studies have shown that activation of GnRH receptor by GnRH analogue causes marked production of inositol phosphates in αT3-1 cells and these effects are resistant to pretreatment of the cells with pertussis toxin implying that the G-protein(s) which couple to the GnRH receptor in these cells are not of the Gi-family. Pertussis toxin-insensitive activation of phosphoinositidase C activity by G-protein coupled receptors is produced by members of the G$_q$-family of G-proteins and GnRH receptor activation of this enzyme in αT3-1 cells has been shown by Hsieh and Martin[10] to be produced by G$_q$ and/or G$_{11}$.

In many circumstances, the maintained exposure of cells to an agonist at a G-protein linked receptor results in a reduction in cellular levels of the receptor[6,7,11,16]. This process is called downregulation and contributes to the desensitization mechanisms which cells can utilize to prevent contiual response to a stimulus. Although it was not widely examined for a number of years, due largely to the lack of suitably sensitive and discriminatory probes, it has now been firmly established that chronic exposure to a G-protein-linked receptor agonist can also frequently result in a downregulation of the G-protein or G-proteins which interact with the receptor. Such effects have been observed for members of each of the G$_s$, G$_i$ and G$_q$ families of G-proteins and in the cases in which the mechanisms have been explored seem to result from enhanced turnover of the G-protein[12,15].

Maintained exposure of αT3-1 cells to the GnRH analogue GnRh analogue results in a sustained, selective downregulation of G$_{q\alpha}$/G$_{11\alpha}$ and this regulation is not altered by PKC activator or inhibitor. Analogous to that the downregulation of Gsα produced by either the activation of an IP prostanoid receptor in neuroblastoma x glioma hybrid NG108-15 cells,[16] or through ADP-ribosylation of Gsa in pituitary GH3 cells[17] or C6 glioma cells[18] by cholera toxin was not as a consequence of increased intracellular cyclic AMP levels. These observations preclude the mechanism of downregulation of the G-proteins being dependent upon activation of second messenger-regulated kinases. However, forskolin has been shown to cause a selective decrease in the half-life of Gsα in S49 mouse lymphoma cells[9], so particular effects may be cell type or G-protein specific. Agonist induced changes in mRNA is well studied contributory mechanism for regulation of the levels of G-protein-coupled receptor proteins.[9] Recent studies have shown that G-protein expression can be transcriptionally regulated in various tissues by agents such as thyroid hormone,[19] somatostatin, vasoactive intestinal peptide[20] and FSH[21]. However the lack of effect of GnRH analogue on steady state levels of $G_{q\alpha}$ and G_{11a} mRNA in αT3-1 cells in the present study indicates that the agonist-mediated reduction in levels of these G-proteins is unlikely to be due to changes in the rate of synthesis, although the possibility that the rate of translation could alter in the absence of any changes in the steady-state levels of mRNA can not be eliminated. These data are similar to our previous observations in chainese hamster ovary (CHO) cells transfected with human muscarinic acetylcholine receptor and β2-adrenergic receptors where chronic treatment with agonists failed to regulate either $G_{q\alpha}/G_{11\alpha}$ or Gsα mRNA levels although agonsit-mediated G-protein down-regulation was observed[15,22].

The two component decay of radiolabel in $G_{q\alpha}/G_{11\alpha}$ observed in the presence of GnRH agonist may imply either that the fast decay rate reflects the fraction of the cellular G-protein which becomes activated upon occupancy of the GnRH receptor while the slower decay rate represents the residual G-protein pool or that a fast initial decay rate occuring upon receptor occupancy is reduced to as lower rate with desensitization of the receptor response. It is difficult to devise experiments which would discriminate between these possibilities. Studies from Levis and Bourne[23] have demonstrated that a mutationally activated variant of Gsα is degraded substantially more rapidly than that the wild type protein, providing further evidence that activation can result in enhanced degradation of G-protein α subunits.

In conclusion, our studies demonstrate that chronic full-scale stimulation of a G-protein coupled receptor by its agonist leads to generation of second messengers and downregulation of the G-protein, a mechanism responsible to dampen the excessive stimulation. The present studies show that GnRH-agonist mediated downregulation of $G_{q/11\alpha}$ protein is neither caused by protein kinase C nor by changes in G-protein mRNA levels, however, the $G_{q\alpha/11\alpha}$ protein turnover is enhanced in agonist treated cells.

Acknowledgements

This work was completed by a grant from the Wellcome Trust.

REFERENCES

1. Eidne, K.A., Seller, R.E., Couper, G., Anderson, L. and Taylor, P.L. (1992) *Mol. Cell. Endocrinol.* 90: R5-R9.
2. Tsutsumi, M., Zhou, W., Millar, R.P., Mellon, P.L., Roberts, J.L., Flanagan, C.A., Dong, K., Gillo, B. and Sealfon S.C. (1992) *Mol. Endocrinology* 6: 1163-1169.
3. Windle, J.J., Weiner, R.I. and Mellon, P.L. (1990) *Mol. Endocrinology* 4: 597-603.
4. Gilman, A.G., Rall, T.W., Nies, A.S. and Taylor, P. (1990) In: *The Pharmacological Basis of Therapeutics.* 8th Ed., pp: 1333-1360.
5. Anderson, L., Milligan, G. and Eidne, K.A. (1993) *J. Endocrinol.* 136: 51-58.
6. Milligan, G. (1993) *Trends Pharmacol. Sci.* 14: 413-418.

7. Milligan, G., Wise, A., MacEwan, D.J., Kennedy, F.R., Lee, T.W., Aide, E.J., Svoboda, P., Shah, B.H. and Mullaney, I. (1995) *Biochem. Soc. Trans.* 23: 166-170.

8. Milligan, G., Shah, B.H., Mullaney, I. and Grassie, M. (1995) *J. Recept. Res.* 15: 253-263.

9. Hadcock, J.R., Port, J.D. and Malbon, C.C. (1991) *J. Biol. Chem.* 266: 11915-11922.

10. Hseih, K.P. and Martin, T.F.J. (1992) *Mol. Endocrinology* 6: 1673-1681.

11. Shah, B.H. and Milligan, G. (1994) *Mol. Pharmacology* 46: 1-7.

12. Shah, B.H., MacEwan, D.J. and Milligan, G. (1995) *Proc. Natl. Acad. Sci.* USA (in press).

13. Lowry, O.H., Rosebrough, N.J., Farr, A.L. and Randall, R.J., (1951) *J. Biol. Chem.* 193: 265-275.

14. Mullaney, I, Dodd, M.W., Buckley, N. and Milligan, G. (1993) *Biochem. J.* 289: 125-131.

15. Mitchell, F.M., Buckley, N.J., and Milligan, G. (1993) *Biochem. J.* 293: 495-499.

16. McKenzie, F.R., and Milligan, G. (1990) *J. Biol. Chem.* 265: 17084-17093.

17. Chang, F.H. and Bourne, H.R. (1989) *J. Biol. Chem.* 264: 5352-5337.

18. Shah, B.H. and Milligan, G. (1994) *J. Neurochemistry* 63: S32A.

19. Rapiejko, P.J., Watkins, D.C., Ros, M. and Malbon, C.C. (1989) *J. Biol. Chem.* 264: 16183-16189.

20. Paulssen, E.J., Paulssen, R.H., Haugen, T.B., Gautvik, K.M., and Gordeladze, J.O. (1991) *Acta Physiol. Scand.* 143: 195-201.

21. Loganzo, F. Jr., Fletcher, P.W. (1993) *Mol. Endocrinol.* 7: 434-440.

22. Mullaney, I., Shah, B.H., Wise, A. and Milligan, G. (1995) *J. Neurochemistry* (in press).

23. Levis, M.J. and Bourne, H.R. (1992) *J. Cell Biol.* 119: 1297-1307.

STRUCTURAL STUDIES OF CARCINOEMBRYONIC ANTIGEN (CEA) AND ANTI-CEA ENGINEERED ANTIBODIES USING MASS SPECTROMETRY

John E. Shively,* Stanley A. Hefta, Kristine Swiderek, Anna Wu, and Terry D. Lee

Divisions of Immunology and Biology
Beckman Research Institute of the City of Hope
Duarte, California 910101

ABSTRACT

We have used mass spectrometry as a structural method to analyze carcinoembryonic antigen (CEA) and other members of the CEA gene family, including engineered anti-CEA antibodies that are use in clinical studies for tumor imaging and therapy. CEA is a highly glycosylated, GPI-anchored, high molecular weight (180 kDa) membrane glycoprotein that is highly expressed in cancers of epithelial cell origin, including colon, lung, breast, and ovary. Since the high level of glycosylation posed serious problems for structural analysis, CEA was deglycosylated with TFMSA, converting all 28- N-glycosylation sites to Asn-GlcNAc derivatives which were easily analyzed by peptide mapping, microsequence analysis, and mass spectrometry. The GPI linkage site was converted to an N-ethanolamine amide which was positively identified by mass spectrometry. Peptide mapping of intact CEA allowed isolation of the majority of the glycopeptides and a mass survey of the structures was performed by SIMS and MS-MS approaches. Structures included high mannose, complex, and polylactosamine. Similar studies were performed on the CEA-related antigen, NCA (non-specific cross reacting antigen or CD66c antigen on granulocytes). The glycosyl units on the N-domain of NCA were shown to contribute to bacterial binding via a lectin site, and when removed by recombinant means, gave a glycoprotein devoid of bacterial binding activity. The glycosylation sites were also identified by peptide mapping and MS or MS-MS analysis. A chimeric anti-CEA antibody used in clinical studies was subjected to peptide mapping and analyzed by LC/ESI-MS, MALDI-TOF, and SIMS, confirming its structure and several posttranslational modifications. A single chain version (sFv) was analyzed by LC/ESI-MS and microsequence analysis, confirming that the correct size protein was expressed and processed by *E.coli*.

---

* Represents person presenting Paper

INTRODUCTION

Carcinoembryonic antigen (CEA) is a highly glycosylated, GPI-anchored, high molecular weight (180 kDa) membrane glycoprotein that is highly expressed in cancers of epithelial cell origin, including colon, lung, breast, and ovary (for reviews see[1] and[2]. CEA has seven immunoglobulin (Ig)-like domains, comprising a 108 amino acid N-domain with no disulfide bonds, and six C2 (constant-like) domains with one disulfide bond per domain. CEA is a member of a multi-gene family located on chromosome 19[3]. Each member is recognizable by high sequence homology, and the characteristic order of an N-domain followed by one or more C2 domains. A simple schematic showing the domain structure and membrane linkage of three members of the gene family are shown in Figure 1. The first member shown, NCA (non-specific crossreacting antigen) received its name based on its known crossreactivity to CEA in immunoassays, a problem frequently encountered with polyclonal and some monoclonal antibodies to CEA. NCA is found in normal lung, colon, and granulocytes (in granulocytes it originally received the designation CD66, and revised to CD66c). Biliary glycoprotein (BGP) was first identified in the liver and bile, but later shown to be in normal colon and granulocytes (originally CD15 in granulocytes and revised to CD66a). CEA and NCA are linked to the membrane by a glycolipid anchor via a glycosylphosphatidylinositol (GPI) moiety, and BGP is a type I transmembrane protein. While the in vivo function of these membrane proteins remains a matter of debate, there is good in vitro evidence that they function as both homotypic cell adhesion molecules[4] and adhesion molecules to bacteria[5]. We have been interested in the structure-function relationships of these molecules and will review some of our findings on CEA and NCA in this report.

Antibodies to CEA have considerable clinical significance as in vitro tests for colon cancer and as in vivo targeting agents for tumor imaging and therapy. We have developed a monoclonal anti-CEA antibody, T84.66[6], which has a high affinity and specificity for CEA. T84.66 is used in the Roche test for CEA, and when radiolabeled with In-111 has been shown to effectively image colon tumors[7]. Since the administration of murine antibodies to man results in an human-anti-mouse response (HAMA), we have engineered a chimeric version of the antibody, cT84.66, with murine variable regions and human constant regions of class

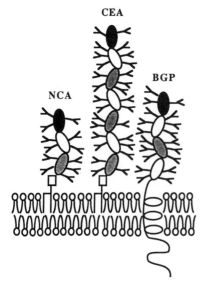

Figure 1. Schematic of CEA-family domains, glycosylation, and membrane anchoring. Solid ovals are 108 amino acid N-domains, resembling immunoglobulin V-regions, but with no disulfide bond. Open and shaded ovals are 80-90 amino acid immunoglobulin C2-like domains, including one disulfide bond. Membrane anchorage shows GPI for NCA and CES, and type I transmembrane for BGP. Glycosylation sites shown as Ys.

IgG-1[8]. In recent studies we have shown good targeting of In-111 labeled cT84.66 to human tumors and have initiated therapy studies with Y-90 labeled cT84.66 (unpublished). Since further improvements in biodistributions and therapeutic ratios (tumor to normal tissue uptake ratios) are expected by further engineering of the antibody, we have generated single chain (sFv) and deleted domain versions of the antibody. In this report we will present approaches to characterize engineered antibodies by mass spectrometry.

Structural Studies on CEA

CEA has 28 N-glycosylation sites. Since it was so difficult to obtain peptide resolution on peptide mapping for CEA, our first approach was to remove the majority

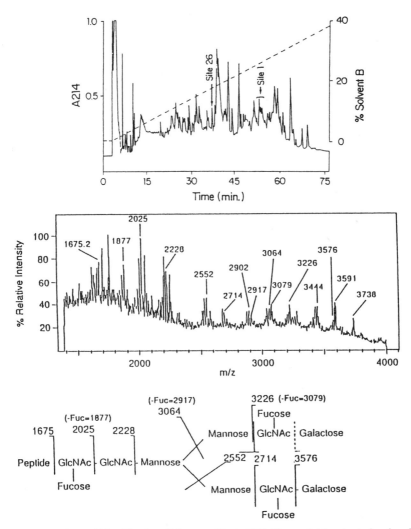

Figure 2. Separation and identification of glycopeptides in CEA. Neuraminidase treated, reduced and S-alkylated CEA was digested with trypsin and chymotrypsin and peptides separated by reversed phase HPLC (upper). Glycopeptides corresponding to sites 1 and 26 are indicated with arrows. The peptides were further purified by chromatography on polyhydroxyethyl aspartamide and reversed phase HPLC. SIMS analysis of the peptide from site 1, together with the proposed structure is shown (bottom).

of glycosylation by treatment with trifluoromethanesulfonic acid (TFMSA). This approach gave simplified peptide maps where each peptide was amenable to microsequence analysis and could be mass analyzed by SIMS[9]. Since TFMSA cleaves O- but not N-glycosyl bonds, each of the glycopeptides was recovered in the Asn-GlcNAc form. We demonstrated that the PTH derivative of Asn-GlcNAc could be identified by the usual chromatography on reversed phase columns, and that all 28 potential N-glycosylation sites were utilized. The C-terminal amino acid Ser-643 was derivatized with ethanolamine in an amide bond[10], a product of the TFMSA removal of the GPI anchor from the C-terminus of CEA. In an attempt to map the full glycosylation patterns for CEA, we have performed peptide mapping on neuraminidase treated and reduced and S-alkylated CEA[11]. In order to obtain the majority of glycopeptides it was essential to digest CEA with both trypsin and chymotrypsin. The peptide map (Figure 2) was extremely complicated, and it was necessary to perform repeated rechromatography on each peak to obtain single species suitable for further analysis. Two glycopeptides (out of the possible total of 29) are indicated. Rechromatography was performed on a C18 and polyhydroxyethyl aspartamide column with a reverse gradient. Figure 4 shows a SIMS analysis for the peptide corresponding to glycosylation site-1 (Asn-70). The sequence of the peptide was confirmed by microsequence analysis (SGREIIPN*ASLLIQ) with a blank cycle observed at the N* site (PTH derivatives of Asn-CHO are not normally extracted into butyl chloride during microsequence analysis). SIMS analysis gave fragments corresponding to the intact peptide (MH+= 3738), and loss of portions of the oligosaccharide chains (as shown), including loss of the entire oligosaccharide (MH+= 1675). This analysis is consistent with, but does not prove the structure of the oligosaccharide moiety. Given that the overall structures of the glycosyl units of CEA are known[12], our conclusions are likely correct, and provide the first site specific glycosylation analysis for CEA. The glycosyl unit at site-26 had a fucosylated polylactosamine branch[11].

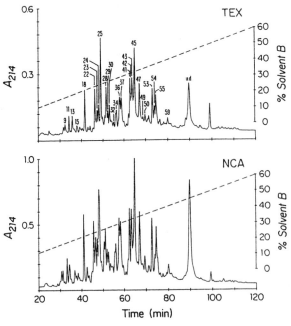

Figure 3. Comparative peptide mapping of NCA-95 (TEX) and NCA-50. The two glycoforms of NCA were treated with TFMSA, digested with chymotrypsin, and peptides separated by reversed phase HPLC. Peptide fractions analyzed by SIMS and microsequence analysis are numbered in the upper trace.

Structural Studies on NCA

The polypeptide portion of NCA has an expected Mr of about 35 kDA, but when NCA is isolated from tumor (or other) cells, a broad band on SDS polyacrylamide gels is observed from about 50-95 kDA. Interestingly, different Mr species can be isolated depending on the source of NCA or isolation method; thus, NCA has been variously reported as molecular species of Mr 50 kDa or 95 kDa. Since each species had identical N-terminal sequences, we hypothesized that the molecular weight heterogeneity was due solely to differences in glycosylation patterns. This was confirmed by deglycosylation of the two species with TFMSA, comparison of their molecular masses, and peptide mapping by mass analysis[13]. Both species are reduced to a mass of about 35 kDa by TFMSA, and give identical chymotryptic peptide maps (Fig. 3) and identical masses of each peptide throughout their entire peptide maps, including each glycosylation site. In addition, they both have a C-terminal ethanolamine, consistent with their anchorage via a GPI moiety. Further studies established that only the 50 kDa species mediated bacterial adhesion that was specifically inhibited by α-methyl mannoside[14].

Mutational analysis of the three N-glycosylation sites in the N-domain of NCA established that sites-1 and 2 accounted for all of the bacterial binding activity of NCA in spite of the presence of an additional 10 glycosylation sites in the remainder of the molecule[15]. The sizes of intact NCA and each of the mutants was determined by MALDI-TOF-MS (Figure 4), a method especially well suited for the analysis of heavily glycosylated glycoproteins. It can be seen that the difference in Mr's are consistent with the loss of about 1500 Da for sites 1 and 2, and about 2000 Da for site 3. The biological activity studies suggested that the first two sites of NCA were unique either in terms of their glycosylation patterns or surrounding protein sequence (i.e., some contribution to binding was contributed by the protein moiety). In order to further study this possibility, intact (wild type) NCA-50 was subjected to chymotryptic mapping as described for CEA. Glycopeptides which were isolated for sites 1 and 2 had SIMS patterns corresponding to high mannose oligosaccharides, a result consistent with their binding to type 1 fimbriae and inhibition with α-methyl mannoside. The SIMS-MS analysis for NCA site-2 is shown in Figure 5. Characteristic fragments are consistent with a branched M3 structure with a fucose on the GlcNAc attached to the Asn-77. Fragments are also observed within the peptide itself.

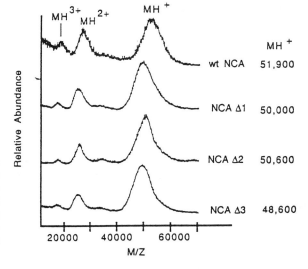

Figure 4. Maldi-Tof analysis of NCA and NCA glycosylation site deleted mutants. Glycosylation sites 1-3 were mutated in the N-domain of NCA and expressed in HeLa cells. The proteins were isolated by affinity chromatography. The samples and masses are shown to the right.

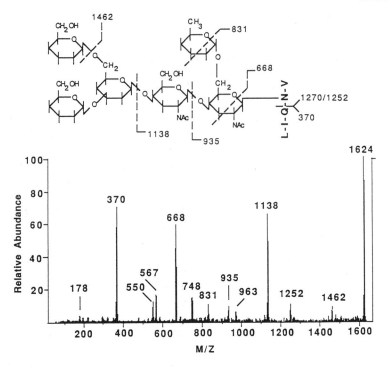

Figure 5. SIMS-MS analysis of glycosylantion site 2 in NCA. The glycopeptide was isolated from a chymotryptic map of reduced and S-alkylated NCA and analyzed by SIMS-MS on a triple quadrupole Finnigan TSQ700 mass spectrometer. The peak corresponding to the intact glycopeptide (1624) was chosen for MS-MS analysis. The postulated fragmentation patter is shown above the mass spectrum.

These studies indicate the utility of SIMS and SIMS-MS analysis for the identification of site specific glycosylation patterns in glycoproteins. In order to be successful, one must be prepared to perform arduous rechromatographies, since the peptide maps are usually complex, and separation by reversed phase HPLC alone is insufficient to resolve multiple glycopeptide species. We have utilized SIMS, MALDI, and ESI (electrospray ionization) for the analyses, and can comment that the SIMS and MALDI spectra are the easiest to interpret. ESI spectra may be difficult to deconvolute due to the simultaneous presence of multiple charging and multiple fragmentations inherent in the oligosaccharide backbone. In many cases it is useful to utilize several techniques to resolve ambiguities.

Structural Analysis of Engineered Antibodies

Radiolabeled (In-111) cT84.66 has been administered to 24 patients with tumors of the colon, breast, and thyroid. In each case specific uptake was observed at the sites of primary, metastatic, or lymph node involvement. In order to verify the primary structure of the intact protein, the antibody was reduced and S-alkylated in situ with vinyl pyridine, run on SDS gels, and the isolated heavy and light chains digested with trypsin and subjected to LC/MS or reversed phase HPLC followed by peak collection and mass analysis. LC/ESI-MS analysis alone was able to account for over 80% of the sequence of the heavy and light chains. The single N-glycosylation site in the CH2 domain of the heavy chain was identified as a peptide with a characteristic mass of 2606. This peptide has been observed by us for a number of IgG-1 antibodies produced either by ascites (murine) or bioreactor (chimeric). It corre-

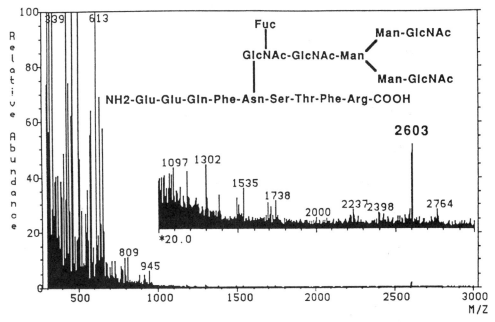

Figure 6. SIMS analysis of a glycopeptide isolated from anti-CEA antibody mT84.66. Reduced and S-alkylated heavy chain was treated with trypsin and the peptides separated by reversed phased HPLC. The glycopeptide was microsequenced to confirm the peptide sequence. SIMS analysis and the postulated peptide-oligosaccharide structure is shown.

sponds to a fucosylated biantennary oligosaccharide with a characteristic fragmentation pattern (Fig. 6). The peptide was identified by all three mass spectrometric methods used in the analysis (Fig. 7). By ESI only the doubly charged ion was observed (m/z= 1302.9), while for SIMS and MALDI only the single charged species were observed. All three methods were sensitive to the level of 1-10 pmoles.

The production of single chain antibodies (sFv) has been advocated for faster blood clearance and better tumor tumor penetration compared to whole antibodies. We have generated two versions of sFv, one with a 14 amino acid linker between V_L and V_H (sFv14), and the other with a 28 amino acid linker (sFv28) between V_H and V_L (the order of the variable regions is not important for activity). The sFv's were expressed in E.coli by secretion into the media and purified by a combination of ion exchange and anti-idiotypic antibody affinity chromatography. The purified sFv's were >90% immunoreactive when radiolabeled and bound to CEA beads. They gave the correct N-terminal amino acid sequences and molecular masses when analyzed by ESI-MS (Fig. 8), demonstrating that they were correctly processed, involving the removal of the bacterial signal peptide. sFv14 gave a mixture of monomers and dimers when analyzed by size exclusion HPLC, whereas sFv28 gave exclusively monomers. Biodistribution studies in nude mice bearing human colon xenografts showed that sFv14 gave better tumor uptake and better tumor to blood ratios compared to sFV28, demonstrating that the bivalent form of engineered antibodies are superior to monovalent forms. We (and others) have also determined that blood clearance is much too rapid for monovalent sFv's. We are now in the process of characterizing bivalent engineered antibodies of intermediate molecular weight (80 kDa) which are expected to exhibit superior biodistribution results.

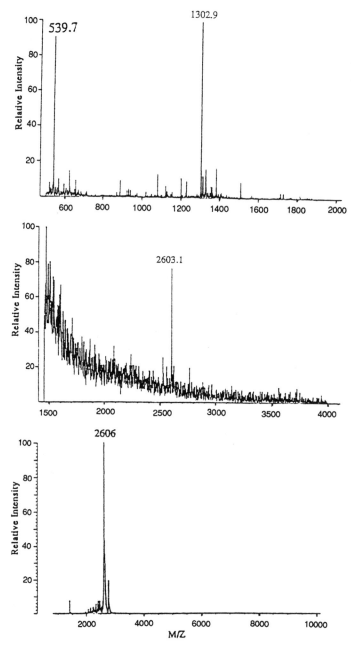

Figure 7. MS analysis of a glycopeptide from cT84.66. The peptide was isolated in the same manner as mT84.66. A. Analysis by ESI-MS. B. Analysis by SIMS. C. Analysis by MALDI-TOF-MS (external calibration with bovine insulin; α-cyano-4hydrozycinnamic acid as matrix).

ACKNOWLEDGMENTS

This research was supported by grants CA 43904 and CA 37808 from the National Cancer Institute.

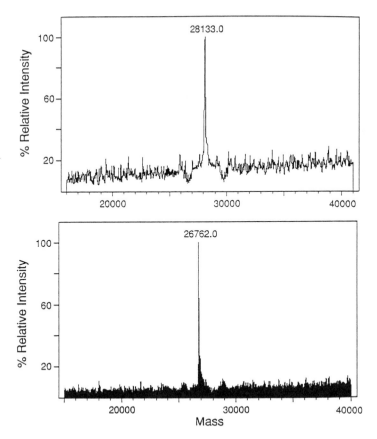

Figure 8. ESI-MS analysis of single chain anti-CEA antibodies. The samples were run on a capillary reversed phased HPLC connected on-line to an ESI source on a Finnigan TSQ700 mass spectrometer. Only the deconvoluted spectra are shown. A. sFv14. B. sFv28.

REFERENCES

1. Shively, J.E. and Beatty, J.D. (1985) *Crit. Rev. Oncol. Hematol.* 2: 355-399.
2. Thomas, P., Toth, A.A., Saini, K.S., Jessup, J.M. and Steele, Jr. G. (1990) *Biochim. Biophys. Acta* 1032: 177-189.
3. Thompson, J., et al., (1992) *Genomics* 12: 761-772.
4. Benchimol, S., Fuks, A., Jothy, S., Beauchemin, N., Shirota, K. and Stanners, C.P. (1989) *Cell* 57: 327-334.
5. Leusch, H.G., Hefta, S.A., Drzeniek, Z., Hummel, K., Markos-Pusztai, Z. and Wagener, C. (1990) *FEBS Lett.* 261: 405-409.
6. Wagener, C., Clark, B.R., Rickard, K.J. and Shively, J.E. (1983) *J. Immun.* 130: 2302-2307.
7. Beatty, J.D., Williams, L.E., Yamauchi, D., Morton, B.A., Hill, L.R., Beatty, B.G., Paxton, R.J., Merchant, B. and Shively, J.E. (1990) *Cancer Res.* 50: 922s-926s.
8. Neumaier, M., Shively, L., Chen, F.-S., Gaida, F.-J., Ilgen, C., Paxton, R.J., Shively, J.E. and Riggs, A.D. (1990) *Cancer Res.* 50: 2128-2134.
9. Paxton, R.J., Mooser, G., Pande, H., Lee, T.D. and Shively, J.E. (1987) *Proc. Natl. Acad. Sci.* USA, 84: 920-924.
10. Hefta, S.A., Hefta, L.J.F., Lee, T.D., Paxton, R.J. and Shively, J.E. (1988) *Proc. Natl. Acad. Sci.* USA, 85: 4648-4652.
11. Swiderek, K.M., Pearson, C.S. and Shively, J.E. (1993) In: *Techniques in Protein Chemistry III.*, (Hogue-Angeletti, R. ed.) pp. 127-134.

12. Yamashita, K., Totani, K., Kuroki, M., Matsuoka, Y., Ueda, I. and Kobata, A. (1987) *Cancer Res.* 47: 3451-3459.
13. Hefta, S.A., Paxton, R.J. and Shively, J.E. (1990) *J. Biol.Chem.* 265: 8618-8626.
14. Sauter, S., Rutherfurd, S., Wagener, C., Shively, J.E. and Hefta, S.A. (1991) *Infection and Immunity* 59: 2485-2493.
15. Sauter, S.L., Rutherfurd, S., Wagener, C., Shively,J.E. and Hefta, S.A.(1993) *J. Biol. Chem.* 268: 8711-8716.

22

SEROTONINERGIC NEURONES IN THE ZONA INCERTA EXERTS INHIBITORY CONTROL ON GONADOTROPHIN RELEASE VIA 5-HT$_{2A/2C}$ RECEPTORS

Arif Siddiqui[*1] and Catherine A. Wilson[2]

[1] Department of Physiology
The Aga Khan University
Karachi 74800, Pakistan
[2] Department of Obstetrics and Gynaecology
St. George's Hospital Medical School
London SW17 0RE, United Kingdom

INTRODUCTION

Gonadotrophin secretion is mainly influenced by distinct neural circuitry in the diencephalon, the components of which communicate through diverse chemical signals [1-3]. Identification of these components and their organization in co-ordinating the episodic and cyclic release patterns of LH has been the subject of intensive investigations for decades. The functional activity of the hypothalamic neurones responsible for gonadotrophin releasing hormone (GnRH) secretion are affected by numerous putative neurotransmitter which stimulate or inhibit GnRH release.

The participation of serotonin (5-HT) in the control of phasic release of luteinizing hormone (LH) appears widely accepted[4-6]. However no unanimity of opinion exists whether 5-HT is stimulatory[7-13] or inhibitory [14-16] on LH release. Administration of steroids can alter 5-HT synthesis, turnover and release [17-20] so it seems likely that at least one mechanism by which steroids exert their feedback effects on gonadotropin release is by altering indoleamine activity. In particular, steroids exert their effects in the preoptic area (POA), suprachiasmatic nucleus (SCN) arcuate/median eminence area (ARC/ME) and ventromedial nucleus (VMN) [18,21] and these areas all contain 5-HT nerve terminals [22-24] as does the zona increta (ZI).[25]

Pharmacological studies supported by histological techniques indicate that there is an extensive serotoninergic innervation close to the dopamine neurones in the ARC, ME, ZI and POA .[25,26] There is some evidence that 5-HT stimulates LH release in the ME and since 5-HT levels in the ZI are inversely correlated with LH release, perhaps it is inhibitory in this

---

[*] Represents person presenting Paper

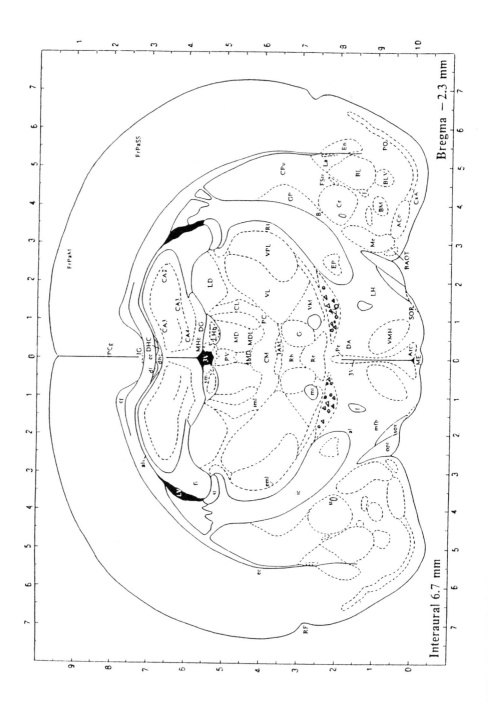

3V third ventricle
ACo ant cortical amygdaloid nu
al ansa lenticularis
alv alveus hippocampus
Arc arcuate hy nu
B cells basal nu of Meynert
BAOT bed nu accessory olf tr
BL basolateral amygdaloid nu
BLV basolateral amygdaloid nu, vent
BM basolateral amygdaloid nu
CA2 field CA2 of Ammon's horn
CA3 field CA3 of Ammon's horn
CA4 field CA4 of Ammon's horn
cc corpus callosum
Ce central amygdaloid nu
eg cingulum
CL centrolateral th nu
CM central med th nu
Cpu caudate putamen
CxA cortex-amygdala transition zone

DA dors hy area
df dors fornix
DG denate gyrus
DHC nu dors hip commissure
dhc dors hip commissure
ec ext capsule
eml ext medullary lamina
En endopiriform nu
EP entopeduncular nu
f fornix
fi fimbria hippocampus
FrPaM frontoparietal cortex, motor area
FrPaSS frontoparietal cortex, somatosensory area
FStr fundus striati
G agelatinosus nu thalamus
GP globus pallidus
I intercalated nuclei amygdala
IAM interanteromedial th nu

ic int capsule
IG induseum griseum
IMD intermodiodorsal th nu
iml int medullary lamina
La lat amygdaloid nu
LD laterodorsal th nu
LH lat hy area
LHb lat habenular nu
LV lat ventricle
MD mediodorsal th nu
MDL mediodorsal th nu. lat
ME median eminence
Me med amygdaloid nu
mfb med forebrain bundle
MHb med habenular nu
mt mammillothalamic tr
opi optic tr
PC paracentral th nu
PCg post cingulate cortex
Pe periventricular hy nu

PO primary olf cortex
PV paraventricular th nu
Re reuniens th nu
RF rhinal fissure
Rb rhomboid th nu
Rt reticular th nu
sm stria medularis thalamus
SOP supraoptic hy nu
retrochiasmatic
sox supraoptic decussation
st stria terminalis
VL ventrolateral th nu
VM ventromedial th nu
VMH ventromedial hy nu
VPL ventroposterior th nu lat
ZI zona incerta

Figure 1. Sites of injection of 2 μg 5-HT into the Zona incerta (ZI) area of rat brain. Circles represent the site of injection of 5-HT and triangles of saline, as estimated histologically by noting the bottom of the needle tract. All injections were made between 14:00 and 15:00 h.

latter site.[27] Identification of different receptor families (5-HT$_1$, 5-HT$_2$, 5-HT$_3$ and 5-HT$_4$)[28,29] and their implication in the control of gonadotropins secretion has further complicated their precise role.[30,31]

In this study we investigated the modulatory role played by the serotoninergic system by noting the effect on LH release by injecting into the ZI of 5-HT itself and 1-(2,5-Dimethoxy-4-iodo-phenyl)-2 aminopropane HCL (DOI), a selective 5-HT 2A/2C agonist. Relatively selective 5-HT$_{1A}$ antagonist WAY 100 135 or 5-HT$_2$ antagonist ritanserin was injected intraperitoneally (i.p.) one hour prior to 4 intra-cerebral injection of 5-HT to see the antagonizing effect of 5-HT antagonists on the inhibitory effect of 5-HT on LH release.

MATERIALS AND METHODS

Adult female Wistar rats weighing 250-300 g were smeared daily and only those showing at least 3 regular 4-day oestrous cycles were used for the study. They were maintained in a fixed lighting system of 12 h light: 12 h dark (lights on at 06:00h). Animals were kept at a constant temperature of 21°C and had food and water ad lib.

Rats were ovariectomized under halothane and nitrous oxide anaesthesia (Halothane; May & Baker Ltd., Daganham). Three weeks later the animals were divided into groups and given either 0.1 ml/rat corn oil subcutaneously (s.c.) or 5μg oestradial benzoate (OB) followed 48 h later by 0.5 mg/rat progesterone (P) s.c. given in 0.1 ml corn oil to induce an LH surge. Between 14:00 and 16:00h rats were anaesthetized with 3 ml/kg Saffan (Glaxo Intervet Ltd, Greenford), i.p. A blood sample of approximately 0.2 ml was obtained by cutting the tip of the tail. Rats were placed in a stereotaxic apparatus (David Kopf Ltd.) and injected with 5-HT creatinine sulphate (Sigma Chemicals, London Ltd., Dorset) (n=16), 5-HT T$_{2A/2C}$ agonist DOI (n=9) at 2 μg/0.4 μl saline/side into the ZI under Saffan anaesthesia. Control group of animals were injected with 0.4 ml saline/ side into the same site (n=10). The stereotaxic parameters taken from the rat brain atlas of Paxinos and Watson [32] were 2.3 mm behind bregma, 0.5 mm lateral from the midline and 7.5 mm below the surface of the brain. To ensure that the effects we are seeing of 5-HT or other agents injected into the ZI rats were decapitated, brains were fixed in formal saline and the site for injection was noted histologically on frozen brain sections (Fig. 1).

Blood samples (0.2 ml) were collected from the tail vein at 30 min. interval post injection for 3 hours. The blood was centrifuged at 400 xg for 10 min at 4°C and the plasma stored at - 20°C until assayed for LH. A group of rats (n=10) also received 2 mg / kg WAY 100 135, a 5-HT$_{1A}$ antagonist i.p. one hour prior to intra-cerebral injection of 5-HT. Control group in this experiment also received 0.9% saline, i.p. (n=11) one hour prior to intra-cerebral injection of 5-HT.

Measurement of Plasma LH

LH was assayed by a modified method of Naftolin and Corker [33] as described by Kendle et al [34]. The sensitivity of the assay was 20 pg/tube against the standard LER-C$_2$-1056 (potency 1.73 times NIH-LH-SI). The intra- and interassay coefficients of variance were 10% and 8.2% respectively.

Histology

Formal saline-fixed brains were frozen and 60μm sections were cut on a sledge microtome. These were stained with thionine blue. Sites of injection were at the lowest point of the needle tract.

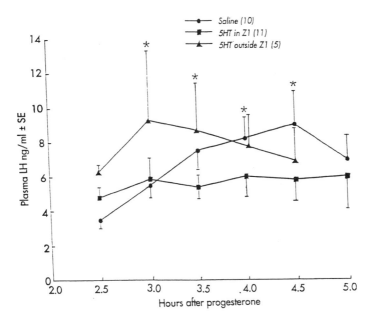

Figure 2. The effect of 2 µg/side 5-HT injected bilaterally into the ZI on plasma LH concentrations. *P<0.05 compared to saline in ZI group.

Statistics

The data pertaining the significance of difference between the concentrations of LH in control and treated groups were made using student's t test.

RESULTS

Based on previous findings that there is an inverse relationship between 5-HT turnover and LH release, it seems possible that 5-HT may have an inhibitory influence on LH release. Thus effect on LH release was noted after removing the negative feedback influence by ovariectomizing the rats and priming with OB 48 hours before and P 2 hours before an injection of 5-HT or saline into the ZI. The steroid regime induces a sure of LH approximately 4 hours after the P thus by injecting the 5-HT between 2-3 hours after the P it was hoped that an inhibitory effect of 5-HT on LH release might be observed. Although all the 5-HT injections were confined to the one hour period an attempt was made to account for the variability in the intervals between the P and 5-HT injections by grouping the results according to the half hour periods after the P treatment. Figure 2 show the mean and standard errors of the results calculated according to the time after P administration. After saline was injected into the ZI, there was a significant rise in LH levels 4-5 hours after the P. This was not significantly reduced by bilateral injection outside the ZI. Bilateral injections of 5-HT in the ZI significantly reduced the rise in LH after (2 hours post 5-HT) usually seen after P (p<0.05). Like 5-HT, the selective 5-HT$_{2A/2C}$ agonist DOI at 2mg/side/rat also inhibited LH rise (p<0.05) seen 4.5-5.0 hours post-saline injected rats (Fig.3).

Since 5-HT inhibited LH release, the possible stimulatory effects of a 5-HT$_{1A}$ antagonist WAY 100 135 and 5HT$_{2/1C}$ antagonist ritanserin was assessed in ovariectomized

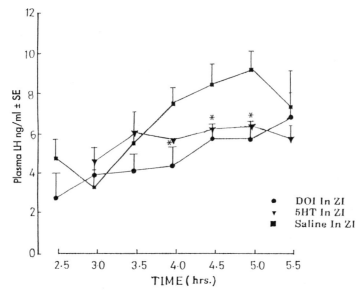

Figure 3. The effect of 2 μg/side 5-HT or DOI injected bilaterally into the ZI on plasma LH concentrations in ovariectomized rats treated with OB+P. *P<0.05 compared to saline in ZI group.

rats primed with OB plus P. 5-HT in these animals were injected into the ZI in a manner similar to that described in previous experiments. In addition they received one of the two 5-HT antagonists mentioned above one hour before the central injection of 5-HT systemically by i.p. injection. Ritanserin was seen to significantly antagonize the 5-HT-induced reduction in LH release and the LH peak was seen to occur between 4-5 hour following P

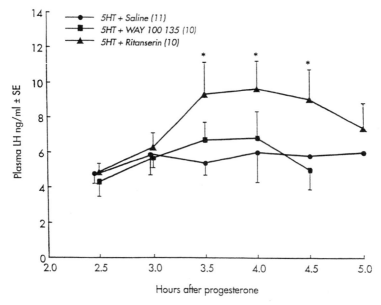

Figure 4. The effect of 5-HT antagonists given systemically on the inhibitory effect of 5-HT in the ZI on LH release. *P<0.05 compared to saline in ZI group.

administration WAY 100 135 ($5HT_{1A}$) antagonist however was unable to alter 5-HT induced inhibition of circulatory LH levels (Fig.4).

DISCUSSION

Several lines of evidence, mainly pharmacological, now support the hypothesis that hypothalamic 5-HT is involved in the release of LH. Manipulation of brain 5-HT activity either by pharmacological agents or selective lesions indicate that 5-HT has a dual effect on GnRH release (see introduction for references). Administration of p-chlorophenyl alanine, an inhibitor of tryptophan hydroxylase, and 5-hydroxytryptophan, an immediate precursor of 5-HT has been reported to stimulate LH surge in steroid-primed ovariectomized rats, while no effect has been observed in intact rats[35]. Thus, one of the contributing factors for these discrepancies may include the steroidal environment of the animals. Furthermore, the existence of two or more sites of action with opposing effects and/or the possibility that stimulation of the subtypes of 5-HT receptors may elicit different responses.

Evidence based on changes in 5-HT turnover indicates that 5-HT exerts a stimulatory effect on LH release in the POA and ME[20,36,37]. On the contrary it is inhibitory to LH release in the ARC[38,39] and in the VMN [40,20]. Alternatively, differential activation of 5-HT receptors may also account for the dual effect on LH release [41,42]. There are studies using 5-HT on female sexual behavior also showing to have stimulatory effect via $5-HT_2$ receptors and inhibitory effect via $5-HT_1$ receptors[43,44].

Previous findings in this laboratory show that 5-HT neurones appear to be part of a network involving GABA and DA neurones. GABA in the ZI exerts an inhibitory effect on LH release by decreasing DA turnover and increasing 5-HT turnover[45,46].

In this report we have investigated the possibility that the effect of 5-HT is exerted at different hypothalamic loci and that a 5-HT system influence gonadotrophin release which is differentially affected by noting the effect of 5-HT injections into the ZI on LH release in a group of ovariectomized rats in which depleted negative feedback control is replaced by injecting 5µg OB (s.c) followed by 0.5mg of P(s.c.).

This model was chosen to see any reduction in relatively high levels of LH. The results shows that 5-HT injected into the ZI did significantly reduce circulating LH levels with respect to saline-injected animals.

Ovariectomized rats do not possess cyclical LH surge. However, chronic administration of 5µg OB induces a daily afternoon surge of LH release. This stimulatory action on LH secretion is known as the positive feedback effect of OB.[47] Treatment of OB-primed rats with progesterone both advances and amplifies the LH surge although P administration by itself is ineffective in term of LH release. Under these experimental conditions 5-HT was injected into the ZI to see if it inhibited this surge. Towards the end of the sample collection period (i.e. 3.0 - 3.5 hours) LH concentrations tend to rise. This may have been due to the Saffan anaesthesia, as this can inhibit LH release,[48] and moreover at the end of the blood sample collection period circulatory levels of Saffan must have been decreasing, in resulting this inhibitory action to be removed and so there is an increase in LH release.

Our finding on reduced concentration of LH in the blood are in conformation with earlier reports where in ovariectomized rats 5µg OB plus 0.5 mg P-induced LH surge has been correlated with the increased 5-HT activity (as assessed by the parallel changes in the concentrations of 5-HT and 5-H1AA were taken as indices of 5-HT neuronal activity) in the ME.[27]

Furthermore, steroids has been reported to increase 5-HT receptors in certain estrogen-concentrating brain nuclei, such as anterior hypothalamus, POA and ARC-ME,[49] which have a well established role in the control of ovulation. Tanaka et al.[50] have suggested that the sensitivities to 5-HT in the levels of hypothalamus and pituitary gland is regulated by 5-HT dependent estradiol synthesis in the ovary and this may be an important key to interpret contradictory effects of 5-HT on ovulation and LH secretion.

It has also been sown that 5-HT neurones appear to be part of a network, involving GABA and DA neurones. GABA in the ZI exerts an inhibitory effect on LH release by decreasing dopamine turnover and increasing 5-HT turnover.[46] The increase in 5-HT activity in the ME is in agreement with Vitale *et al* [51] who in intact females showed increased 5-HT turnover in the ME at the time of the preovulatory surge with a concomitant reduction in 5-HT activity.

This study also investigate the types of 5-HT receptors involved in its inhibitory influences on LH releases. For this purpose on 5-HT2A/2C receptors agonist DOI, was injected into the ZI in rats pre-treated with 5μg OB plus 0.5 mg P. Therefore a plasma LH surge would occur and the agonist DOI would be expected to inhibit that if the effect is mediated by a selective 2A/2C subtype of 5-HT receptor. The result of this experiment indicate that the inhibitory effect of 5-HT in the ZI is exerted by $5\text{-HT}_{2A/2C}$ if not any other post-synaptic receptor/s.

Since 5-HT inhibited LH release, the possible stimulatory effect of selective 5-HT antagonists WAY 100 135 and ritanserin was assessed in ovariectomized rats primed with OB plus P. Fig. shows that ritanserin very significantly antagonized the 5-HT induced reduction in LH release and the LH peak was seen to occur between 4-5 hours following P administration. WAY 100 135 (5-HT_{1A} antagonist) however was unable to alter 5-HT-induced inhibition of circulating LH concentrations.

CONCLUSION

In summary, our results indicate that the intra-cerebral infusion of 5-HT into the ZI in OB plus P-primed ovariectomized female rats inhibits the LH release.

Serotoninergic inhibitory influence on steroid-induced LH release is mediated by 5-HT receptors in the ZI since the subtype 2A/2C action of 5-HT is mimicked by DOI and antagonized by ritanserin. Further studies are however needed to screen and elucidate the effects of all other serotoninergic pharmacological agents in the ZI on LH release.

Acknowledgments

The authors would like to thank the Aga Khan University for the support provided to A.S. We are also indebted to National Hormone and Pituitary Programme, NIH, NIDDK, Bethesda, MD 20892, USA for providing rat LH RIA kits used in this study. Excellent secretarial assistance of Mr. Hilary F. Fernandes is also gratefully acknowledged.

REFERENCES

1. Wilson, C.A. (1979) In: *Finn, Oxford reviews of reproductive biology*. vol. I. pp. 373-383, Oxford University Press, Oxford.
2. Barraclough, C.A. and Wise, P.M. (1982) *Endocr. Rev.* 3: 91-119.
3. Kalra, S.P. and Kalra, P.S. (1983) *Endocrine Rev.* 4: 311-351.
4. Weiner, R.I. and Ganong, W.F. (1978) *Physiol. Rev.* 58 : 905-976.
5. Biegon, A., Bercovitz, H. and Samuel, D. (1980) *Brain Res.* 187 : 221-225.
6. Marko, M. and Fluckiger, E. (1980) *Neuroendocrinology* 30: 228-231.
7. Hery, M., Laplante, E. and Kordon, C. (1978) *Endocrinology* 102:1019-1025.
8. Meyer, D.C. (1978) *Endocrinology* 103 : 1067-1074.

9. Coen, C.W. and Mackinnon, P.C.B. (1979) *J. Endocr.* 82:105-113.

10. Walker, R.R. (1980) *Life Sci.* 27 : 1063 - 1068.

11. Ruzsas, C., Limonta, P. and Martin, L. (1982) *J. Endocr.* 94: 83-89.

12. Iyengar, S. and Rabii, J. (1983) *Brain Res. Bull* 10: 339-343.

13. Vitale, M.L., Villar, M.J.,Chiochio, S.R. and Tramezzani, J.R. (1986) *J. Endocr.* 111: 309-315.

14. Kamberi, I.A., Mical, R.S. and Porter, J.C. (1970) *Endocrinology* 87: 1-12.

15. Arendash, G.W. and Gallo, R.V. (1978) *Endocrinology* 102: 1199-1206.

16. Charli, J.L., Rotsztejn, W.H., Patton, E. and Kordon, C. (1978) *Neuroscience* 10: 159-163.

17. Cone, R.I., Davis, G.A. and Goy, R.W. (1981) *Brain Res. Bull* 7: 639-644.

18. Hery, M., Faudon, M., Dusticier, G. and Hery, F. (1982) *J. Endocr.* 94: 157-166.

19. Meyer, D.C. and Eadens, D.J. (1985) *Brain Res. Bull.* 15: 283-286.

20. Johnson, M.D. and Crowley, W.R. (1986) *Endocrinology* 118: 1180-86.

21. Goodman, R.L. (1978) *Endocrinology*, 102: 142-150.

22. Kiss, A., Culman, J., Kvetansky, R. and Palkovits, M. (1981) *Endocr. Exp.* 15: 219-228.

23. Jennes, L., Beckman, W.C., Stumf, W.E. and Grzanna, R. (1982) *Exp. Brain Res.* 46:331-338.

24. Kiss, J. and Halasz, B. (1985) *Neuroscience* 14:69-78.

25. Bosler, O., Jon, T.H. and Beaudet, A. (1984) *Neurosci. Lett.* 48: 279-285.

26. Steinbusch, H.W.M. (1981) *Neuroscience* 6:557-618.

27. James, M.D., Hole, D.R. and Wilson, C.A. (1989) *Neuroendocrinology*, 49:561-569.

28. Bradley, P.B., Engel, G., Feniuk, W., Fozard, J.R., Humphrey, P.P.A., Middlemiss, D.N., Mylecharane, E.J., Richardson, B.P. and Saxena, P.R. (1986) *Neuropharmacology* 25: 563-576.

29. Frazer, A., Maayani, S. and Wolfe, B.B. (1990) *Ann. Rev. Pharmacol. Toxicol.* 30: 307-48.

30. Di Sciullo, M.T. (1990) *Neuroendocrinology* 127: 567-572.

31. Jorgensen, H.U., Knigge, and Warberg, J. (1992) *Neuroendocrinology* 55:336-343.

32. Paxinos, G. and Watson, C. (1992) In: *The rat brain in stereotaxic coordinates,* Academic Press, New York.

33. Naftolin, F. and Corker, C.S. (1971) In : *Radioimmunoassay methods,* (Kirham, K.E. and Hunter, W.M. eds.) pp. 641-645, Churchill Livingstone, Edinburgh.

34. Kendle, K.E., Paterson, J.R. and Wilson, C.A. (1978) *J. Reprod. Fertil.* 53: 363-368.

35. Coen, C.W., Franklin, M., Laynes, R.W. and Mackinnon, P.C.B. (1980) *J. Endocr.* 87: 195-201.

36. King, T.S., Steger, R.W. and Morgan, W.N. (1986) *Neuroendocrinology* 42: 344-350.

37. Vitale, M.L., Villar, M.J., Chiochio, S.R. and Tramezzani, J.H. (1987) *Neuroendocrinology*: 46: 252-257.

38. Foreman, M.M. and Moss, R.L. (1978) *Horm. Behav.* 10: 97-106.

39. Gallo, R.V. (1980) *Neuroendocrinology* 30: 122-131.

40. Frankfurt, M,, Renner, K., Azmitia, E. and Luine, V. (1985) *Brain Res.*340: 127-133.

41. Johnson, J.H. and Sanders, K. (1987) *Anat. Res.* 218: 67A-68A.

42. Lenahan, S.E., Seibel, H.R. and Johnson, J.A. (1986) *Neuroendocrinlogy* 44: 89-94.

43. Ahlenius, S., Fernandez-Guasti, A., Hjorth, S. and Larsson, K. (1986) *Eur. J. Pharmacol.* 129: 361 - 363.

44. Mendelson, S.D. and Gorzalka, B.B. (1986) *Physiol. Behav.* 37: 345-351.

45. Mackenzie, F.J., Hunter, A.J., Daly, C. and Wilson, C.A. (1984) *Neuroendocrinology* 39: 289-295.

46. Wilson, C.A., James, M.D. and Leigh, A.J. (1990) *Neuroendocrinology* 52: 354-360.

47. Caligaris, L. and Taleisnik, S.(1974) *J. Endocr.* 62:25-33.

48. Dyer, R.G. and Mansfield, S. (1984) *J. Endocr.* 102: 27-31.

49. Biegon, A., Fischette, C.T., Rainbow, T.C. and McEwen, B.S. (1982) *Neuroendocrinolgy* 35: 287-291.

50. Tanaka, E.; Baba, N., Toshida, K. and Suzuki, K. (1993) *Life Sci.* 53: 563-570.

51. Vitale, M.L., Parisi, M.N., Chiochio, S.R. and Tramezzani, J.H. (1984) *Neuroendocrinolgy* 39: 136-141.

23

STRUCTURE ELUCIDATION OF HUMAN LENS PROTEINS BY MASS SPECTROMETRY

David L. Smith,[*] Peiping Lin, Anders Lund, and Jean B. Smith

Department of Chemistry
University of Nebraska-Lincoln
Lincoln, Nebraska, 68588-0304

ABSTRACT

Development of fast atom bombardment and electrospray ionization mass spectrometry have greatly facilitated investigations of proteins and peptides. The extraordinary accuracy of the molecular weights determined by these techniques permit proteins and their modifications to be identified with much less ambiguity than was possible with previously available techniques. Combining fast atom bombardment mass spectrometry with directly-coupled HPLC and with collision-induced dissociation mass spectrometry has further extended the information that can be obtained about the primary structures of proteins. We describe the use of these techniques in investigations into the structure of the proteins present in the human eye lens. Modifications of the lens proteins associated with age are of particular interest because lens proteins do not turnover; i.e., the lens of an old person contains even the proteins that were present in the fetus. It is also an important area of investigation because elucidation of the structure of old lens proteins may lead to an understanding of the mechanisms of cataract formation, the most common cause of blindness, worldwide.

Delivery of the first commercial mass spectrometer, manufactured by Consolidated Electronics Corporation, in 1942 is a benchmark for analytical mass spectrometry. During the subsequent forty years, mass spectrometry became a principal tool for identification and quantification of volatile organic materials. Peptides, even those with molecular weights less than 500, are not volatile because they are highly polar, and required chemical derivatization before mass spectrometric analysis[1]. In 1980, Barber *et al.* described the first use of fast atom bombardment mass spectrometry (FABMS) for analysis of involatile substances[2]. During the ensuing 15 years, FABMS has proved extremely useful for determining the molecular weights of peptides with as many as 40-50 residues[3-5]. The utility of FABMS has been enhanced by joining this ionization method with directly-coupled HPLC[6,7] and with collision-induced dissociation (CID) mass spectrometry.[8,9]

---

[*] Represents person presenting Paper

The notion that molecules with an excess of positive or negative charge in solution might be analyzed directly from solution by mass spectrometry had been proposed, but not clearly demonstrated, prior to 1988 when Fenn *et al.* demonstrated the reality of this wishful-thinking[10]. Specifically, they showed that analytically useful ions could be produced when a solution is sprayed from a small needle into a strong electric field. During the past several years, this approach has become known as electrospray ionization mass spectrometry (ESIMS), and is now one of the most important analytical tools available for determining the primary structures of peptides and proteins[11-13]. The molecular weights of peptides and proteins to M_R 75,000 are routinely determined with error less than 0.01-0.05% using commercial instrumentation. With this accuracy, it is now possible to determine the molecular weights of proteins with M_R 20,000 (typical of proteins isolated from the eye lens) with an uncertainty of approximately ±2 Da[14,15].

In protein structure studies, ESIMS is used most often to determine the molecular weights of peptides and proteins. The use of this information is limited only by the investigator's imagination. For example, when the molecular weight of a protein matches, within a few Daltons, the molecular weight calculated from the amino acid sequence of a proposed protein, it may be concluded that the protein has been tentatively identified. If the found molecular weight differs slightly from the expected molecular weight, posttranslational modifications may be considered. For example, acetylation of the N-terminus of a protein, a common *in vivo* posttranslational modification, increases the molecular weight by 42 Da. This increase is easily detected by ESIMS analysis of most proteins. Furthermore, analysis of a mixture of mono-acetylated and non-acetylated forms of a protein with M_R 50,000 gives separated peaks in the ESI mass spectrum. When compared with traditional methods of determining the molecular weights of proteins, such as sodium dodecyl sulfate polyacrylamide gel electrophoresis (SDS-PAGE), ESIMS has vastly higher accuracy, substantially better resolution, and detection limits somewhat lower than those typically achieved by silver staining techniques.

The high performance of ESIMS makes it particularly useful for investigations of the primary structures of complex mixtures of proteins, such as structural proteins extracted from human eye lenses. This tissue, whose function is to focus images on the retina of the eye, is approximately 37% structural protein. The eye lens is unusual because it continues to grow throughout the life of an individual by adding more layers of cells, and because the protein never turns-over. As a result, copies of the lens proteins present in the inner portion (the nucleus), of the lens, were present at birth. Identification of which gene products are actually expressed, as well as posttranslational modifications that occur during the life of an individual, are important for advancing our understanding of the natural *in vivo* aging of lens proteins and of the possible causes of cataract. The physiology and biochemistry of the lens have been described in detail elsewhere[16,17].

Approximately 15 different genes code for the most abundant structural proteins found in the lens. Fractionation of lens extracts by gel filtration chromatography (Sephadex G-200) may be used to isolate the three general classes, α-, β-, and γ-crystallins, of lens proteins[18,19]. The α-crystallins consist of two gene products (αA and αB), which aggregate to form large amorphous complexes with an average molecular weight of 500,000 Da. The β-crystallins include at least 8 different gene products, which aggregate to form complexes with average molecular weights of 80,000 to 150,000 Da. The γ-crystallins consist of 6 different gene products and do not aggregate under normal conditions. All of the crystallin monomers have molecular weights in the 20-30 kDa range. The N-termini of all the α- and β-crystallins and γs-crystallin are acetylated.

Because individual copies of each protein are as old as the cell in which they were formed, and those in the center of the lens are as old as the person, one can imagine that a lens from an elderly person has crystallins that have undergone a variety of modifications.

For example, modifications that have been found in the human eye lens include phosphorylation,[20-22] C-terminal truncation,[14,23] deamidation,[22,24] and disulfide bonding.[22,25,26] Oxidation of Met, Trp, and Cys have also been proposed as likely modifications. Understanding the aging of proteins, as well as chemical reactions causing or accompanying cataract, poses a significant analytical challenge. One would like to know which modifications occur, and specifically on which residue in a particular gene product. Furthermore, one would like to know the extent to which these modifications are present, as well as the point in lens development at which they occurred.

Our general approach to identifying crystallins present in human eye lenses, including their modified forms, starts with their separation into α-, β-, and γ- fractions by gel filtration chromatography. Constituents of these fractions are further purified by a combination of ion exchange and reversed phase high performance liquid chromatography (HPLC). Although the final products of this isolation procedure may contain only one gene product, different modifications (phosphorylation, deamidation, truncation) may be included, and these modifications may be located at different sites along the peptide backbone. These semi-pure samples are analyzed by ESIMS to determine the molecular weights of the constituent proteins, and to assess the purity of the sample. The molecular weights may be used to identify the protein as belonging to a particular gene, and to tentatively identify the existence of posttranslational modifications. The sample can also be fragmented into peptides whose molecular weights are determined either by ESIMS or FABMS. This molecular weight information may be used to relate large but incomplete protein fragments to specific genes, and to identify specific posttranslational modifications. Peptides that cannot be identified by systematically searching proposed sequences of proteins believed to be present in the sample[27] may be sequenced by CID mass spectrometry. This technique is also useful for determining which residues are modified. The purpose of this communication is to describe how we have used mass spectrometry to investigate the primary structure of human lens crystallins.

RESULTS AND DISCUSSION

Directly-Coupled HPLC ESIMS

Several different ion source designs have been used to effect ESI.[28-30] In all of these designs, the sample is dissolved in an aqueous solution, which is sprayed from a needle at atmospheric pressure. A potential of several kilovolts, applied between the needle and a counter electrode, assists in forming small, highly charged droplets. These droplets are carried by the gas flow into a vacuum chamber where desolvation takes place. Analyte ions, sometimes highly charged, are ejected from the droplets and analyzed by a mass spectrometer to determine their mass-to-charge ratios.

A particularly useful feature of ESIMS is that analyte ions are sampled directly from solution, facilitating the use of mass spectrometry to directly monitor the effluent from a liquid chromatograph. Since the flow rates required for ESIMS are normally in the 1-100 μl/min range, microbore or capillary HPLC is most suitable. The HPLC ESIMS system developed in our laboratory for analyzing lens proteins is illustrated in Fig. 1. Two different systems are used, depending on the requirements of the experiment. The system illustrated in the upper part of Fig. 1 consists of an injector (Rheodyne model 8125), a C8 reversed phase column (50 x 1 mm), and a post-column flow splitter (Alltech 01-0165). Highest resolution from this type of column is obtained with a flow rate of 50 μl/min. The post-column splitter is used to direct a small portion of the effluent to the ESI mass spectrometer which is used to determine the

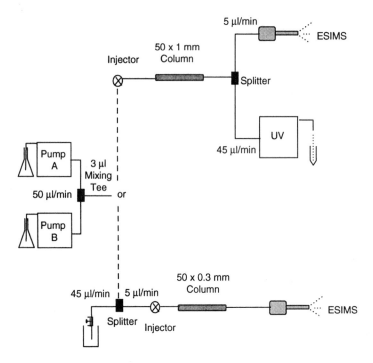

Figure 1. Microbore (upper) and capillary (lower) reversed-phase HPLC systems directly-coupled to an ESI mass spectrometer. The A and B mobile phases were H_2O and $CH_3CN:H_2O$ (4:1). Both contained 0.1% trifluoroacetic acid. The microbore and capillary columns were purchased from Applied Biosystems and LC Packings, respectively. A VG Fisons Platform II quadrupole mass spectrometer was used to obtain the ESI mass spectra presented here.

molecular weight of each component in the sample. A major portion of the flow is sent to a UV detector and fraction collector. This approach is useful because a major portion of the sample is recovered and can be used in other experiments. For example, proteins may be proteolytically fragmented into peptides whose molecular weights are determined by a subsequent HPLC ESIMS analysis. Since the volume of chromatographic peaks from a 1 mm column is small, the volume of the UV detector cell should not exceed approximately 1 µl to avoid a loss in chromatographic resolution. Highest sensitivity is achieved by using a smaller, capillary column, as illustrated in the lower part of Fig. 1. Decreasing the column diameter from 1 mm to 0.3 mm requires lowering the flow rate from 50 µl/min to 5 µl/min. This flow is ideal for most ESIMS systems. Therefore, post-column splitting of the effluent is not required.

Relatively standard HPLC pumping systems suitable for making gradients with high pressure mixing can be used for either type of column. The Gilson pumping systems used in our laboratory work well at flow rates as low as 50 µl/min, and can be connected directly to the microbore system. It is important, however, to use a low-volume mixing tee (Upchurch U-466). It is not practical to create a gradient flow of 5 µl/min with these pumps, and probably not with any commercial pumping system. Gradient elution using such very low flow rates can be achieved, however, by running the pump at a higher flow rate, and splitting the flow prior to the injector. The split ratio is selected by clamp pressure on the waste side outlet.

Figure 2. Directly-coupled microbore HPLC ESIMS analysis of human αA-crystallin, which was extracted from a single human eye lens and isolated by Sephadex G-200 gel filtration chromatography. The sample was desalted by off-line reversed-phase HPLC prior to analysis by HPLC ESIMS. Approximately 0.2 nmoles of αA and αB-crystallin were injected in a volume of 2 μl. Proteins were eluted directly into the ESI mass spectrometer using a gradient of 5% B for 10 min., to 45% B at 15 min., and to 60% B at 45 min. The mass spectrometer was tuned to a resolution of 1,000 (50% valley), and scanned repetitively through the 800-1500 m/z (mass-to-charge) range. Each scan required a total time of 10 sec. The UV absorbance (0.02 aufs) was determined at 214 nm (Applied Biosystems model 757).

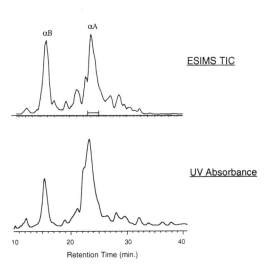

Applications of HPLC ESIMS to Structure Elucidation of Proteins

Alpha-crystallins are composed of two gene products, αA and αB, which are present at approximately the same levels in the human eye lens. These proteins, which co-elute as large, amorphous aggregates in gel filtration chromatography, have 173 and 175 residues, respectively. Separation of αA- and αB-crystallin by reversed-phase microbore HPLC

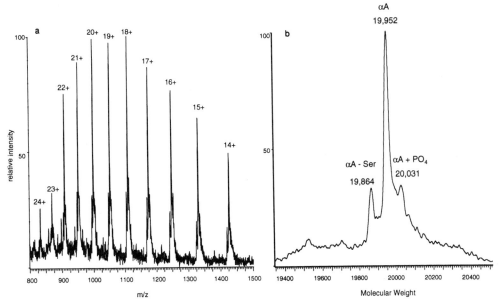

Figure 3. Determination of the molecular weights of proteins eluting in the time interval 22-24 min. This time interval coincides with the elution of αA-crystallin. (a) Mass spectral data (intensity vs. m/z) obtained by summing scans 155-167, and subtracting scans 143-148 and 175-180. (b) Reconstructed ESI mass spectrum (intensity vs. molecular weight) indicating the molecular weights of the proteins eluting during this time interval.

(Fig. 1 upper) is illustrated in Fig. 2. The total ion current plot (TIC) is made by plotting the sum of all ion signals in each scan vs. the scan number (or vs. the elution time). This presentation of data, as well as plots of specific m/z intervals, can be made following the actual analysis because all mass spectral data are stored for off-line processing. The UV absorbance chromatogram (Fig. 2) is similar to the TIC chromatogram, indicating that the chromatographic resolution is similar for both detectors. Differences in peak heights reflect the difference in sensitivity of the two detectors for various proteins. The two large peaks correspond to αB- and αA-crystallin. These peaks are broad, probably due to unusual physical properties of the α-crystallins. When the same system is used to analyze peptides, as well as several other proteins, the peaks are much narrower (e.g., see Fig. 4).

Although analysis by HPLC ESIMS requires less than an hour to acquire the data, determining the molecular weights of all components in the sample may require several additional hours, despite use of high speed computers. Results for analysis of one interval (indicated in Fig. 2) in the elution chromatogram is illustrated in Fig. 3. The ESI mass spectrum (Fig. 3a) has many prominent peaks, each corresponding to a multiply charged form of the protein. Since this is a plot of intensity vs. m/z, ions with the same molecular weight but different charge appear as discrete peaks in the mass spectrum. The maximum number of positive charges on a protein that are likely to be found in the ESI mass spectrum can usually be predicted from the sum of the basic residues (Lys, Arg, and His) in the protein. For a relatively uncomplicated mass spectrum, as illustrated in Fig. 3a, data analysis proceeds by assuming that adjacent major peaks are formed from the same protein with one more (moving down in m/z) proton. Computer programs use this approach to determine exact values for mass and charge (m and z) for each peak in a series. The net charge on ions corresponding to the major peaks is indicated in Fig. 3a. The mass and charge of these ions is used to determine the molecular weight of the protein (M_R 19,952), as presented in the reconstructed mass spectrum in Fig. 3b. If several proteins co-elute, there will be a corresponding number of series of peaks.

The reconstructed mass spectrum is useful for determining the number of components in a sample (elution interval) and their molecular weights. From the data presented in Fig. 3b, it is apparent that there are three components with molecular weights of 19,864 and 19,952 and 20,031. Although several additional minor components are undoubtedly present, they only contribute to the baseline noise. The molecular weight of the major component agrees very well with the molecular weight of monoacetylated αA-crystallin (found 19,952; calculated 19,951.7). The two smaller peaks correspond to des-Ser (found 19,864; calculated 19,864.6) and monophosphorylated (found 20,031; calculated 20,031.7) forms of αA-crystallin. Since other proteins, as well as other modifications to αA-crystallin could possibly have the same molecular weights, without additional data, these assignments must be considered to be tentative. For αA-crystallin, results of previous peptide mapping studies have confirmed these assignments[22]. From these data, it is apparent that the molecular weights of lens proteins can be determined with very high accuracy by directly-coupled HPLC ESIMS. Furthermore, this molecular weight information may be used to tentatively identify many of the common posttranslational modifications. It should be noted however, that it is generally not possible to detect deamidation or one disulfide bond in proteins with M_R greater than 20,000 because these modifications change the molecular weight of the protein by only 1 or 2 Da, respectively.

Electrospray ionization mass spectrometry and SDS-PAGE are both techniques for determining the molecular weights of proteins. The molecular weights determined by SDS-PAGE frequently have errors as large as 10%, whereas molecular weights determined by ESIMS normally have errors less than 0.05%. Although the precision of SDS-PAGE may be high, the accuracy depends on the shape of the protein, the effects of which cannot be predicted accurately. The increased accuracy of ESIMS is essential for identifying possible posttranslational modifications. It is also interesting to compare the detection limits of ESIMS and SDS-PAGE. Although approximately 10 pmoles of αA-crystallin were used to obtain the ESI mass spectrum presented in Fig. 3, satisfactory results are routinely obtained for 200 fmoles of well-behaved proteins. The ESIMS response for different proteins is not understood at this time, but it is likely affected by such factors as the net charge on a protein in solution, the tendency of a protein to aggregate, and the solubility of the protein in the electrospray solution. The detection limit of ESIMS is generally much better than SDS-PAGE, even when silver staining is used to visualize the bands. Use of radioactively labeled proteins separated by SDS-PAGE, permits detection of much smaller amounts.

Determining the primary structures of proteins is complicated by the observation that most samples isolated from natural sources are mixtures, as illustrated in Fig. 3. ESIMS and SDS-PAGE each have advantages and disadvantages for analysis of protein mixtures. The high mass resolution of ESIMS is particularly valuable when analyzing mixtures. If one considers the peak widths at half-height for αA-crystallin, and its des-Ser form in Fig. 3b (a mass difference of 87 Da), it is evident that constituents whose molecular weights differ by as little as 0.1% could have been detected using these experimental conditions. When similar measurements are made with a double-focusing high resolution mass spectrometer, the resolution may be improved by as much as a factor of 10. Because migration in SDS-PAGE depends on the shape of the proteins, it is not possible to specify a universal figure for the resolution of this technique. However, from a survey of published literature and our own experience, one can expect to separate proteins only if their molecular weights differ by approximately 5-10%. It follows that ESIMS has much better resolution, and therefore, is more useful than SDS-PAGE for analyzing mixtures of proteins with similar molecular weights. On the other hand, very complex mixtures of proteins with widely different molecular weights are better analyzed by SDS-PAGE than ESIMS. With ESIMS, each protein in the mixture gives a series of peaks (Fig. 3a) resulting in very complex mass spectra that are difficult to interpret. However, each protein in a complex mixtures of proteins can be detected by SDS-PAGE as long as the molecular weights of the constituent proteins differ by more than 1000-2000. Detection of minor components in mixtures is also difficult by ESIMS because it may not be possible to distinguish peaks due to the minor components from chemical noise in the mass spectrum. However, if the molecular weight of the minor component is very different from the major components, the gel may be purposely over-loaded so that the minor component can be detected. Hopefully, it is evident from this discussion that molecular weight information derived from ESIMS and SDS-PAGE is complementary, and that one will often benefit from using both techniques.

Application of HPLC ESIMS to Proteolytic Digests of Proteins

Additional structural information may be obtained from the molecular weights of proteolytic digests of proteins, which may be used to confirm the identity and to determine the approximate location of modifications consistent with the molecular weight of the intact protein. For example, one would expect to find a C-terminal peptide in a tryptic digest of the αA-crystallin sample described above that has a molecular weight 87 Da less than expected, corresponding to the loss of Ser from the C-terminus of αA-crystallin. Likewise, if the +80 Da peak in Fig. 3b (M_R 20,031) is due to phosphorylation, one should find a tryptic peptide whose molecular weight is 80 Da higher than expected. Identification of the parent protein of an extensively modified protein is another important use of the molecular weights of the proteolytic peptides. If the protein has undergone substantial processing since formation (e.g., backbone cleavage, glycosylation, deamidation), the molecular weight of the intact protein alone may not be sufficient for rigorous identification of the protein. However, most of its peptides may have molecular weights consistent with its amino acid sequence. It is now possible to identify proteins by comparing the molecular weights found in a tryptic digest with peptide molecular weights calculated for proteins whose sequences are stored in a large data base. The higher the number of matching peptides, the higher the probability that the origin of the protein has been identified. Peptides whose molecular weights do not match those of peptides from the data bank may be analyzed further to determine whether they represent modified forms of the protein, or whether they may be due to other proteins contaminating the sample.

Molecular weight maps of proteolytic digests of proteins may be made by digesting the protein, separating and desalting the peptides by reversed-phase HPLC, and collecting and analyzing each fraction by FABMS or ESIMS. Directly-coupled HPLC ESIMS is a faster and more sensitive method for determining molecular weights. Use of the capillary system illustrated in Fig. 1b gives the highest sensitivity. The microbore system (Fig. 1a) is particularly useful when further experiments are required to identify selected peptides, that is, peptides whose molecular weights do not match the peptides expected to be produced by the enzyme and protein used in that particular digestion. Deciding which peptides match is facilitated by using highly specific proteases to fragment the protein. For example, one would expect most of the peptides in a tryptic digest to have either Lys or Arg as the C-terminal residue. Likewise, most of the peptides in an Asp-N digest have Asp on the N-terminus. Identification of peptides in chymotryptic or peptic digests is more difficult because these proteases frequently cleave at any of several different sites, and their cleavage is not highly predictable. Consequently, there may be several portions of the protein that could yield a peptide of any given molecular weight and unambiguous identification requires further data, such as the CID analysis described below.

Application of microbore HPLC ESIMS to analyze the tryptic digest of αB-crystallin is illustrated in Figs. 4 and 5. It is important to note that the same HPLC column, mobile phases, and ESI mass spectrometer used to analyze the mixture of α-crystallin proteins can be used to obtain peptide molecular weight maps of proteolytic digests of these proteins. The ESIMS total ion current display and UV absorbance of a tryptic digest of 100 pmoles of αB-crystallin are given in Fig. 4. Because the chromatographic peaks for the peptides are substantially narrower than for the proteins (Fig. 2), it is possible to critically compare the

Chromatograms of tryptic digest of αB-Crystallin

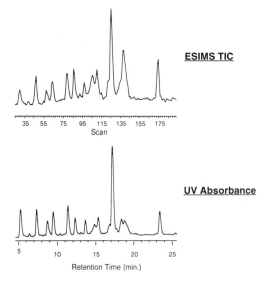

Figure 4. Directly-coupled microbore HPLC ESIMS analysis of a tryptic digest of human αB-crystallin. The αB-crystallin was isolated from a lens extract by gel filtration and reversed-phase HPLC. Approximately 100 pmoles of a tryptic digest of αB-crystallin were injected in a volume of 2 μl. Peptides were eluted into the ESI mass spectrometer using a gradient linear from 2%-50% B over 25 min. The mass spectrometer was tuned to a resolution of 1000 (50% valley), and scanned repetitively through the 300 to 1600 m/z range. Each scan required a total time of 7 sec. The UV absorbance was determined at 214 nm.

ESIMS Spectra of αB-Crystallin Peptide Map

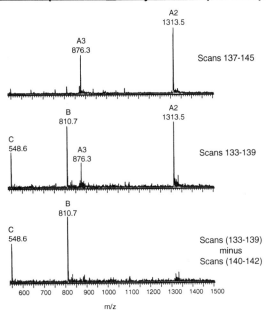

Figure 5. ESI mass spectra of peptides in a tryptic digest of αB-crystallin. The top trace includes fractions eluting at 18.5-19.5 min. (scans 137-145). The two peaks can be attributed to the doubly (m/z 1313.5) and triply (m/z 876.3) charged forms of peptide 124-149 (MH⁺ 2624.4). The middle trace includes the fractions eluting at 18.0-18.5 min (scans 133-139). Peaks at m/z 548.6 and 810.7 correspond to segments 15-18 and 12-18, respectively, of αB-crystallin. The third trace was obtained by subtracting scans 140-142 from scans 133-139.

chromatographic resolution obtained by the ESIMS and UV detectors. The results in Fig. 4 indicate that the resolution of the mass spectrometer is only slightly less than that of the UV detector, which may be due to the much longer tube required to transmit the peptides to the mass spectrometer. The molecular weights of peptides present in each of the chromatographic peaks can be determined from the appropriate ESI mass spectra. The ESI mass spectra given in Fig. 5 show how proper background subtraction can enhance the spectra of minor components. The top trace in Fig. 5 is the spectrum for the large fraction eluting in scans 137-145. The major peaks have been identified as peptide 124-149, doubly charged (MH_2^{+2} 1313.5) and triply charged (MH_3^{+3} 876.3). The middle trace shows the spectrum for the preceding small shoulder in the chromatogram (scans 133-139) which contains peaks corresponding to two chymotryptic cleavages, peptide 12-18 (MH⁺ 810.7) and peptide 15-18 (MH⁺ 548.6), but the spectrum is still dominated by the masses belonging to peptide 124-149. After subtraction of scans 140-142, the spectrum in the bottom trace display the minor chymotryptic peptides more clearly. Note that only 10% of the sample was used for ESIMS analysis. The remaining 90% (90 pmoles) was collected and saved for further analysis.

Peptide Sequencing by CID MS/MS

When the molecular weights of the protein and its proteolytic fragments are not adequate for positive identification of the protein, collision induced dissociation mass spectrometry/mass spectrometry (CID MS/MS) may be used to sequence a peptide, and in so doing, locate specific sites of modification[8,9,31]. This method is implemented by generating a beam of molecular ions of a peptide of interest, either by FABMS or ESIMS, energetically exciting the molecular ions through collisions with an unreactive gas, and using a second stage of mass analysis to determine the m/z's of fragments formed by dissociation of the excited molecular ions. Just as fragment ions in the electron ionization mass spectra of small volatile materials can be mentally fitted together to give likely structures of the starting

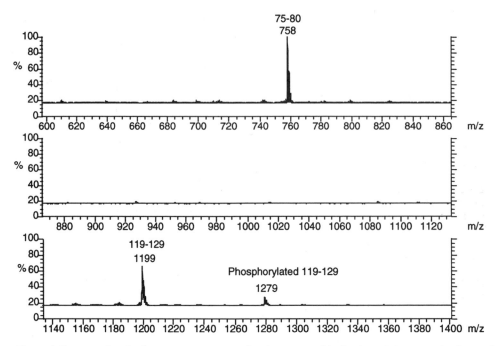

Figure 6. Fast atom bombardment mass spectrum of a chromatographic fraction of chymotryptic digest of αA-crystallin. The peaks at m/z 1199 and 1297 are due to the non-phosphorylated and phosphorylated forms, respectively, of the 119-129 segment of αA-crystallin. The mass spectrum was acquired using a VG Fisons Autospec mass spectrometer. Also shown is another peak at m/z 758, which corresponds to the 75-80 segment of αA-crystallin. These three peptides co-elute on most reversed-phase HPLC systems.

material, peptide fragment ion mass spectra formed by CID MS/MS can be used to deduce the amino acid sequence of the peptide. Although the amino acid sequences of proteins are more easily determined from the sequence of the appropriate DNA segment or by sequencing the protein by Edman or related chemistry, CID MS/MS complements these techniques by being able to sequence peptides that are not amenable to these techniques. CID MS/MS can be used to sequence peptides in which the N-terminus is blocked, or that contain modifications or unusual amino acid constituents.

We have found CID MS/MS useful for our studies of lens proteins. The FAB mass spectrum of one HPLC fraction, which contains the 119-129 segment of αA-crystallin isolated from a chymotryptic digest, is given in Fig. 6. The peak at m/z 1199 corresponds to the MH[+] ion expected for the unmodified 119-129 segment of αA-crystallin. This assignment was verified by sequencing the peptide using the CID MS/MS spectrum presented in Fig. 7. Results of the CID MS/MS spectrum are summarized in Fig. 7b, following a standard nomenclature system[8,32]. After collision with neutral gas molecules, the protonated molecule ions of peptides (MH[+]) may have sufficient vibrational and rotational energy to undergo fragmentation. Detailed studies of CID mass spectra of many reference peptides have led to formulation of a series of expected fragmentation reactions. To determine the amino acid sequence of a peptide, one is most interested in fragments formed by cleaving the backbone of the peptide, with the excess charge remaining on either the N- or C-terminus. Fragment ions comprising the A, B, and C series, formed by cleavage of the backbone with the charge remaining on the N-terminus, are illustrated in Fig. 7b. The C-terminal fragment of many

a **MS/MS of Peptide 119-129**

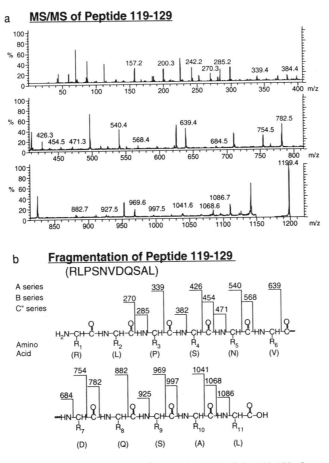

b **Fragmentation of Peptide 119-129**
(RLPSNVDQSAL)

Figure 7. Fragments formed by CID of the MH$^+$ ion (m/z 1199) of the 119-129 chymotryptic fragment of αA-crystallin (a) and their assembly to indicate the amino acid sequence of this peptide (b). Data acquired using a VG Fisons' Autospec high resolution mass spectrometer equipped with an orthogonal acceleration time-of-flight mass analyzer. The molecular ion (MH$^+$ 1199) was selected by the first stage of mass analysis, decelerated to 800 eV of energy, fragmented through collisions with xenon, and analyzed by the time-of-flight analyzer. The m/z of the parent ion, as well as the m/z's of its fragments, were determined to within 0.2 Da.

peptides, especially those with a basic residue (Lys or Arg) on the C-terminus, may retain the positive charge. Then the peptide fragments form the X, Y, and Z series ions[8,32]. Although the fragments identified in Fig. 7 may not be sufficient to sequence this peptide, they are sufficient to identify unequivocally the peptide as the 119-129 segment of αA-crystallin. In general, tryptic peptides are better for sequencing peptides because, with either Lys or Arg at their C-terminus, these peptides usually give fragments attributable to both ends of the peptide. The 119-129 does not have Lys or Arg on the C-terminus, and therefore, does not give intense fragment ions derived from its C-terminus. By also including fragments formed from cleavage of side chains, known as the "d, w, and v "series[8], all of the major peaks in the CID mass spectrum can be attributed to fragmentation processes characteristic of protonated peptides.

The extreme usefulness of CID MS/MS for identifying modified peptides is further illustrated in the analysis of peak (MH$^+$ 1279) in the spectrum (Fig. 6). Finding a peptide with MH$^+$ 80 Da greater than that of the 119-129 segment, is consistent with the +80 Da

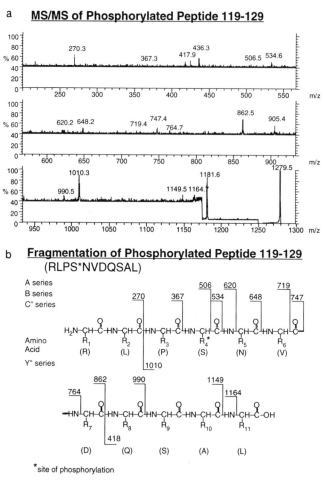

a   **MS/MS of Phosphorylated Peptide 119-129**

b   **Fragmentation of Phosphorylated Peptide 119-129**
    (RLPS*NVDQSAL)

*site of phosphorylation

Figure 8. Fragments formed by CID of the MH$^+$ ion (m/z 1297) of the phosphorylated form of the 119-129 chymotryptic fragment of αA-crystallin (a) and their assembly to indicate that Ser 122 is phosphorylated (b). Experimental conditions are the same as those used to acquire data for Fig. 7.

peak found in the ESI mass spectrum of the intact protein, (Fig. 3b) suggesting that one residue in the 119-129 segment is phosphorylated. However, the molecular weight of the peptide is insufficient for unambiguous identification of the peptide because there are 3 possible explanations for a peptide of this molecular weight. It may be peptide 119-129 phosphorylated at either Ser 122 or Ser 127, or it may possibly be peptide 140-157, although not an expected chymotryptic peptide, which also has an MH$^+$ 1279. Loss of 98 from the parent ion, corresponding to loss of a phosphate group, to give the large peak at m/z 1181, is diagnostic for peptides phosphorylated on Ser and confirms identification of the peptide as phosphorylated peptide 119-129. This CID spectrum (Fig. 8) has fewer fragment peaks with intensities above the background than the CID spectrum for MH$^+$ 1199 because the intensity of the parent in Fig. 6 is much lower. Although this must be regarded as a "weak" spectrum, it is evident that the fragmentation is consistent only with phosphorylation at Ser 122.

Acknowledgments

This research has been supported by the National Institutes of Health through grants RO1 EY07609 and RO1 GM40384, and by the Nebraska Center for Mass Spectrometry.

REFERENCES

1. Khorana, H.G., Gerber, G.E., Herlihy, W.C., Gray, C.P., Anderegg, R.J., Nihei, K. and Biemann, K. (1979) *Proc. Natl. Acad. Sci. USA* 76: 5046-5050
2. Barber, M., Bordoli, R.S., Sedgwick, R.D. and Tyler, R.N. (1981) *J. Chem. Soc. Chem. Commun.* 325-327.
3. Naylor, S., Findeis, A.F., Gibson, B.W. and Williams, D.H. (1986) *J. Am. Chem. Soc.* 108: 6359-6363.
4. Carr, S.A., Hemling, M.E., Bean, M.F. and Roberts, G.D. (1991) *Anal. Chem.* 63: 2802-2824.
5. Fenselau, C. (1991) *Annu. Rev. Biophys. Biophys. Chem.* 20: 205-220.
6. Caprioli, R.M., Fan, T. and Cottrell, J.S. (1986) *Anal. Chem.* 58: 2949-2954.
7. Coutant, J.E., Chen, T.-M. and Ackermann, B.L. (1990) *J. Chromatogr.* 529: 265-275.
8. Bieman, K. (1990) In: *Methods in Enzymology,* (McCloskey, J.A. ed.) Vol. 193: pp. 455-479 Academic Press, San Diego, CA.
9. Hunt, D.F., Yates, J.R., Shabanwitz, J., Winston, S. and Hauer, C.R. (1986) *Proc. Natl. Acad. Sci. USA* 83: 6233-6237.
10. Fenn, J.B., Mann, M., Meng, C.K., Wong, S.F. and Whitehouse, C.M. (1989) *Science* 246: 64-71.
11. Smith, R.D., Loo, J.A., Ogarzalek, R.R., Loo, Busman, M. and Udseth, H.R. (1991) *Mass Spectrom. Rev.* 10: 359-451.
12. Chait, B.T. and Kent, S.B.H. (1992) *Science* 257: 1885-1893.
13. Covey, t.r., Huang, e.c. and Henion, J.D. (1991) *Anal. Chem.* 63: 1193-1200.
14. Smith, J.B., Thevenon-Emeric, G., Smith, D.L. and Green, B. (1991) *Anal. Biochem.* 193: 118-124.
15. Smith, J.B., Sun, Y., Smith, KD.L. and Green, B. (1992) *Protein Sci.* 1: 601-608.
16. Davson, H. (ed.) (1984) *The Eye,* Academic Press. London.
17. Berman, E. (1991) *Biochemistry of the Eye,* Plenum Press, New York.
18. Asselberg, F.A.M., Koopmans, M., Venrooij, W.J. van and Bloemendal, J. (1979) *Exp. Eye Res.* 28: 223-228.
19. Smith, J.B., Miesbauer, L.R., Leeds, J., Smith, D.L., Loo, J.A., Smith, R.D. and Edmonds, C.G. (1991) *Int. J. Mass Spectrom. Ion Proc.* 111: 229-245.
20. Chiesa, R., Gawinowicz-Kolks, M.A. and Spector, A. (1987) *J. Biol. Chem.* 262: 1438-1441.
21. Voorter, C.E.M., de Haard-Hoekman, W.A., Roersma, E.S., Meyer, H.E., Bloemendal, H. and de Jong, W.W. (1989) *FEBS Lett.* 259: 50-52.
22. Miesbauer, L., Zhou, X., Yang, Z., Yang, Z., Sun, Y., Smith, D.L. and Smith, J.B. ((1994) *J. Biol. Chem.* 269: 12494-12502.
23. de Jong, W.W., van Kleef, F.S.M. and Bloemendal, H. (1974) *Eur. J. Biochem.* 48: 271-276.
24. Groenen, P.J.T.A., van Dongen, M.J.P., Voorter, C.E.M., Bloemendal, H. and de Jong, W.W. (1993) *FEBS Lett.* 322: 69-72.
25. Lou, M.F. and Dickerson, J.E. (1992) Jr., *Exp. Eye Res.* 55: 889-896.
26. Kamei, A. (1993) *Biol. Pharm. Bull.* 16: 870-875.
27. Zhou, Z. and Smith, D.L. (1990) *J. Protein Chem.* 9: 523-532.
28. Fenn, J.B., Mann, M., Meng, C.K. and Wong, S.F. (1990) in: M.L. Gross, Ed., *Mass Spectrom. Rev.* 9: 37-70.
29. Chowdhury, S.K., Katta, V. and Chait, B.T. (1990) *Rapid Commun. Mass Spectrom.* 4: 81-87.
30. Briuns, A.P., Covey, T.R. and Henion, H.D. (1987) *Anal. Chem.* 59: 2642-2646.
31. Biemann, K. and Scoble, H.A. (1987) *Science* 237: 992-998.
32. Roepstorff, P. and Fohlman, J. (1984) *Biomed. Mass Spectrom.* 11: 601.

24

ISOLATION, CHARACTERIZATION AND AMINO ACID SEQUENCE OF THE N-TERMINAL FUNCTIONAL UNIT FROM THE *Rapana thomasiana grosse* (MARINE SNAIL, GASTROPOD) HEMOCYANIN

Stanka Stoeva,[1] Krasimira Idakieva,[2] Wolfgang Voelter,[*1] and Nicolay Genov[2]

[1] Department of Physical Biochemistry
Institute of Physiological Chemistry
University of Tübingen
Hoppe-Seyler Straße 4, D-72076 Tübingen, Germany
[2] Institute of Organic Chemistry
Bulgarian Academy of Sciences
Sofia 1113, Bulgaria

ABSTRACT

The amino-terminal functional unit Rta (domain a) of the *Rapana* hemocyanin "heavy" structural subunit was obtained after limited trypsinolysis of the whole polypeptide chain. Mass-spectrometric analysis showed a molecular mass of 49698 daltons for the electrophoretically homogeneous fragment. 33 amino acid residues were sequenced directly from the N-terminus of Rta which allowed the location of the domain in the polypeptide chain of the subunit. Physico-chemical parameters were determined by absorption and fluorescence spectroscopy and circular dichroism. Comparison with the respective parameters of the whole *Rapana* hemocyanin showed that after the isolation of the functional unit, the polypeptide backbone, binuclear active site and capability of binding oxygen are identical to the native hemocyanin. The domain was digested with endoproteinases LysC and AspN and BrCN. 12 peptides were isolated and about 290 residues sequenced. The obtained structures were aligned and compared with the sequences of other molluscan hemocyanin functional units.

---

[*] Represents person presenting Paper

INTRODUCTION

Hemocyanins are oxygen-carrying copper proteins suspended freely in the hemolymph of invertebrates. Their binuclear active sites contain two copper ions, ligated to histidyl residues from the polypeptide chain. Arthropod hemocyanins occur as multiples of hexameric aggregates of 75 kDa minimal subunits, each containing one oxygen-binding site[1]. Molluscan hemocyanins are composed of large subunits with a molecular mass of 250 to 450 kDa which are arranged as a cylindrical 4 million Da decamer in cephalopods and as a 8 million Da di-decamer in gastropods. The later results from a face-to-face assembly of two decamers[2,3]. The elongated subunit is organized as a string of globular domains of ca 50 kDa which are quite different immunologically. Each domain carries a binuclear copper active site and is capable of reversible binding one dioxygen molecule. Amino acid sequences of domains from several molluscan species have been elucidated[4-8]. Most of the information about the structure and function of molluscan hemocyanins has been obtained from studies on dioxygen carriers isolated from *Helix pomatia,*[9-12] *Marisa cornuarietis*[12-14] and *Octopus dofleini*[8,15,16]. However, the microstructure of the amino-terminal functional unit from molluscan hemocyanin has not so far been reported. Intensive studies on the hemocyanin of the Californian giant keyhole limpet *Megatura crenulata*, KLH, have also been performed.[17-20] KLH is widely used in research and clinics as an immune stimulant and for immunotherapy of bladder cancer[21,22] and renal cell carcinoma.[23]

Rapana thomasiana grosse (gastropod) is a marine snail living along the west coast of the Black Sea, in the Yellow Sea and in the East China Sea. At neutral pH the associated native *Rapana* hemocyanin appears in the electron micrographs as cylinders with 300 Å diameter and 380 Å height. In contrast to the other gastropod hemocyanins, the central hole of the cylinders formed by the *Rapana* hemocyanin is filled with a protein material[24]. At pH values above 9 the aggregates dissociate into two subunits designated RHSS1 and RHSS2, with a molecular mass of 250 and 450 kDa, respectively. The dissociation and reassociation is reversible, depending on the absence or presence of calcium ions. The two subunits have been sequenced from the amino-terminus and a homology with the N-terminal region of structural subunits or functional units of other molluscan hemocyanins has been observed.[25]

This study was undertaken in order to further characterize the *Rapana* hemocyanin. We report here results of the limited trypsinolysis of the heavy chain, RHSS2 and the isolation of the amino-terminal functional unit with preserved structure, binuclear active site and oxygen-binding affinity. A special reference to the amino acid sequence and to some physicochemical properties of this fragment is made. The microstructural data are compared with those of other molluscan hemocyanins.

MATERIALS AND METHODS

Preparation of Rapana Hemocyanin and Isolation of the Structural Subunits

Rapana thomasiana specimens were caught near the Bulgarian coast of the Black Sea (Varna). The *Rapana thomasiana* hemolymph and hemocyanin was isolated as described previously.[24] The two structural subunits, RHSS1 and RHSS2, were purified by ion-exchange chromatography of the dissociated material on DEAE-Sepharose CL-6B according to the procedure given in Idakieva *et al.*[25]

Tryptic Hydrolysis of the Heavy Chain, RHSS2, of the Rapana Hemocyanin and Fractionation of the Fragments

The heavy structural subunit of *Rapana thomasiana* hemocyanin, dissolved in 100 mM ammonium hydrogen carbonate, containing 1mM EDTA, pH 8.2, was treated with TPCK-trypsin in the same buffer at an enzyme/substrate ratio of 1/400 (w/w) for 20 min at 20°C. At the end of the reaction, the enzyme was inactivated by phenylmethanesulfonyl fluoride. The reaction mixture was fractionated on a Sephadex G-100 (119x2cm) column, equilibrated and eluted with 100 mM ammonium acetate buffer, pH 8.2, containing 0.02% NaN$_3$.

SDS-Polyacrylamide Gel Electrophoresis (SDS-PAGE)

SDS-polyacrylamide gel electrophoresis was carried out according to Laemmli[26] using a 10% gel. For the molecular mass determination the following protein markers were used: albumin (66.25kDa), a subunit of catalase (58kDa), ovalbumin (42.7kDa) and carboanhydrase (30kDa).

Spectroscopic Measurements

Absorption spectra were recorded with a Shimadzu recording spectrophotometer, model 3000. Fluorescence measurements were performed with a Perkin Elmer model LS5 spectrofluorimeter, equipped with a thermostatically controlled assembly and a data station model 3600. The optical absorbance of the solutions was lower than 0.05 at the excitation wavelength to avoid inner filter effects. Circular dichroism was measured with a Roussel Jouan Dichrographe III instrument. The data were expressed in terms of mean ellipticity using a specific absorption coefficient at 280nm ε_{280}=6.88 x 10^4 M^{-1} cm^{-1}. The molecular mass of the functional unit was determined using a matrix-assisted laser desorption mass spectrometer (Lasermat, Finnigan MAT, Bremen, Germany).

Automatic Amino Acid Sequence Analysis

Amino acid sequence analysis was performed using an Applied Biosystems sequencer model 473A (Weiterstadt, Germany).

RESULTS AND DISCUSSION

The "heavy" subunit RHSS2 of *Rapana thomassiana* hemocyanin was split with TPCK-trypsin into fragments and the reaction mixture was fractionated by gel-filtration on Sephadex G-100, yielding two well separated peaks T$_1$ and T$_2$ (Fig.1). SDS-PAGE showed that fraction T$_1$ was heterogeneous and contained several fragments with different electrophoretic mobilities, whereas T$_2$ was characterized by one major band and small amounts of contaminating products. Fraction T$_2$ was purified by HPLC (Fig.2), and the isolated material was used for further investigations.

It was analyzed by SDS-PAGE under reducing and non-reducing conditions and in both cases a single band was observed. The electrophoretic band of T$_2$ (designated later Rta) was located between those of the catalase subunit (58 kDa) and ovalbumin (42.7 kDa) (Fig. 3) and corresponded to molecular mass of 52 kDa. A value of 49698 daltons was obtained by LDMS (Fig. 4). The fragment was localized in the polypeptide chain of the

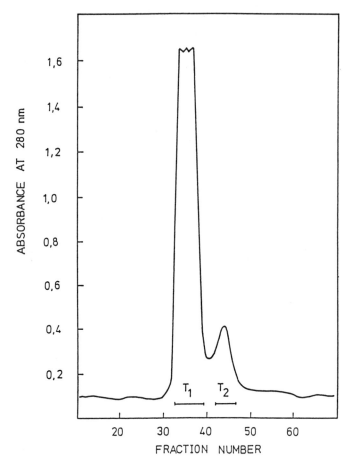

Figure 1. Chromatography on Sephadex G-100 of a tryptic hydrolysate of the "heavy" structural subunit from the *Rapana thomasiana* hemocyanin. The column (119 x 2 cm) was equilibrated and eluted with 0.05 M borate/HCl buffer, pH 8.2, containing 1 mM EDTA. Fractions were pooled as indicated. Fraction T_1 is a mixture of protein fragments; fraction T_2 contains the amino-terminal functional unit (domain).

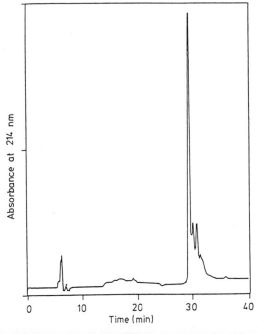

Figure 2. HPLC chromatogram of the purified N-terminal functional unit (domain) isolated from the tryptic hydrolysate of the *Rapana thomasiana* hemocyanin "heavy" structural subunit. Only the material from the major fraction was used for the further investigations. Column: Nucleosil 7C18, 250 x 10 mm. Eluents: A) 0.058% trifluoroacetic acid; B) 80% acetonitrile in solution A). Gradient: 5-100% B/40 min. Flow rate: 2.5 ml/min.

Figure 3. SDS-polyacrylamide gel electrophoresis on 10% gel of the purified N-terminal functional unit (domain) from the *Rapana thomasiana* hemocyanin "heavy" structural subunit. Lanes 1, 2, 3 and 8, molecular mass markers. 1: subunit of catalase, 58 kDa; 2: albumin, 66.25 kDa; 3: carboanhydrase, 30 kDa; 8: ovalbumin, 42.7 kDa. Lanes 5 and 7: the N-terminal unit Rta (domain a) in the absence of 2-mercaptoethanol; lanes 4 and 6: the N-terminal unit in the presence of 50 mM 2-mercaptoethanol.

1   2   3   4   5   6   7   8

"heavy" structural subunit by amino acid sequence analysis of the first 33 residues from the N-terminus. The amino-terminal sequence of the fragment was identical with that of the intact "heavy" structural subunit RHSS2. This domain was designated Rta according to the nomenclature accepted for the molluscan hemocyanin functional units.[8] The prefix Rt signifies *Rapana thomasiana* and the letter a characterizes that the unit represents the N-terminal domain.

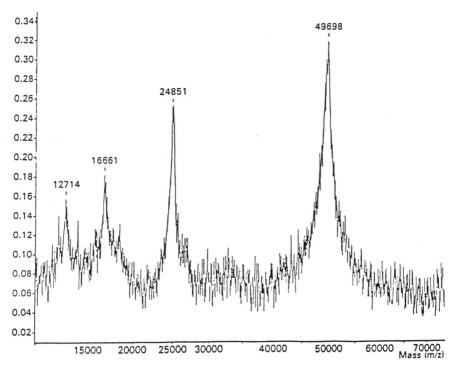

Figure 4. Mass-spectrometric analysis of the N-terminal functional unit (domain) Rta of the *Rapana thomasiana* hemocyanin "heavy" structural subunit. A molecular mass of 49698 daltons is determined for the unit.

Table 1. Physico-chemical parameters of the N-terminal functional unit (domain) Rta
of the *Rapana thomasiana* hemocyanin "heavy" structural subunit and
of the whole native hemocyanin

Parameters	Functional unit Rta in oxy-form	Oxy-hemocyanin of *Rapana thomasiana**
Absorption	280 (71000)	280 (68800)
$\lambda_{max}(\varepsilon)$	345 (22000)	345 (19200)
A_{345}/A_{280}	0.31	0.28
Circular dichroism $\lambda_{max}[\theta]$	208 (-6830) 222 (-6360) 253 (125) 280 (41) 287 (35) 345 (-186)	208 (-7130) 222 (-6850) 253 (130) 280 (99) 287 (76) 345 (-149)
Fluorescence	331 (295)	330 (295)
$\lambda_{max}(\lambda_{ex})$	330 (280)	330 (280)

*Data from Boteva *et al.*, 1991.

Table I presents a summary of spectroscopic parameters of Rta and the native *Rapana* hemocyanin. Besides the absorption band at 280 nm, due to the aromatic chromophores, a band with a maximum at 345 nm was observed which is characteristic for oxyhemocyanins and results from transitions of the oxygen-copper-protein active site. The presence of the band at 345 nm proves that on proteolysis the functional groups are preserved and the binuclear active site is conserved in Rta. The spectroscopic ratio 345nm/280nm is informative for the oxygen saturation of the hemocyanin molecule[27] and can surve as a measure for the oxygen-binding at the copper site. The value calculated for the fragment, 0.31, is practically the same (even slightly higher) as that determined for the whole *Rapana* hemocyanin (Table I). This observation characterizes the isolated fragment Rta as a fuctional unit with preserved capability of binding dioxygen with the same affinity as the native hemocyanin. Evidently, the experimental conditions used lead to limited proteolysis of the "heavy" chain without disturbing the oxygen binding properties of the Rta active site.

The far-ultraviolet (UV) CD spectrum of the functional unit which reflects the backbone conformation of the protein molecule, is characterized by negative bands at 208 and 222 nm connected mainly with the presence of a-helical structure. The β-sheet also contributes to the ellipticity around 222 nm, though to a lesser extent than the α-helix. The far-UV dichroic spectra of Rta and the native *Rapana* hemocyanin are essentially identical suggesting a closely similar folding of the polypeptide chains of the isolated functional unit and the units (domains) constituting the native hemocyanin. The differences in the intensity of the bands at 208 and 222 nm for the unit and for the whole hemocyanin are 4 and 7%, respectively, and are practically in the region of the experimental errors. The CD data showed that little or no conformational change occurs in the polypeptide backbone as a result of the limited trypsinolysis and the functional unit Rta preserved the native backbone conformation.

The near-UV CD spectrum of Rta and that of the *Rapana* hemocyanin are qualitatively identical. In the region 250 - 300 nm these spectra reflect an asymmetric environment for the aromatic (tyrosyl and tryptophyl) amino acid residues and give information on the conformation and flexibility of their side chains. However, the tryptophyl residues are mainly responsible for dichroic transitions above 280 nm. The intensities of the positive bands at 280 and 287 nm in the CD spectrum of Rta are reduced in comparison with those of the respective bands in the spectrum of the *Rapana* hemocyanin (Table I). This suggests that minor changes occur in the environment of the aromatic chromophores of the functional unit as a result of its separation from the long polypeptide chain. Probably, the whole structural subunit creates a specific environment for aromatic residues located in the N-terminal domain which is somewhat different from that after the separation of Rta as an individual unit. The negative CD band at 345 nm in the spectrum of the functional unit should be attributed to the complex between the copper atoms from the binuclear active site and dioxygen molecule. The intensity of this band is somewhat higher than that of the respective band in the spectrum of the native oxyhemocyanin (Table I). This result confirms the conclusion that the oxygen-binding capability of the N-terminal functional unit is completely preserved. The intensity of the dichroic band at 253 nm in the two spectra is identical (Table I).

The most important feature to be noted in the comparison of the UV and CD spectra of Rta and *Rapana* hemocyanin is that the secondary structure and the native form of the oxygen-binding binuclear copper site of the N-terminal domain are conserved in the separated functional unit and only minor changes occur in the environment of the aromatic chromophores.

Fluorescence spectroscopy is a sensitive method for studying protein conformation in solution and changes in the conformation. The aromatic residues of the unit Rta and those of the *Rapana* hemocyanin are potential fluorophores. Upon excitation at 295 nm, where at least 93% of the incident light is absorbed by the indole groups, the fluorescence spectra of the functional unit and that of the whole hemocyanin were characterized by a maximum (λ_{max}) at 331 ± 1 nm (Table I). This emission maximum position is typical for tryptophyl residues buried in a nonpolar environment. Under excitation at 280 nm both, phenolic and indole groups, absorb. Again, an emission maximum at 330 ± 1 nm was observed in both cases (Table I) which can be explained by an efficient nonradiative energy transfer from tyrosyl to tryptophyl side chains[28]. The coincidence of the emission maximum position of Rta and that of native *Rapana* hemocyanin after excitation at 295 or 280 nm is indicative of similarity (or identity) in the microenvironment of the tryptophyl residues in the two proteins. This confirms the conclusion from the CD data that Rta preserved its native structure.

All these spectroscopic parameters showed that little happens to the structural integrity of the functional unit upon the limited trypsinolysis of the whole *Rapana* hemocyanin "heavy" structural subunit.

The partial primary structure of the N-terminal functional unit of the *Rapana* hemocyanin "heavy" chain, consisting of about 400 amino acid residues, is presented in Figure 5. It was mainly determined by analysis of the Lys C peptides (LC1-LC13) (Table II). From the BrCN and Asp N digests, two peptides, CN X and AN, respectively, could also be isolated and sequenced (Table II). The digests were fractionated by reversed-phase HPLC and the fragment sequences were determined by automatic Edman degradation on an Applied Biosystems 473A protein sequencer. The alignment of the first Lys C peptides could be achieved on the basis of the N-terminal sequence of the polypeptide chain (33 steps). All other peptides were located by homology with the known primary structures of domains of

```
                  1                    10                        20
Rta:                          S L L R K N V D S L T E E E I
Hd:           D A V T V A S H V R K D L D T L T A G E I
Oe:               E G N E Y L V R K N V E R L S L S E M
Of: I P P S K Q D A D I D I P L N H I R R N V E S L D E R D I
Hg:           D I H T T A V A G V G V R K D V T R L T V S E T
Og:               E A V R G T I I R K N V N S L T P S D I
Sh:           D H D T E T L I S L N V N S L S P S E I

                      30                40        46          50
Rta: L T L Q S V L R E L E N D Q T E H G F Q T L A S F H G S P P L
Hd:  E S L R S A F L D I Q Q D H T - - - Y E N I A S F H G K P G L
Oe:  N S L I H A F R R M Q K D K S S D G F E A I A S F H A L P P L
Of:  Q N L M A A L T R V K K D E S D H G F Q T I A S Y H G S T - L
Hg:  E N L R E A L R R I K A D N G S D G F Q S I A S F H G S P P G
Og:  K E L R D A M A K V Q A D T S D N G Y Q K I A S Y H G I P - L
Sh:  K N L R D A L V A V Q A D K S G N G Y Q K I A S Y H G M P - L

         53          60    65    67    70      74        80
Rta: - - - - - - E V A C S Q H G M A S F P Q W H R I I T K Q
Hd:  C Q H - - E G H K V A C S V S G M P T F P S W H R L Y V E Q
Oe:  C P R P T A K H R H A C C L H G M A T F P H W H R L Y V V Q
Of:  C P S P E E - P K Y A C C L H G M P V F P H W H R V Y L L H
Hg:  C E H - - E N H S V A C S I G G M A N F P Q W H R L Y V K Q
Og:  S C H Y E N G T A Y A C C Q H G M V T F P N W H R L L I K Q
Sh:  S C H Y P N G T A F A C C Q H G M V T F P H W H R L Y M K Q

                        90                100              110
Rta: M E A A L M G H G A H L G M P Y W D W T T S F T K - - - -
Hd:  V E E A L L D H G S S V A V P Y F D W I S P I Q K L P D L I
Oe:  F E Q A L H R H G A T V G V P Y W D W T R P I S K I P D F I
Of:  F E D S M R R H G S S V A T P Y W D W T Q P G T K L P R L L
Hg:  W E D A L T A Q G A K I G I P Y W D W T T A F T E L P A L V
Og:  M E D A L V A K G S H V G I P Y W D W T T T F S H L P V L V
Sh:  M E D A M K A K G A K I G I P Y W D W T T T F S H L P F L V

                                    120                      130
Rta: - - - - - - - - - - - - - - - - - -
Hd:  S K A T Y Y N S R E Q R F D P N P F F S G K V A - - G E D A V T
Oe:  A S E K Y S D P F T K I E V Y N P F N H G H I S L I S E D T T T
Of:  A D S D Y Y D A W T D N V I E N P F L R G Y I T - - S E D T Y T
Hg:  T E E - - - - - - - - - V D N P F H H G T I Y - - - N G E I T
Og:  T E E K - - - - - - - - D N S F H H A H I D - - V A N T D T
Sh:  T E P K - - - - - - - - N N P F H H G Y I D - - V A D T K T

                          140                150              160
Rta: - - - - - - - - - - - - - - - - - -
Hd:  T R D P Q P E L F N N N - - - - - - Y F Y E Q A L Y A L E Q
Oe:  K R E V S E Y L F E H P A L G K Q T W L F D N I A L A L E Q
Of:  V R D F K P E L F E I - G G G E G S T L Y Q Q V L L M L E Q
Hg:  T R A P R D K L F N D P E F G K E S F F Y R Q V L L A L E Q
Og:  T R S P R A Q L F D D P E K G D K S F F Y R Q I A L A L E Q
Sh:  T R N P R P Q L F D D P E Q G D Q S F F Y R Q I A F A L E Q

                         170     174     178   180            190
Rta: - - - - - - - - - - - - - - - - - - S
Hd:  D N F D D F E I Q F E V L H N A L H S W L G G H A K Y S F S
Oe:  T D Y C D F E I Q L E I A H N A I H S W I G G H A K Y S F S
Of:  E D Y C D F E V Q F E V V H N S I H Y L V G G H Q K Y A M S
Hg:  T D Y C D F E V Q Y E I S H N A I H S W T G G Q S P Y G M S
Og:  G D F C D F E I Q F E I G H N A I H S W V G G S S P Y G M S
Sh:  R D F C D F E I Q F E M G H N A I H S W V G G S S P Y G M S

                       200       205     210                  220
Rta: T L E Y T A Y D P L F I H H S N V D R L W A I W Q E L Q K
Hd:  S L D Y T A F D P V F F L H A N T D R L W A I W Q E L Q R
Oe:  H L H Y A A Y D P I F Y L H S N V D R L W V I W Q E L Q K
Of:  S L V Y S S F D P I F Y V H S M V D R L W A I W Q A L Q E
Hg:  T L E Y T A Y D P L F L L H H S N V D R Q F A I W Q A L Q K
Og:  T L H Y T S Y D P L F Y L H H S N T D R I W S V W Q A L Q K
Sh:  T L H Y T S Y D P L F Y L H H S N T D R I W A I W Q A L Q K
```

```
                  230                    240                         250
Rta: - - - - - - - - - - E A I V T L R A P L A P L - F K K - - - - -
Hd:  Y R G L P Y N E A D C A I N L M R K P L Q P F Q D K K L - N P R
Oe:  L R G L N A Y E S H C A L E L M K V P L K P F S F G A P Y N L N
Of:  H R H L P F D K A Y C A L E Q L S F P M K P F V W E A - - N P N
Hg:  F R G L P Y N S A N C A I Q L L H Q P M R P F S D A D - - N V N
Og:  Y R G L P Y N T A N C E I N K L V K P L K P F L N D T - - N P N
Sh:  Y R G L P Y N S A N C E I N K L K K P M M P F S S D D - - N P N
                  260                    270                         280
Rta: - - - - - - - - - - S L F S Y L - Q L G Y T Y D T L T L N G M
Hd:  N I T N I Y S R P A D T F D Y R N H F H Y E Y D T L E L N H Q
Oe:  D L T T K L S K P E D M F R Y K D N F H Y E Y D I L D I N S M
Of:  L N T R A A S T P Q H L F D Y - N K L G Y K Y D D L E F H G M
Hg:  P V T R T N S R A R D V F N Y D - R L N Y Q Y D D L N F H G L
Og:  A V T K A H S T G A T S F D Y H - K L G Y D Y D N L N F H G M
Sh:  E V T K A H S T G T K - - - - - - - - - - - - - - - - - - -
                  290                    300                         310
Rta: T I S Q L - - - - - - - - - E E - D R H F A N F M L R D I D S S
Hd:  T V P Q L E N L L K R R Q - E Y G R V F A G F L I H N N G L S
Oe:  S I N Q I E S S Y I R H Q K D H D R V F A G F L L S G F G S S
Of:  N I D Q L E N A I H K T Q N - K D R V F A S F L L F G I K T S
Hg:  S I S E L N D V L - - - - - - - A R I F A E F L L H G I G A S
Og:  T I P E L E E H L K E I Q H E - D R V F A G F L L R T I G Q S
Sh:  - - - - - - - - H L N K I Q - E K D R V C A G F L L R A I G Q S
                  320                              330
Rta: A D V T F D L C - K D E H - D F - - - - - - A G T F A V L G G P
Hd:  A D V T V Y V C V P S G P K G K N D C N H K A G V F S V L G G E
Oe:  A Y A T F E I C I E G G E C H E G - - - - - S H F A V L G G S
Of:  A D V H L K L C - K D E D C E D A - - - - - G V V F V L G G D
Hg:  A D V T F D L C D S H D H C E F A - - - - - G T F A I L G G P
Og:  A D V N F D V C T K D G E C T F G - - - - - G T F C I L G G E
Sh:  A D V N F D I C R K D G E C K F G - - - - - G T F C V L G G Q
                  340                    350                         360
Rta: L E M P W S F D R L F K Y D X T - - - - - - V H L F Q T L A A - F
Hd:  L E M P F T F D R L Y K L Q I T D T I K Q L G L K V N N A A S Y
Oe:  T E M P W A F D R L Y K I G I T D V L S D M H L A F D S - - A F
Of:  N E M P R P F D R T Y K M D I T N V L H K M H I P L E D - - L Y
Hg:  L E H P W A F D R L F K Y D V T D V F S K L H L R P D S - - E Y
Og:  H E M P W A F D R L F K Y D I T T S L K H L R L D A H D - - D F
Sh:  H E M A W A F D R L F L Y D I S R T L L Q L R L D A H D - - D F
                  370                              380                390
Rta: H G E P S Q T A L R A L V Y S V R G T L L R S Q I L G K P Y I
Hd:  Q L K V E I K - - - - - - - A V P G T L L D P H I L P D P S I
Oe:  T I K T K I - - - - - - - V A Q N G T E L P A S I L P E A T V
Of:  V H G S T I H L E V K I E S S V D G K V L D S S S L P V P S M
Hg:  H F N I H I - - - - - - - V S V N G T E L D S H I I R S P T V
Og:  D I K V T I K - - - - - - - G I D G H V L S N K Y L S P P T V
Sh:  D V K V T I - - - - - - - M G I D G K S L P T T L L P P P T I
                  400
Rta: S H R P A H G F T X P V I H K K
Hd:  I F E P G T K E R
Oe:  I R I P P S K Q D A
Of:  I Y V P A - K E F T K E I G K
Hg:  Q F V P G V K D Y Y E K
Og:  - F L A P A K T T H
Sh:  L F K P G T G T Q L T R
```

Figure 5. Alingnment of complete or almost complete sequences for functional units of molluscan hemocyanins. The sequences are from domains a of *Rapana thomasiana*; e, f, g of *Octopus dofleini*; h of *Sepia officinalis;* d and g of *Helix pomatia*. Boxes show identical and similar residues shared by all seven sequences. The following groups of amino acid residues are considered to be similar: E and D; N and Q; S and T; S and C; M, L, I and V; F, Y and W; H, K and R. Numbering refers to the sequence of Odg.

Table 2. Amino acid sequences of the peptides obtained from Rta after
cleavage with LysC proteinase

Peptide	RT	Pos.Nr	Sequence
LC 1	-	6-10	SLLRK
LC2	58,92	11-58	NVDSLTEEEILTLQTVLRELENDQTEHGFQTLASFHG SPPL
LC3	41,99	59-79	EVACSQHGMASFPQWHRIITK
LC4	55,65	80-105	QMEAALMGHGAHLGMPYWDWTTSFTK
LC5 CN V	-	106-220 190-...	STLEYTAYDPLFFIHHSNVDRLWAIWQELQK
LC6	35,93	230-247	EAIVTLRAPRPQLFKK
LC8	55,65	263-271	FNYRNFAGYK
LC9 CN X	-	263-271 281-...	SVLQYXSSXLXLQK
LC10	51,24	294-320	EEDRHFANFMLRGIGSSADVTFDLCK
LC11	55,61	321-348	DEHDFAGTFAVLGGPLEMPWSFDRLFK
LC12 AN	- 49,14	349-364 350-373	DXTVHLFQTLAAFHGEPSQTALR
LC13	38,2	388-416	PYISHRPAHGFTXPVIHKK

gastropodan or cephalopodan hemocyanins (Fig.5). 290 of ca 400 amino acid residues (72,5%) of Rta could be sequenced and positioned. Comparison of the Rta partial sequence with those of other domains constituting subunits of molluscan hemocyanins revealed 31% of structural homology (Fig. 5). The degree of homology is higher, 44 - 60%, if isofunctional residues are also taken into consideration. It should be mentioned that the domains which are compared in Fig. 5 occupy different positions in the whole subunits of the respective molluscan hemocyanin. A much higher degree of homology should be expected if the functional unit Rta is compared with domains with the same position in the respective subunits. However, the amino acid sequence of the N-terminal functional unit from molluscan hemocyanin has not so far been reported.

In Fig. 5 the alignment of the sequences around the copper B binding site in molluscan hemocyanins can be seen. It is obvious that histidines-174, -178, and -205 serve as ligands for the copper ion in the copper B site in molluscan hemocyanins similar to this in arthropodan hemocyanins[4,7]. The question what the ligands of copper A may be in molluscan hemocyanins remains open. Lang[7] suggested His-46, His-53, and His-74 as candidates. However, in contrast to His-46 and His-74, His-53 is not conserved in all molluscan hemocyanin sequences (Fig. 5). Because there are no other conserved histidine residues in molluscan hemocyanins, according to Lang and van Holde,[8] following possibilities for copper A ligands could be considered:

I. The third ligand for copper A may be a histidine which may not be conserved between individual units of molluscan hemocyanins.

II. The third side chain involved in copper A binding may be another kind of amino acid residue. Indeed, Met-67 is conserved in all molluscan hemocyanin functional units . As a close neighbour of His-65, which has been found in copper binding in tyrosinases[8], it could be substituting as a copper A ligand in molluscan hemocyanins. Glu-82 is conserved in all molluscan sequences and is also known to serve as a metal ligand in metalloproteins, but has not yet been observed as copper ligand in proteins.

III. The third possibility is that there are only two copper A ligands, both of them histidines. Results from ligand substitution reactions can be interpreted to mean the presence of only two ligands for the copper A site in molluscan hemocyanins.

Further investigations on the microstructure of the unit Rta with respect of the copper A binding site as well as on the composition of fraction T_1 from the tryptic hydrolysate of the *Rapana* hemocyanin "heavy" structural subunit are now in progress.

Acknowledgment

We express our gratitude to the Volkswagen Stiftung (Hannover, Germany) for the financial support by the research grant No I/68 703.

REFERENCES

1. Salvato, B. and Beltramini, M. (1990) *Life Chem. Rep.* 8: 1-47.
2. van Holde, K.E. and Miller, K.I. (1982) *Q. Rev. Biophys.* 15: 1-129.
3. van Holde, K.E., Miller, K.I. and Lang, W.H. (1992) In: *Adv. Comp. Environ. Physiol* (Mangum C.P. ed.). 13: 257-300, Springer Heidelberg
4. Drexel, R., Siegmund, S., Schneider, H.-J., Linzen, B., Gielens, C., Preaux, G., Lontie R., Kellermann, J. and Lottspeich, F. (1987) *Biol. Chem. Hoppe-Seyler* 368: 617-635.
5. Xin, X.-Q., Gielens, C., Witters, R. and Preaux, G. (1990) In: *Invertebrate Dioxygen Carriers* (Preaux G. and Lontie R. ed.), pp. 113-117. Leuven University Press, Leuven.
6. Declercq L., Witters R. and Preaux G. (1990) In: *Invertebrate Dioxygen Carriers* (Preaux G. and Lontie R. ed.), pp. 131-134. Leuven University Press, Leuven.
7. Lang, W.H. (1988) *Biochemistry* 27: 7276-7282.
8. Lang, W.H. and van Holde, K.E. (1991) *Proc. Natl. Acad. Sci. U.S.A.* 88: 244-248.
9. Gielens, C., Verschueren, L.J., Preaux, G. and Lontie, R. (1980) *Eur. J. Biochem.* 103: 463-470.
10. Torensma, R., van der Laan, J.M., van Bruggen, E.F.J., Gielens, C., van Paemel, L., Verschueren, L.J. and Lontie, R. (1980) *FEBS Lett.* 115: 213-215.
11. De Sadeleer, J., Gielens, C., Preaux, G. and Lontie, R. (1983) In: *Structure and Function of Invertebrate Respiratory Proteins. Life Chem. Rep. Suppl.* 1, (Wood, E.J. ed.) pp 133-134.
12. Dijk, J., Brouwer, M., Coert, A. and Gruber, M. (1970) *Biochim. Biophys. Acta* 221: 467 - 479.
13. Herskovits, T.T., Blake, P.A., Gonzalez, J.A., Hamilton, M.G. and Wall, J.S. (1989) *Comp. Biochem. Physiol.* 94B: 415-421.
14. Herskovits, T.T., Gonzalez, J.A. and Hamilton, M.G. (1991) *Comp. Biochem. Physiol.* 98B: 271-278.
15. Lamy, J., Leclerc, M., Sizaret, P.-Y., Lamy, J., Miller, K.I., McParland, R. and van Holde, K.E. (1987) *Biochemistry* 26: 3509-3518.
16. Miller, K.I., van Holde, K.A., Toumadye, A., Johnson, W.C. Jr. and Lamy, J. (1988) *Biochemistry* 27: 7282-7288.

17. Weigle, W.O. (1964) *Immunochemistry*, 1 295-302.

18. Maier, B., Rzepka, R. and Melchers, I. (1989) *Immunobiology* 179: 68-85.

19. Markl, J., el Din, M.N., Winter-Simanowski, S. and Simanowski, U.A. (1991) *Naturwissenschaften* 78: 30-31.

20. Markl, J., Savel-Niemann, A., Wegener-Strake, A., Süling, M., Schneider, A., Gebauer, W. and Harris, J.R. (1991) *Naturwissenschaften* 78: 512-514.

21. Jurincic, C.D., Engelmann, U., Gasch, J. and Klippel, K.F. (1988) *J. Urol.* 139: 723-726.

22. Recker, F. and Rübber, H. (1989) *Urol. Int.* 44: 77-80.

23. Bichler, K.-H., Kleinknecht, S. and Strohmaier, W. L. (1990) *Urol. Int.* 45: 269-283.

24. Boteva, R., Severov, S., Genov, N., Beltramini, M., Filippi, B., Ricchelli, F., Tallandini, L., Pallhuber, M.M., Tognon, G. and Salvato, B. (1991) *Comp. Biochem. Physiol.* 100B: 493-501.

25. Idakieva, K., Severov, S., Svendsen, I., Genov, N., Stoeva, S., Beltramini, M., Tognon, G., Di Muro, P. and Salvato, B. (1993) *Comp. Biochem. Physiol.* 106B: 53-59.

26. Laemmli, U.K. (1970) *Nature* 227: 680-685.

27. Gielens, C., Preaux, G. and Lontie, R. (1975) *Eur. J. Biochem.* 60: 271-280.

28. Stryer, L. (1978) *Ann. Rev. Biochem.* 47: 819-846.

29. Finotto, M., Witters, R., Gielens, C. and Preaux, G. (1988) *Arch. Intern. Physiol. Biochim.* 96: B157.

25

POTENTIAL APPLICATIONS OF PLANT GLYCOHYDROLASES FOR OLIGOSACCHARIDE SYNTHESIS

Jisnuson Svasti,[*][1,2] Chantragan Srisomsap,[1] Rudee Surarit,[1,3]
Em-on Benjavongkulchai,[4] Wipa Suginta,[2] Sauvarat Khunyoshyeng,[1]
Voraratt Champattanachai,[1] Sirinun Nilwarangkoon,[2] and
Saravud Rungvirayudx[2]

[1] Laboratory of Biochemistry
   Chulabhorn Research Institute
[2] Department of Biochemistry
   Faculty of Science, Mahidol University
[3] Department of Physiology and Biochemistry
   Faculty of Dentistry, Mahidol University
[4] Department of Biochemistry
   Faculty of Dentistry
   Chulalongkorn University
   Bangkok, Thailand

ABSTRACT

Reversal of hydrolytic enzymes is an alternative approach for oligosaccharide synthesis. Thus, we have screened for the hydrolytic activity of nine glycohydrolase enzymes from the seeds of some fifty Thai plant species from 17 families, with a view to selecting suitable enzymes to reverse. Good sources of α-D-mannosidase, β-D-glucosidase, β-D-fucosidase, and β-D-galactosidase were found. Some have been purified including a β-D-glucosidase/β-D-fucosidase enzyme from Thai Rosewood (*Dalbergia cochinchinensis* Pierre) and a β-D-galactosidase from Thai Jute (*Hibiscus sabdariffa* L.var *altissima)*, and their kinetic and other properties have been investigated. Synthesis of oligosaccharides by reversal of hydrolytic enzymes was studied by incubation with high concentrations of monosaccharide substrates for 1-20 days at elevated temperature, followed by analysis of products by h.p.l.c. Typically, disaccharide and trisaccharide products could be readily detected confirming synthetic capability. In general, total synthesis at equilibrium was higher (about 35-75%) for α-D-mannosidase than for β-D-glucosidase (15-25%) or for β-D-galactosidase (10-15%).

---

[*] Represents person presenting Paper

Protein Structure–Function Relationship, Edited by Zaidi
and Smith, Plenum Press, New York, 1996

INTRODUCTION

Oligosaccharides perform many useful functions in many biological processes, such as cellular recognition, secretion and clearance of glycoproteins, and modulation of cell growth and differentiation[1]. Chemical synthesis of oligosaccharides is a rather cumbersome process, requiring many protection and deprotection steps, because monosaccharides contain several hydroxyl groups. Enzymatic synthesis is another alternative, which should have some potential due to the inherent stereospecifity and catalytic efficiency of enzyme molecules[2,3]. One approach is to use the enzymes that catalyse synthetic reactions in living systems, such as the glycosyl transferases, but often these enzymes require expensive substrates and are present in lower amounts. We have therefore recently become interested in the possibility of synthesising oligosaccharides by reversing the action of hydrolytic enzymes, such as the glycohydrolases, which are more readily abundant. Typically conditions favouring synthesis would include high monosaccharide concentrations, low water activity, and high temperature. Moreover, since previous studies on enzymatic synthesis of oligosaccharides has utilised commercially available enzymes[3-6], we initiated our studies by searching for new glycohydrolases from Thai plants. This paper describes the results of our search for new enzymes, the purification and properties of some novel glycohydrolases, and some initial studies of enzyme reversal.

MATERIALS AND METHODS

Seed Extraction. Plant seeds were kindly provided by the Field Crops Research Institute, Department of Agriculture, Ministry of Agriculture, Thailand, the ASEAN-Canada Forest Tree Seed Center, Muaklek, Saraburi, Thailand, and the Department of Forestry, Ministry of Agriculture, Thailand. Seeds were surface-sterilized by sodium hypochlorite, imbibed overnight in distilled water and homogenized in 2 vols. of 0.05 M sodium acetate buffer, pH 5.0 containing 1 mM phenylmethylsulfonylfluoride (PMSF) and 5% (w/v) polyvinyl-polypyrrolidone (PVPP). The homogenate was filtered through miracloth and centrifuged at 12,000 g for 30 min, stirred with 25 % (w/v) Dowex 2-X8 for 1 h, and centrifuged again at 12,000 g for 30 min to yield the crude extract.

Assay of Glycohydrolase Activity. The standard reaction mixture[7] contained 50 μl suitably diluted enzyme, 100 μl 0.01 M p-NP-glycoside, 850 μl 0.10 M sodium acetate buffer, pH 4.0 or pH 5.0. The reaction was incubated at 30°C for 10 min and stopped by addition of 2 M Na_2CO_3. p-Nitrophenol released was quantitated spectrophotometrically at 400 nm. One unit of enzyme is defined as the amount of enzyme releasing 1 μmol p-nitrophenol per min at 30°C. Optimum temperature and pH of purified enzyme was studied in the same manner except that pH was varied from 2.5 to 7.5 and temperature varied from 10°C to 80°C. K_m and V_{max} were determined by measurement of activity at varying substrate concentrations, followed by non-linear regression (Michaelis-Menten plot) or linear regression (Lineweaver-Burk plot) using the Enzfitter computer program (Elsevier Biosoft, Cambridge, U.K.). Hydrolysis of β-lactose was measured by following the release of glucose using a glucose oxidase kit, while hydrolysis of glucose disaccharides was followed by h.p.l.c.

Fractionation of Glycohydrolases. Plant crude extracts were generally fractionated with 35-75% ammonium sulphate, and purified by standard techniques such as DEAE-cellulose chromatography, gel filtration on Sephadex, preparative isoelectric focusing (Bio-Rad

Table 1. Selected Data on Glycohydrolase Enzyme Levels in Some Thai Plant Seeds
(μmol/min/g seed)

Botanical Name	α-D-Man	β-D-Glc-NAc	α-D-Gal	β-D-Gal	β-D-Fuc	β-D-Glc
F.Fabaceae (Leguminoceae)						
Glycine max Mer.	0.53	NS	0.21	0.17	NS	NS
Vigna radiata (L.) Wilzcek	0.43	0.11	0.15	0.18	NS	NS
Vigna sinensis L. Saviex Hassk	0.33	NS	0.12	0.28	NS	NS
F.Malvaceae						
Hibiscus cannabinus L.	0.49	0.15	0.27	0.32	NS	NS
Hibiscus sabdariffa L.var altissima	0.82	0.35	0.39	0.92	0.17	0.21
Hibiscus sabdariffa L.var sabdariffa	1.20	0.40	0.40	0.63	0.11	NS
F.Mimosaceae						
Acacia auriculaeformis Cunn.	0.94	0.63	0.32	0.15	NS	1.06
Acacia catechu Willd.	0.56	0.56	0.61	NS	NS	NS
Albizia procera Benth.	1.24	0.14	NS	NS	NS	NS
F.Papilionaceae (Leguminoceae)						
Dalbergia cochinchinensis Pierre.	0.31	0.29	0.17	0.32	26.26	17.40
Dalbergia nigrescens Kurz.	0.42	NS	NS	NS	2.28	1.18
Gliricidia sepium Steud.	0.50	0.16	0.61	0.28	1.06	NS
Sesbania grandiflora (L.) Poir.	0.37	NS	0.20	0.15	ND	NS
F.Tiliaceae						
Corchorus capsularis L.	0.50	0.48	0.20	0.74	0.47	0.30
Corchorus olitorius L.	0.30	0.33	0.18	0.53	0.20	0.12

* ND = Not determined N.S. = Not significant (<0.100 μmol/min/ g seed)

rotofor), or affinity chromatography (e.g. Lactosyl-Sepharose). Purity was checked by electrophoresis on polyacrylamide gels under both native conditions (with activity staining using 4-methyl-umberriferrylglycosides and protein staining with Coomassie blue) and under denaturing conditions (with sodium dodecyl sulphate).

Test of Oligosaccharide Synthesis. Oligosaccharide synthesis by reversal of hydrolytic action was performed by incubating enzyme with the suitable monosaccharide at high concentration (30-80% (w/w)) at optimal pH (~ pH 5) and elevated temperature (50° - 60° C) for 1-14 days. Products were analyzed in a Waters 625 LC h.p.l.c., typically employing an Aminex HPX-87C column at 85° C with refractive index detector (Waters 410), but improved separations could be obtained on a Dionex Carbopac PA-1 column, eluted with 75 mM NaOH and connected to a pulsed amperometric detector (Waters 464) with gold electrode.

RESULTS

Screening of Glycohydrolase Enzymes in Thai Plant Seeds

The hydrolytic activity of nine glycosidase enzymes were tested in the crude extracts from the seeds of some fifty Thai plant species from 17 families. Selected data (Table I) show that six major enzymes were found: α-D-mannosidase, β-D-N-acetylglucosaminidase, α-D-galactosidase, β-D-galactosidase, β-D-fucosidase and β-D-glucosidase. α-D-Mannosidase was the most frequently found enzyme, present in most species of plants, and good sources of this enzyme included *Albizzia procera* Benth, *Hibiscus sabdariffa spp.* and *Acacia catechu* Willd. The *Hibiscus* spp. and *Corcorus* spp. also contained moderate levels of β-galactosidase. However, most interestingly, the levels of β-D-glucosidase and β-D-fucosidase found in *Dalbergia cochinchinensis* Pierre were more than ten-fold higher than those of any other glycohydrolase. Accordingly, our initial studies have focused on the purification and study of β-glucosidase/β-fucosidase from *Dalbergia cochinchinensis* Pierre, and α-mannosidase and β-galactosidase from various species including *Hibiscus sabdariffa* var altissima (Thai Jute*)*, *Hibiscus sabdariffa* var. sabdariffa (Rosella*)* and *Albizzia procera* Benth.

β-Glucosidase/β-Fucosidase from *Dalbergia cochinchinensis* Pierre

Hydrolytic activity towards p-NP-β-D-glucoside and p-NP-β-D-fucoside were co-purified from the seeds of *Dalbergia cochinchinensis* Pierre by 35%-75% ammonium sulphate fractionation, preparative isoelectric focusing and Sephadex G-150 chromatogaphy (Table I). Both hydrolytic activities were found in the same peak in both isoelectric focusing and in gel filtration, corresponding to pI of 5.6 and molecular weight of 330,000 respectively. The final product, obtained in ~ 30 % yield, gave a single band for both activity and protein in non-denaturing gels and also gave one with molecular weight 66,000 on SDS-PAGE. These results suggest that both β-D-glucosidase and β-D-fucosidase activities are due to the same enzyme.

Optimum pH for hydrolysis of both p-NP-β-D-glucoside and p-NP-β-D-fucoside were pH 5.0. Moreover, the purified enzyme also showed activity towards p-NP-β-D-galactoside and p-NP-α-L-arabinoside, which have the same trans equatorial configuration with respect to oxygen at C_1, C_2 and C_3. Electrophoresis on native gels stained with the four methyl-umberriferyl-glycosides gave a single band at the same position with all four substrates suggesting that all four activities are due to the same molecular species.

The kinetic properties of the Thai Rosewood enzyme were also studied using different nitrophenyl-glycosides (Table II). Interestingly, p-NP-β-D-glucoside has both a higher K_m and a higher V_{max} compared to p-NP-β-D-fucoside, raising the question whether the enzyme should be designated a β-D-glucosidase or β-D-fucosidase. Nevertheless, the substantially higher V_{max}/K_m ratio for p-NP-β-D-fucoside compared to the other substrates suggests that the enzyme may be more properly designated a β-D-fucosidase. In terms of natural substrates, we have been able to show that the enzyme can hydrolyse the glucose disaccharides sophorose [β1-2], lamaribiose [β1-3] and gentiobiose [β1-6], but shows much slower release of glucose from cellobiose [β1-4], cellotriose, and cellotetraose. However, the enzyme seems unable to hydrolyse the cyanogenic glucoside linamarin or the [β1-3] glucose polymer laminarin. Further studies will be necessary in identifying the natural substrates present in Thai Rosewood, and defining the biological role of the β-D-glucosidase/β-D-fucosidase enzyme.

The effect of various compounds was tested on the hydrolysis of p-NP-β-D-glucoside and p-NP-β-D-fucoside (Table III). δ-Gluconolactone at 1mM and 0.1 mM caused > 90% inhibition and <50% inhibition respectively, which is not surprising since reactions of

Table 2. Kinetic Properties of Thai Rosewood β-Glucosidase/ β-Fucosidase

Substrate	K_m mM	V_{max} U/ml	V_{max}/K_m
p-NP-β-D-glucoside	5.39 ± 0.09	13.49 ± 0.09	2.28
p-NP-β-D-fucoside	0.54 ± 0.03	6.64 ± 0.10	12.3
p-NP-β-D-galactoside	4.50 ± 0.64	1.69 ± 0.02	0.38
p-NP-α-L-arabinoside	1.01 ± 0.03	0.26 ± 0.01	0.28
o-NP-β-D-glucoside	1.45 ± 0.14	2.39 ± 0.08	1.65
o-NP-β-D-fucoside	0.72 ± 0.04	1.32 ± 0.02	1.83

Table 3. Effect of Substances on Thai Rosewood β-Glucosidase/ β-Fucosidase

Substance	Final conc	% Activity remaining	
		β-D-Glucosidase	β-D-Fucosidase
Control	-	100.0	100.0
FeSO$_4$	1mM	112.9	99.9
FeCl$_3$	1mM	104.2	113.6
CuSO$_4$	1mM	74.3	86.3
CaCl$_2$	1mM	102.1	93.8
MnCl$_2$	1mM	94.8	91.5
KCN	1mM	106.7	104.5
ZnSO$_4$	1mM	111.2	107.6
MgCl$_2$	1mM	111.3	112.1
HgCl$_2$	1mM	*0.3*	*3.1*
HgCl$_2$	10 μM	*9.2*	*10.8*
HgCl$_2$	0.1 μM	*10.7*	*9.4*
Iodoacetate	1mM	84.2	98.3
NaF	1mM	104.4	91.8
CdOAc	1mM	81.1	99.3
EDTA	1mM	93.9	99.3
p-CMB	1mM	*11.7*	*10.2*
p-CMB	10 μM	*15.2*	*13.2*
p-CMB	1 μM	*25.2*	*23.7*
Gluconolactone	1mM	*8.8*	*6.8*
Gluconolactone	0.1 mM	*50.3*	*69.8*

Effect of substances were tested on the hydrolytic activity towards
pNP-β-D-glucoside (2mM) or pNP-β- D-fucoside (1mM) in 0.1 M sodium
acetate buffer, pH 5.0 at 30°C. p-CMB = p-chloromercuribenzoate.

glycohydrolases often involve lactones as transition state analogs[8]. However, much stronger inhibition was observed with the mercuric compounds, where HgCl$_2$ at 10^{-7} M caused 90% inhibition and p-chloromercuribenzoate (p-CMB) at 10^{-6} M caused 75% inhibition. Although mercuric compounds generally react with sulphydryl groups, it is also possible that they may chelate catalytically active aspartate or glutamate residues. In this connection, it has been shown that the activity of Thai Rosewood β-D-glucosidase/β-D-fucosidase can be abolished by reaction with 1 molecule of conduritol B epoxide per active site, suggesting that acidic amino acid(s) are essential for catalysis (R. Surarit, personal communication). Further studies are in progress in determining the primary structure of the enzyme, and defining the amino acids involved in the active site.

The Thai Rosewood β-D-glucosidase/β-D-fucosidase enzyme may be reversed by incubation with high concentrations of monosaccharides at elevated temperature. Best yields were obtained with D-glucose as substrate. Thus after incubation with 50% (w/w) D-glucose as substrate, disaccharides and trisaccharides (with total combined yields of 15-25%) could be readily detected by h.p.l.c. using an Aminex HPX-87C column with refractive index detection (Figure 1A). Much poorer yields were obtained with 50% (w/w) D-fucose as substrate, but novel oligosaccharides could also be produced with mixtures of D-glucose and D-fucose. These and other studies are described elsewhere[9].

β-Galactosidase and α-Mannosidase from Plants

Further studies were carried out on the *Hibiscus sabdariffa spp.* since they contained relatively high levels of β-D-galactosidase and α-D-mannosidase (Table I). Thus, β-D-galactosi-

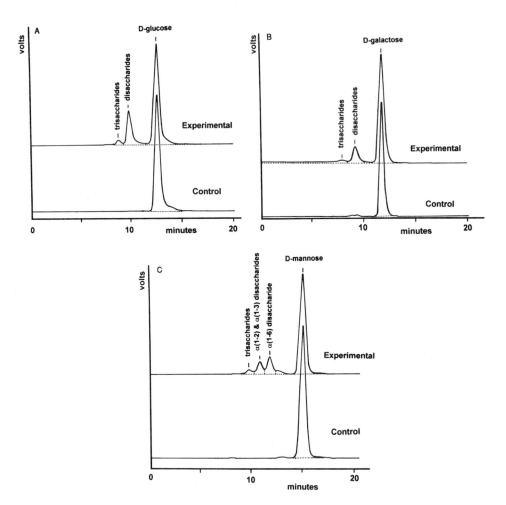

Figure 1. Fractionation of the oligosaccharide synthesis products obtained by reverse hydrolysis of glycohydrolases from various plant seeds. H.p.l.c separation was performed on an Aminex HPX87-C column at 85°C using a Waters 625LC instrument equipped with refractive index detector. The stated monosaccharide at 50% (w/w) was incubated at pH 5, 50°C for 7 days in the presence of enzyme (*Experimental*) or in the absence of enzyme (*Control*). *A*: β-glucosidase/ β-fucosidase from *Dalbergia cochinchinensis* Pierre incubated with D-glucose; *B*: β-galactosidase from *Hibiscus sabdariffa* var altissima incubated with D-galactose; *C*: α-mannosidase from *Albizzia procera* Benth. incubated with D-mannose.

dase was purified from Thai Jute (*Hibiscus sabdariffa* var altissima) by ammonium sulphate fractionation, DEAE-cellulose chromatography (to remove α-D-mannosidase), Sephadex G-100 gel filtration (to remove α-D-galactosidase), Lactosyl-Sepharose affinity chromatography, and a second DEAE-cellulose chromatography step. The final product was purified 870 fold with a yield of 13 %, and gave a single band of molecular weight 66,000 on SDS-PAGE. The purified enzyme had an pH optimum of 4.0 and a temperature optimum of 55°C similar to other plant β-galactosidases[10,11]. Kinetic studies (Table IV) indicate that Thai Jute β-galactosidase showed

Table 4. Kinetic Properties of Thai Jute β-Galactosidase

Substrate	K_m (mM)	V_{max} (nmol/min)	V_{max}/K_m
p-NP-β-D-Galactoside	0.80±0.02	63.5±0.4	79.38
o-NP-β-D-Galactoside	12.8±0.51	16.9±0.3	1.32
β-Lactose	84.7±1.28	6.0±0.4	0.07

Reactions were performed in 0.1 M sodium acetate buffer, 4.0 at 30°C ;
values are mean ±S.D.

preference for p-NP-β-D-galactoside as the best substrate, followed by o-NP-β-D-galactoside, with β-lactose being a rather poor substrate. The enzyme also showed 98% inhibition by 1 mM HgCl$_2$ and 95% inhibition by 5 mM galactano-1,4-lactone.

Oligosaccharide synthesis studies were also performed by incubation of Thai Jute β-D-galactosidase with 33% (w/w) D-galactose. Disaccharide products were also obtained (Fig. 1B), but yields tended to be rather low (10-15%). Accordingly, studies were also carried

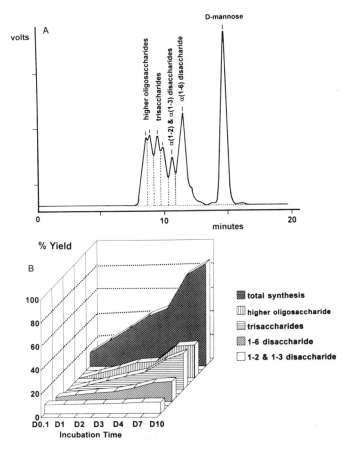

Figure 2. Oligosaccharide synthesis by commercial α-mannosidase from Jack Bean (*Canavalis ensiformis*) incubated with 50% (w/w) D-mannose at pH 4, 50°C for various times. Reaction products were analysed h.p.l.c. on an Aminex HPX87-C column as described in Fig.1. *A*: H.p.l.c. profile at day 7; *B*: percent yield of each product as a function of time.

on the synthetic capability of α-D-mannosidase, which is found in *Hibiscus sabdariffa spp.* and also in many other species, such as *Albizzia procera* Benth. In general, synthesis yields with α-mannosidases (Fig. 1C) tended to be rather good (over 30%), with trisaccharides and usually also higher oligosaccharides being produced. However, α-mannosidases tend to be more difficult to purify, and attempts are still being made to purify α-mannosidases from *Hibiscus sabdariffa spp.* and from *Albizzia procera* Benth. to homogeneity, so as to permit more detailed studies of their properties. Meanwhile, some synthesis studies have been carried out with commercial α-mannosidase from Jack Bean (*Canavalis ensiformis*). Incubation of this enzyme with 50% (w/w) D-mannose at 50°C, gave excellent yields of synthesis (Figures 2A and 2B). The results obtained not only showed high levels of total synthesis (60-80%), but also high levels of oligosaccharides larger than trisaccharides, comparable or greater to those previously reported using 83 % (w/w) mannose at 75°C[12]. Further studies will be required to separate and characterise the structure of these larger oligosaccharides.

CONCLUSIONS

This paper has described the identification of several glycohydrolase enzymes in the seeds of plants indigenous to Thailand. Some have been purified and studied in some detail in terms of kinetic and other properties, most notably a β-glucosidase/ β-fucosidase from Thai Rosewood, while others have proved to be more difficult to purify. Studies have also shown that under suitable conditions of high monosaccharide concentration and elevated temperature, the hydrolytic reactions of glycohydrolases can be reversed leading to a net synthesis of oligosaccharides. High monosaccharide concentrations favour reversal of the hydrolysis, not only through the law of mass action, but they also tend to lower the water activity, since in a 50% (w/w) monosaccharide solution, the ratio of the concentration of -OH groups from the sugar to the concentration of -OH groups from water molecules will be as high as 1:2. Moreover, high sugar concentrations help to stabilise enzymes, so that prolonged incubations at elevated temperatures (usually 50°C) were possible. Overall, α-D-mannosidases tend to show higher levels of synthesis (30-75%), while β-D-glucosidase/ β-D-fucosidase shows more moderate levels of synthesis (20-35%), while β-D-galactosidase shows the poorest synthesis (10-15%). While disaccharide products may be identified in a relatively simple manner by comparison with commercial standards, the characterisation of larger oligosaccharides will require their separation in preparative amounts to enable structural determination. Although much more progress is clearly required, the present results suggest that a study of the glycohydrolases and their reversal offers an interesting alternative approach for the synthesis of oligosaccharides.

Acknowledgments

This research was supported by the Chulabhorn Research Institute, the National Research Council of Thailand, and the Thai-UK Biotechnology Programme.

REFERENCES

1. Welply, J.K. (1989) *Trends Biotechnol.* 7: 5-10.
2. Nilsson, K.G.I. (1988) *Trends Biotechnol.* 6: 256-264.
3. Bucke, C. and Rastall, R.A. (1990) *Chemistry in Britain* pp. 675-678
4. Nilsson, K.G.I. (1987) *Carbohydrate Res.* 167: 90-103.
5. Ajisaka, K., Nishida, H. and Fujimoto, H. (1987) *Biotechnology Letters* 9: 387-392.

6. Rastall, R.A., Rees, N.M., Wait, R., Adlard, M.W. and Bucke, C. (1993) *Enzyme Microb. Technol.* 14: 53-57.

7. Montreuil, J., Bouquelet, S., Debray, H., Fournet, B., Spik, G., and Strecker, G. (1986) In: *Glycoproteins* (Chaplin, M.F. and Kennedy, F., eds.) pp 143-204, IRL Press, Oxford.

8. Esen, A. (ed.) (1993) β-*Glucosidases: biochemistry and molecular biology*. American Chemical Society, Washington D.C.

9. Srisomsap, C. Khunyoshyeng, S., Boonpuan, K., Sawangareetrakul, P., Surarit, R. and Svasti, J. (1995) Poster abstract, *Proceedings, Fourth International Symposium on Protein Structure Function Relationship*, Karachi, Pakistan, 20-25 January 1995.

10. Steers, E., Jr. and Cuatrecasas, P. (1974) *Methods Enzymol.* 34: 350-358.

11. Dey, P.M. (1984) In: *Adv. Enzymol.* (Meister, A., ed.) Vol. 56: pp 141-249, John Wiley & Sons Inc., New York.

12. Johannsson, E., Hedbys, L., Larsson, P.-E., Mosbach, K., Gunnarsson, A., and Svensson, S. (1986) *Biotechnol. Letters* 8: 421-424.

26

TWO-DIMENSIONAL ELECTROPHORESIS AND AUTOMATED MICROSEQUENCING

Introduction and Overview

Brigitte Wittmann-Liebold[*][1] and Peter Jungblut[2]

[1] Department of Protein Chemistry
Max-Delbrück-Centrum für Molekulare Medizin
Robert-Rössle-Str. 10, D-13125 Berlin-Buch, Germany
[2] Wittmann Institute of Technology and Analysis of Biomolecules
Potsdamer Str.18a, D-14513 Teltow, Germany

INTRODUCTION

In organelles, cells and tissues of all organisms complex functions and metabolic reactions are to be maintained by proteins within and at the surface of each compartment. The number of proteins occuring in a biological compartment corresponds to the complexity of the functions it has to fulfill. Ribosomes as relatively high specialized compartments contain in the case of E.coli 54 different proteins, whereas estimations for the number of expressed genes in a typical human cell are in the range of about 5000. The human genome contains about 50000-100000 genes. For the registration and characterization of the components of these complex systems high resolution methods are necessary. The idea to characterize all components of a biological compartment became reality for the ribosomes when methods became available for the resolution of about 50 different proteins. These methods were first two-dimensional electrophoresis (2-DE)[1] and later also HPLC[2]. Today, improved techniques of 2-DE play an important role to investigate the functional part of the genes, the proteins, within the human genome project. Furthermore changes in protein composition may be elucidated by subtractive analyses in different biological situations for example diseases or during differentiation.

RESOLUTION

Since the first description of a 2-DE protein separation by Smithies and Poulik[3] the resolution of 2-DE was increased from 15 protein spots up to about 10000 protein spots in

---

[*] Represents person presenting Paper

1995 as described by Klose[4]. One milestone in this development was the introduction of polyacrylamide into 2-D separation by Raymond[5]. Another milestone was the combination of isoelectric focusing (IEF) with sodium dodecylsulfate polyacrylamide gel electrophoresis (SDS-PAGE) in 1975[6-8]. IEF as well as SDS-PAGE are high resolution electrophoretic methods separating up to 100 components per gel and the combination of these methods resulted already in the first attempts in a resolution of about thousand proteins. Attempts to perform native 2-DE with a high resolution comparable with 2-DE under denaturating conditions resulted in 2-DE patterns with low reproducibility, smear and not very distinct protein spots. A prerequisite to achieve high resolutions for cells and tissues is to reduce all S-S bridges by addition of reducing agents like dithiothreitol (70 mM) and to cleave protein complexes completely into their polypeptide chains by preparation in 9 M urea and 3 % SDS.

Major factors for the further increase in resolution were the increase of sensitivity, the use of complex ampholyte mixtures, the decrease of the thickness of the gels down to 1.5-0.5 mm thick gels and an increase of the size of the gels up to 30 x 40 cm.

REPRODUCIBILITY

Whereas such an outstanding resolution as obtained by 2-DE may not nearly be obtained by other protein analytical methods, reproducibility is a main problem for 2-DE separations. Causes are that the technological stage of development is not comparable with HPLC, the complexity of the separation problem is much higher than in HPLC separations, the gel matrix is flexible in all directions, cathodic drift effects may occur, if there are any impurities in the gel system of IEF, and the handling of the gels needs experience.

The reproducibility of protein spot silver staining intensity showed coefficients of variability in the range of 15 % with a maximum value for one spot of 35%. These data were obtained for a large gel 2-DE technique using carrier ampholytes in the first dimension, comparing 8 gels of one protein sample. The gels were produced by a student with one month 2-DE experience. Comparable data of Immobiline-2-DE gels are missing.

Reproducibility was improved by simplification of the procedure, for example omission of the stacking gel and the use of a continuous separation gel in the second dimension and use of ready-made gel solutions in IEF and SDS-PAGE, or in the case of immobiline gels[9] by the introduction of ready-made gels. First attempts were started for an automatization of the whole 2-DE procedure.[10]

Under optimal conditions the reproducibility of protein spot position is not a main problem. For eight gels the coefficients of variability of the distance between two spots were in the range of 2% with a maximum value of 3.4%. Visual and computer-assisted assignments of spots from different gels are in most cases unequivocal. Even inter-laboratory comparisons are successful[11] by comparing the identified spots of human heart 2-DE patterns produced in Berlin, Munich and London.[12]

SENSITIVITY

Proteins are detected on the gels by Copper staining, Coomassie Brilliant Blue R, Coomassie Brilliant Blue G, silver staining or by radiochemical methods with increasing sensitivity. The detection limit for Coomassie Brilliant Blue R is in the range of 0.1 μg and that of silver staining in the range of 1-10 ng. The sensitivity of the detection by radioactive labelling depends on the obtained specific radioactivity of the proteins. Proteins occuring in the cell with only 7 molecules per cell could be visualized on fluorograms of gels.[13]

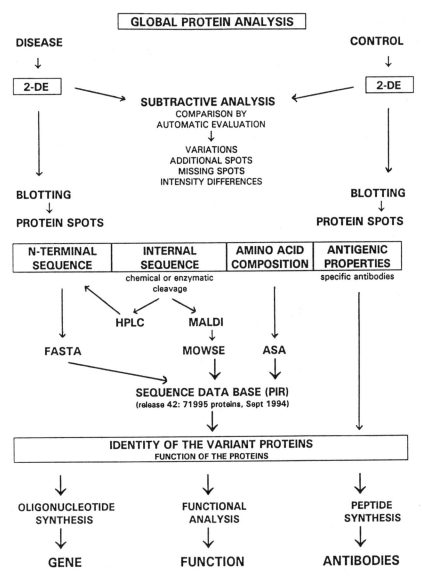

Figure 1. Overview of the proceeding for a global protein analysis. PIR, Protein Information Resource.

Proteins may be visualized on membranes like polyvinylidenedifluoride membranes or modified glass fiber membranes by Poinceau S, amido black, Coomassie Blue R, sulphorhodamine, or with much higher sensitivity by immunostaining.

SUBTRACTIVE ANALYSIS

In contrast to the classical proceeding, where the needle is looked for in the haycock, subtractive analyses[14] allow the hypothesis-free elucidation of biological changes-associated protein variations for example disease-associated proteins in a global approach as shown in Fig. 1. These proteins can be proteinchemically analysed by sequencing, amino acid analysis or by mass spectrometry and identified by protein database searches[15-19]. A tumor-associated protein could be identified as aldosereductase[18]. From 315 amino acids 112 amino acids could be sequenced and from these data not only the protein name but also the right isoform, an embryonic one of aldose reductase could be assigned to the variant protein. For dilated cardiomyopathy four proteins were found disease-associated changed in spot intensity on 2-DE gels: kreatine kinase, malate dehydrogenase, alpha-crystallin B, and heat-shock protein 27.[19,20]

EVALUATION OF 2-DE GELS

2-DE gels may be evaluated visually or automatically with the help of a computer. Visual evaluation is very precise, differences in protein spot staining intensity of 10 % may be detected reliably[21] and visually finding of corresponding spots in different gels is much easier than automatically matching. It is not dependent on expensive hard and software, but it becomes more and more laborious the more protein spots are investigated and the larger the gels are. For the comparison of 20 gels with the size of 30 x 40 cm it is nearly impossible to compare the intensity of a single spot in the whole data set, whereas by automatic evaluation the interesting spot may be zoomed out and compared on twenty windows on the monitor at the same time allowing to control the calculated intensity differences. Highest gel quality, optimized spot detection parameters and the possibility of interactive changes are necessary to use automatic evaluation. The process may be divided into five steps: digitalization of the data after scanning, background and smear corrections, spot detection, quantification and comparison of different gels (matching). Different automatic evaluation systems are commercially obtainable. The sensitivity of the spot detection can not be increased by the change from visual to automatic evaluation without inclusion of artefacts[22].

Criteria for the efficiency of an automated evaluation system are the local resolution, the dynamic range of optical density, time for each software procedure, quality of the spot detection and matching procedure, possibilities of interactive operations and the price.

2-DE DATABASES

The immense amount of data obtained by subtractive analyses, identification and characterization of proteins on 2-DE gels are stored in protein 2-DE databases, which are published once a year in Electrophoresis and some of them are accessible in the Internet using the World Wide Web technology. For example a 2-DE database of human heart proteins was built up containing 3239 proteins from which 66 were annotated until now[11]. Annotation criteria used in this database are dilated cardiomyopathy-associated proteins, N-terminal sequence, internal sequence, amino acid composition, protein name, molecular mass, isoelectric point, and reference for identification. Despite the fact that the myocardial proteins of Baker et al. 1992 were prepared by a different method and also another 2-DE gel system and gel size was used, 40 of the proteins characterized in this database could be easily transferred to our database.

INTRODUCTION

The human body contains proteins and peptides to 17%. These biopolymers are of general interest since they are involved in all essential processes in the living cell, such as transport within the body fluids and transfer of molecules through membranes, signal transduction, the transcription and translation of the genetic information and the immune responses. Proteins (enzymes) and peptides (hormones) control and regulate the biosynthetic pathways. Needless to say that the correlation of their structure and function is of great importance for an understanding of all these processes on the molecular level.

Proteins are complicated structures, they form three-dimensional folds of the linear polypeptide chain in a manner which is specific and unique to each of the various proteins. The specific function is fully imprinted in the amino acid sequence, yet how the folding of the linear sequence into the higher order structure is accomplished is still one of the great obstacles in biochemistry. Firstly, it is necessary to investigate the primary structure (amino acid sequence), secondly, its secondary structural elements (alpha helices, extended structures, turns or coil) have to be examined and thirdly, the protein's fold to the specific higher order structure (the tertiary structure) has to be studied. Finally, its interactions with other molecules (quarternary structure) have to be known. Hence, the first approach towards structure elucidation is to determine the polypeptide's amino acid sequence. Although X-ray structure analysis of protein crystals or NMR-studies on proteins in solutions became extraordinary powerful tools in determining its tertiary structure this would not be possible without knowing the amino acid sequence. Furthermore, many proteins are constituents of complexes composed from proteins and RNA or DNA, they may be composed from or bound to lipids and glycosides. Any type of modifications, such as phosphorylation, methylation, acetylation or others add to the character of the protein or might alter its properties. Therefore, it is not only important to know the sequence and tertiary fold of the protein but also ligands, cofactors and other molecules which constitute a complex and to evaluate which are the binding sites and binding forces. From this it becomes clear that it is not simple to determine the structure of a protein and to explain its functional role on the molecular level. Many complicated methods must be applied to study the implications in the biological processes. The methods span a wide variety of techniques, such as physical, physico-chemical, chemical and biological methods.

PRIMARY STRUCTURE ANALYSIS

Several approaches are in use to determine the polypeptide's primary structure (see Table 1). Partial amino acid sequencing leads to valuable sequence information which can be used for localization and sequence determination of the protein's gene and the final deduction of the amino acid sequence from the gene.

However, it should be pointed out that this information is not sufficient to fully describe the primary structure of a protein! Whether it is really starting and ending at the deduced sequence, e.g. with or without expression of a N-terminal methionine, and whether the reading frame is correctly assigned may not be certain, further modifications, e.g. the sites of glycosylations or phosphorylations cannot be deduced from the nucleotide sequence. The protein has to be isolated and purified, and its primary structure must be characterized by direct protein investigation in combination with mass spectrometry.

In recent years mass spectrometry became a valuable tool in protein and peptide structure analysis. As for direct amino acid sequencing the proteins must be isolated to high purity. Depending on the strategy applied, a different purity grade must be achieved for successful mass spectrometry of the polypeptides. Salts usually disturb largely the measurements, which is one of the severe constraints in this technique.

THE EDMAN CHEMISTRY

Most commonly microsequencing of proteins and peptides is done in automates based on the Edman chemistry in so-called sequencers[23]. The sequences are established by

Table 1. Stategies for Sequence Determination of Peptides and Proteins

1. Genetic approach	Partial N-terminal or internal amino acid sequencing
	Selection of suitable oligonucleotide probes
	Hybridization with the genomic DNA
	Localization of the protein's gene
	Cloning and sequencing of the gene
	Deduction of the complete amino acid sequence from the gene sequence
	Check of the correct open reading frame
2. Sequence analysis by direct amino acid sequencing	N-terminal sequence analysis
	Protein cleavages to generate peptide fragments for internal sequencing
	Purification of peptides
	N-terminal sequence analysis of the peptides
	Additional fragmentation employing other enzymes
	Separation and sequence analysis of these sets of peptides
	Sequence alignment
	Determination of S-S-bridges
	Check for modifications
3. Sequence analysis by mass spectrometry	Determination of the molecular masses by MALDI MS (proteins and peptides)
	Mass determination of peptide mixtures by ES-MS
	Sequence analysis of small peptides (FAB-MS/MS)

a stepwise chemical procedure degrading the amino acids of the polypeptide chain successively from the amino-terminal end. Sequencers perform the degradation with higher better repetitive yields and with higher sensitivity than is possible by any of the manual procedures. This is due to optimal exclusion of oxygen, good preservation of all chemicals in the machine, long durability and performance of all sequencer parts.[24]

The Edman chemistry consists of three different reactions, namely the coupling of the Edman reagent phenylisothiocyanate (PITC) to the alpha-amino group(s) of the protein, the cleavage of the first residue from the remaining intact polypeptide chain as anilinothiazolinone (ATZ-amino acid) and the conversion of the ATZ-amino acid into the more stable phenylthiohydantoin derivative (PTH-amino acid) within the converter[25]. All these reactions are performed in a fully automatic mode and the cycles are repeated many times in order to read extended sequences. The released PTH-amino acids are identified on-line by chromatography in a microbore-HPLC-system[26]. Typically, between 20 to 200 picomole of sample are applied to the reactor of the sequencer and depending on the amount, the quality of the sample and the individual sequence about 50 residues can be identified. Under optimal technical conditions the machine can reach 96 to 99% repetitive yields which means that at cycle 50 about 20% of the correct amino acid residue in the 50th position can be detected together with up to 80% of the previous PTH-amino acids (the so-called overlap). The sequencers are equiped with a comfortable software allowing for a quantitative determination of each PTH-amino acid derivative in comparison to a standard PTH-amino acid mixture. Furthermore, the software package enables substraction of background and/or the previous cycle.

Most sequencers in use employ aqueous trimethylamine as base for the coupling with PITC dissolved in heptane. After coupling the excess of reagent and base have to be removed completely by repeatedly washing with organic solvents, e.g. heptane and ethylacetate. The cleavage is done in anhydrous TFA in order to avoid any hydrolytic cleavage of internal peptide bonds of the polypeptide chain. This would have a drastic influence on the further degradation of the sample since new N-terminal groups may be raised, e.g. in TFA containing water. This causes severe background contaminations by newly released other PTH-amino acids. The thiazolinone is transferred into the converter with butylchloride and the conversion is done in 25% aqueous TFA. Thereafter, the PTH-amino acid is dissolved in a suitable solvent mixture for automatic injection into the column of the HPLC system. Sequencers are equipped with a gradient HPLC system or employ isocratic elution of the PTH-amono acid. Both techniques have advantages and disadvantages.

Important for a high performance of the degradation is the use of high quality reagents and solvents (sequencer grade chemicals), which must be freed of aldehydes, peroxides and radicals. Without this care the amino-terminal amino acids are blocked during the cycles and the degradation may be stopped.

The construction of highly sensitive machines[27,28] have over come some of these problems and have provided various possibilities for detection of the released amino acid derivatives.

REFERENCES

1. Kaltschmidt, E., and Wittmann, H.G. (1970) Anal.Biochem. 36: 401-412.
2. Kamp, R.M., Bosserhoff, A., Kamp, A. and Wittmann-Liebold, B. (1984) J.Chromatogr. 317: 181-192.
3. Smithies, O. and Poulik, M.D. (1956) Nature, 177: 1033.
4. Klose, J. (1995) Electrophoresis 16: (in press).
5. Raymond, S. (1964) Annals N.Y. Acad. Sci. 121: 350-365.
6. O'Farrell, P.H. (1975) J.Biol.Chem. 250: 4007-4021.
7. Klose, J. (1975) Humangenetik 26: 211-234.
8. Scheele, G.A. (1975) J.Biol.Chem. 250: 5375-5385.
9. Görg, A., Postel, W. and Günther, S. (1988) Electrophoresis, 9: 531-546.
10. Nishine, T., Nakamuro, S., Hazama, M., and Nokihara, K. (1991) Anal.Sci. 7: 285-288.
11. Jungblut, P., Otto, A., Zeindl-Eberhart, E., Pleißner, K.-P., Knecht, M., Regitz-Zagrosek, V., Fleck, E., and Wittmann-Liebold, B. (1994) Electrophoresis 15: 685-707.
12. Baker, C.S., Corbett, J.M., May, A.J., Yacoub, M.H. and Dunn, M.J. (1992) Electrophoresis, 13: 723-726.
13. Jungblut, P., Prehm, J. and Klose, J. (1987) Biol.Chem. Hoppe-Seyler 367: 439.
14. Aebersold, R. and Laevitt, J.(1990) Electrophoresis 11: 517-527.
15. Vandekerckhove, J., Bauw, G., Puype, M., Van Damme, J. and Van Montagu, M. (1985) Eur.J.Biochem. 152: 9-19.
16. Aebersold, R.H., Teplow, D.B., Hood, L.E. and Kent, S.B. (1986) J. Biol.Chem. 261: 4229-4238.
17. Pappin, D.J.C., Hojrup, P. and Bleasby, A.J. (1993) Current Biology, 3: 327-332.
18. Zeindl-Eberhart, E., Jungblut, P., Otto, A. and Rabes, H.M. (1994) J.Biol.Chem. 269: 14589-14594.
19. Knecht, M., Regitz-Zagrosek, V., Pleissner, K.-P., Jungblut, P., Hildebrandt, A. and Fleck, E. (1994) Europ. Heart J. 15: Supplement D: 37-44.
20. Otto, A., Benndorf, R., Wittmann-Liebold, B. and Jungblut, P. (1994) J.Prot.Chem. 13: 478-480.
21. Jungblut, P., Schneider, W. and Klose, J. (1984) In: Electrophoresis'84, (Neuhoff, V. ed.), pp. 301-303, Verlag Chemie GmbH, Weinheim.
22. Prehm, J., Jungblut, P. and Klose, J. (1987) Electrophoresis 8: 562-572.
23. Edman, P. and Begg, G. (1967) Eur. J. Biochem. 1, 80.
24. Wittmann-Liebold, B. (1992) Pure and Appl. Chem. 64: 537-543.
25. Wittmann-Liebold, B., Graffunder, H. and Kohls, H. (1976) Anal. Biochem. 75, 621-633.
26. Wittmann-Liebold, B. and Ashman, K. (1985) In: Methods of Protein Chemistry, (Tschesche, H. ed.), pp. 303, Walter deGruyter Verlag, Berlin.
27. Wittmann-Liebold, B. (1986) In: Methods in Protein Microcharacterization, (Shively, J.E. ed.), pp. 249-277, The Humana Press, Inc.
28. Hewick, R.M., Hunkapiller, M.W., Hood, L.E. and Dreyer, W.J. (1981) J. Biol. Chem. 256: 7990-7997.

Posters

CHARACTERIZATION OF PROTEINS FROM MANGROVE
Avicenna marina

Atiya Abbasi,[1] Zafar H. Zaidi,[1] Syed Abid Ali,[1] S. M. Saifullah,[2] and Saima Saeed[1]

[1] HEJ Research Institute of Chemistry
University of Karachi, Karachi, Pakistan
[2] Department of Botany
University of Karachi, Karachi, Pakistan

Mangroves, the marine forests of tropical and sub tropical inter-tidal regions are woody plants with associated flora and fauna that utilize a coastal saline depositional land with typical anaerobic soil.

The economic importance of mangroves have long been recognized. They serve as feed substances, timber or medicine on the one hand and as reservoir and refuge for animal and plant species on the other. Today most of the world shrimp fisheries are related to mangrove ecosystems. Additionally they may serve as marine parks and mariculture ponds.

The mangrove forests of Pakistan are among the largest ones lying on the coastal borders of Sind and Baluchistan. Atleast eight species have been identified and all are invariably present in Sind whereas five species have been found in Baluchistan. Among the identified species *Avicennia marina* (formerly known as *A. officinalis*) constitutes for approximately 95% of the forest population. Despite the economic importance associated with these marine forests and the well known problem of salinity in Pakistan very little attempt has been made for utilization of these natural resources.

We have undertaken some preliminary studies on the proteins of *Avicenna marina*. Leaves were extracted with Tris buffer pH 7.2 and 8M urea. The extracts were analysed for total proteins as well and amino acids. The crude extracts were also analysed using native and SDS polyacrylamide gel electrophoresis, isoelectric focussing and gel filtration. Results will be discussed.

---

*Represents person who presented the poster.

INVESTIGATION OF MODE OF INTERACTION OF MUSCARINIC ACETYLCHOLINE RECEPTORS WITH G-PROTEINS

Shahram Khademi and Ramin Adli

Department of Medicinal Chemistry
Shaheed Beheshti University of Medical Sciences, Tehran, Iran

The muscarinic acetylcholine receptors (mAchR) mediate a variety of cellular responses through the action of G-proteins. The cloning and functional expression of five mammalian mAchR genes has revealed that M1, M3 and M5 primary coupled to stimulation of phosphoinositide (PI) turnover, whereas M2 and M4 are strongly linked to inhibition of adenylate cyclase. There are peptides which could activate G-proteins without receptor activation, such as GF-P14 (P14) and Mastoparan (MP). The investigation of structural similarity between these peptides and various mAchR, could provide valuable information regarding the mode of receptor/G-protein interaction.

In the present study, the amino acid sequences of mAchRs has been aligned with P14 and MP by FASTA program using PAM-250 matrix.

The results of this study indicate that P14 has significant similarity to C-terminal third cytoplasmic loop (i3) and second cytoplasmic loop (i2) of M1, M3 and M5 receptors, while P14 shows only a similarity to N-terminal of i3 in M2, M4 receptors. On the other hand, MP shows a similarity to both C-terminal and N-terminal of i3 as well as i2 and C-terminal domain of all mAchRs. Studies have demonstrated that the third cytoplasmic loop of the mAchR is important for the receptor/G-protein interaction. This study shows that N-terminal and C-terminal of i3 of mAchRs are responsible for selective coupling with different effector systems, probably through interaction with distinct G-proteins.

PLASMA MEMBRANE GLYCOPROTEINS OF RABBIT CORNEAL EPITHELIAL CELLS: Potential Role of Their Oligosaccharides with Terminal Galactose Residues in Cell-Cell and Cell-Matrix Interactions

N. Ahmed,[1,2] S. I. Ahmad,[1] M. B. Raizman,[1] and N. Panjwani[1]

[1] New England Eye Center
Tufts University School of Medicine
Boston, Massachusetts
[2] Department of Biochemistry
University of Karachi, Karachi, Pakistan

Plasma membrane glycoproteins of corneal epithelium are believed to play a role in corneal epithelial cell migration, mitosis and wound healing by influencing cell-cell and cell-matrix interactions and serving as receptors for growth factors. We have previously demonstrated that levels of two plasma membrane sialoglycoproteins (28 and 21K) of corneal epithelium are elevated during cell migration in culture. The purpose of the present study was to establish whether levels of corneal epithelial cell surface glycoproteins with terminal galactose residues are altered during cell migration and mitosis in culture. The plasma membrane glycoproteins of five preparations of sparse (S) and confluent (C) corneal epithelial cell cultures were radiolabeled with NaB^3H_4 after treatment with galactose oxidase (37°C for 1 hour). Appropriate controls were also labeled with NaB^3H_4 only omitting the treatment with galactose oxidase. The radiolabeled glycoproteins were extracted in 0.5% CHAPS buffer containing various protease inhibitors. The solubilized glycoproteins were analyzed by sodium dodecyl sulfate polyacrylamide gel electrophoresis and fluorography. Analyses of fluorography patterns of S-cells and C-cells revealed 10 components ranging in apparent molecular weight from 28K to 220K. Comparison of the S-cells and C-cells patterns revealed that: (1) one component (MW 220K) was present in at least 20 times greater amount in C-cells compared to S-cells; (2) two components (33K and 28K) were detected only in S-cells; and (3) one component (42K) was present in at least 4 times greater amount in S-cells compared to C-cells. No radioactive components were detected in the fluorographs of cells treated with NaB^3H_4 along. This study has demonstrated that relative amounts of four plasma membrane glycoproteins of corneal epithelium are altered during cell migration and mitosis in culture. Further characterization of these glycoproteins should contribute to a better understanding of the factors which mediate corneal epithelial cell-cell and cell-matrix interactions.

ACKNOWLEDGEMENT

Sponsored by NIH grants EY07088 and EY09349.

SUBUNIT HETEROGENEITY IN ARTHROPOD HEMOCYANINS:
Studies on Scorpion (*Buthus sindicus and Androctonus australis*) Hemocyanins

Syed Abid Ali, Zafar H. Zaidi, and Atiya Abbasi

HEJ Research Institute of Chemistry
University of Karachi-75270 Pakistan

Hemocyanins are large multisubunit, extracellular, copper containing glycoproteins which perform the important function of oxygen transport in many molluscs and arthropods. Arthropod hemocyanin is built from discrete number of cubic hexameric building block (6 x1) the number of hexamers being species specific. All arthropod hemocyanins have been found to be hetrogenous and 3-15 different polypeptide chains can be distinguished. Subunit heterogeneity is a characterstic feature of all arthropod hemocyanins not only at the level of primary structure but also in terms of antigenic specificity, electrophoretic mobility, chromatographic behaviour and oxygen binding properties.

Scorpions are regarded as the oldest terrestrial arthropods. Scorpion *Androctonus australis* and *Buthus sindicus* belong to the family Buthidae. We have studied the hemocyanins of these animals.

Hemolymph was analyzed by polyacrylamide gel electrophoresis, isoelectric focussing and high performance ion-exchange chromatography. N-terminal sequence and amino acid composition of purified subunits Bsin 1,2 and 3 is presented. Results of these studies are discussed with special reference to sequence similarities and subunit hetereogeniety in scorpion and other arthropod representatives.

PROTEIN PROFILES OF RESISTANT AND SUSCEPTIBLE VARIETIES OF LEAF RUST INFECTED WHEAT

Amin Uddin,[1] M. I. Khan,[1] and Zafar H. Zaidi[2]

[1] Department of Botany
University of Karachi, Karachi, Pakistan
[2] H.E.J. Research Institute of Chemistry
University of Karachi, Karachi, Pakistan

Seed proteins play a significant role in the growth and the vigor of plant is dependent on the seed proteins. After the onset of germination great biochemical changes take place in the biochemical constituents of the seed. In this study the protein profiles were screened by using SDS-PAGE which revealed a significant pattern in the upper most region of the gel among the resistant varieties. The number of bands are also found to be increased in the resistant varieties.

KINETICS OF FORMATION AND ARCHITECTURE OF FIBRIN NETWORK STRUCTURE USING CRUDE VENOM FROM *Vipera russelli*

Abid Azhar, Fatima S. Ausat, and Fizza Ahmad

Coagulation and Haemostasis Research Unit
Department of Biochemistry
University of Karachi, Karachi, Pakistan

Many snake venoms possess enzymes that can cleave off one or other fibrinopeptide preferentially or exclusively. These can serve as excellent probes for studying the effects of cleavage of each fibrinopeptide separately. A number of these enzymes also possess proteolytic activities of sufficiently narrow specificity that allow their use in investigation of a wide variety of structure-function relationship of fibrinogen and fibrin.

In the present investigation, crude extract of venom from local snake, *Vipera russelli*, was used to study the kinetics of formation and characteristics of human plasma fibrin networks. There was a dose dependent increase in the turbidity of networks indicating an increase in the thickness of the fibres. This effect was profound when the networks were developed with 25 mM $CaCl_2$ in the clotting mixture. The venom seems to act through thrombin generation. The alterations in the network characteristics may be attributed to increased concentration of monomers in protofibril network and enhanced end-to-end polymerization.

SEPARATION AND CHARACTERIZATION OF HUMAN ERYTHROCYTE ASPARTIC PROTEINASES

Muhammad Kamran Azim* and Zafar H. Zaidi

HEJ Research Institute of Chemistry
University of Karachi, Karachi, Pakistan

Human erythrocytes contain many types of proteolytic systems including calcium activated neutral proteinases (CALPAINs), aspartic proteinases such as cathepsin E, high molecular weight multicatalytic proteosome and serine type proteases etc. These proteolytic enzymes are either membrane associated or cytosolic. Erythrocyte proteases actively degrade globin chains, submembranous cytoskeletal proteins, transmembrane proteins and many cytosolic proteins.

We have attempted to study the aspartic proteinases of human erythrocytes. Enzymes were partially purified using salt fractionation, gel filtration and ion exchange chromatograghy. The molecular mass of the enzyme(s) was determined by SDS-PAGE and purity by isoelectric focusing. Further work for purifying and characterizing the enzyme(s) is in progress and will be discussed.

STUDIES ON LACTATE DEHYDROGENASE IN VARIOUS TISSUES OF *Rana tigrina*

Shamsa Akhtar Baqai

Armed Forces Medical College
Rawalpindi, Pakistan

Specific activity of the enzyme, lactate dehydrogenase (E.C. 1.1.1.27) was studied in the brain, heart, skeletal muscle, liver, kidney and spleen of *Rana tigrina* in summer during monsoons and winter during hibernation. Polyacrylamide gel electrophoresis was done to determine the tissue-specific LDH isoenzyme patterns. Findings were correlated with normal physiology and changes in specific activity detected. Molecular weight was determined by gel filtration column chromatography. Structure-function relationship and unique control are discussed.

INFLUENCE OF ALUMINUM ION ON PROTEOLYTIC HYDROLYSIS

R.-A. Boigegrain,[1] B. Pavoni,[2] and M. -A. Coletti-Previero[1]

[1] Centre de Recherche INSERM
   70 Rue de Navacelles, Montpellier, France
[2] Dp. Scienze Ambientali
   Universita di Venezia, Italy

Aluminum is ubiquitous, being the fifth most common element in earth's croust. Animals and plants have limited exposure to this element mainly because of the insoluble nature of most aluminum compounds. However, increased levels of aluminum were found in the brain plaques of patients suffering from Alzheimer disease. A hypothesis was that it plays a dismetabolic role on the proteolytic activity of the brain which regulates the processings of the β-amyloid precursor protein. A study on the influence of aluminum ion on the *in vitro* activity of different proteases is presented.

Kinetic constants were measured in a buffer solution containing aluminum ion at concentrations from 0 to 0.01 mM at pHs compatible with its solubility (from 4.5 to 6.5). Whereas no variations were observed for Km, kcats resulted significantly altered in presence of aluminum for chymotrypsin, trypsin and pepsin, if low molecular weight substrates were used. When, however, more complex substrates were digested, no significant effect of aluminum was detectable. The results suggest that aluminum ion, when present, did not interact with the enzymes but with the substrates that undergo a much easier digestion by trypsin and chymotrypsin and a much lower one by pepsin.

Surprisingly enough, the HIV protease, which belongs to the same family as pepsin (aspartic protease), showed a significant increase in its activity toward the undecapeptide substrate:

His-Lys-Ala-Arg-Val-Leu($-\rho-NO_2$)Phe-Glu-Ala-Nleu-Ser-NH_2.

PROTEASE INHIBITORS FROM INSECT HAEMOLYMPH

R. A. Boigegrain,[1] P. Paroutaud,[2] M. A. Coletti-Previero,[1] S. Amir,[3]
Z. A. Malik,[3] and T. Razzaki[3]

[1] Centre de Recherche INSERM
70 Rue de Navacelles, Montpellier, France
[2] Perkin Elmer, Roissy, France
[3] Department of Microbiology, University of Karachi, Pakistan

Insects lack the complicated immunoglobin system of higher animals to distinguish between self and non-self; they fight intruding substances by producing anti-bacterial peptides and by engulfing the "foreigner" in melanine produced biosynthetically by phenoloxidase. The production of this key enzyme is under the control of proteolytic events. Accidental activation of the system is avoided, under normal conditions, by the presence of protease inhibitors in the haemolymph.

A protease peptide inhibitor was isolated from haemolymph of the desert locust *Schistocerca gregaria*, it was characterized and sequenced. The peptide is relatively short, with three disulfide bridges and strong inhibitory activity toward α-chymotrypsin and human leukocyte elastase.

When compared with the peptide inhibitors, isolated from *Locusta migratoria*, it showed a high degree of similarity, suggesting that these molecules perform an important task in protecting the insects. They could assume responsibility for one or the other (or both) of the following functions: prevent uncontrolled proteolysis and/or protect proteins of insect fluids from foreign proteolytic enzymes.

The ultimate aim of this research is to conceive new and specific bioinsecticides, which would have the double advantage to unharm the useful species and to protect the environment.

GIBBS FREE ENERGIES ANALYSIS FOR INTERACTION OF WEGION HEMOGLOBIN AND SURFACTANTS

A. K. Bordbar,[2] Z. H. Zaidi,[3] Atiya Abbasi[3] and A. A. Moosavi-Movahedi[1]

[1] Institute of Biochemistry and Biophysics
University of Tehran, Tehran, Iran
[2] Department of Chemistry, Tarbiat-Modares University, Tehran, Iran
[3] H.E.J. Research Institute of Chemistry, University of Karachi, Pakistan

The binding of sodium n-undecyl sulfate with wegion hemoglobin was studied at pH 3.2 and 10 by equilibrium dialysis at 27°C. It is well established that the interaction between ionic surfactant and proteins involves initially by electrostatic binding and followed by more extensive hydrophobic binding. On the basis of this fact, the binding data have been analysed in terms of two interacting sets of binding sites, by fitting the data to Hill equation for more than one term. The Gibb's free energies of interaction, which were determined by using the theoretical model of Wyman binding potential, were resoluted to electrostatic Gibbs and hydrophobic Gibb's free energies.

SINGLE CELL PROTEIN PRODUCTION BY *Penicillium expansum* PRETREATED RICE HUSK AS A SUBSTRATE

M. Umar Dahot, M. Yakoub Khan, and M. Yousuf Khan

Enzyme and Fermentation Biotechnology Research Laboratory
Department of Biochemistry, Institute of Chemistry
University of Sindh, Jamshoro, Pakistan

Rice husk was degraded chemically ($H_2SO_4/HClO_4/$ $NaOH/NH_4OH$) and enzymatically (cellulase) to fermentable sugars and these were used as a substrate for the biosynthesis of single cell protein by *Penicillium expansum*. Fungal mycelial biomass obtained by *Penicillium expansum* grown on H_2SO_4 treated rice husk medium supplemented with sucrose showed higher protein content than $H_2SO_4/HClO_4/NaOH/NH_4OH$ and cellulase treated rice husk. Fungal cells grown under rice husk and rice husk incorporated with pure sugars were analyzed for amino acids content and data indicates that all essential amino acids are present. This fungal mycelial protein biomass was supplemented with poultry feed. It was noted that growth rate is more effective (27.59%) in body weight gain of chicks feed on basal diet supplemented with 10% single cell protein in comparison to chicks feed on normal diet.

DETERMINATION OF SELENOMETHIONINE BY MICROWAVE HYDROLYSIS AND ION-EXCHANGE CHROMATOGRAPHY

Gerald Grubler,[1] Stanka Stoeva,[2] Hans Zimmermann,[1] Kirstin Pfeiffer,[2] Arie Kant,[1] and Wolfgang Voelter[2]

[1] Eppendorf/Biotronik, Maintal, Germany
[2] Abteilung für Physik-alische Biochemie
Universität Tübingen, Tübingen, Germany

Selenomethionine (SeMet) in food plants is one of the protein type and can be incorporated in all types of animal proteins. SeMet has many functions. It can be metabolized in the human body into selenocysteine, an essential component of glutathione peroxidase. Selenium yeast (containing about 30% SeMet) is an effective agent for selenium supplementation and is likely a potential anticarcinogen. For these reasons, the determination of SeMet in biological materials is of significant interest. Applying an Eppendorf/Biotronik LC 3000 amino acid analyser, the detection limit for SeMet is about 10 nmol/ml approaching values, found for standard amino acids. Further experiments showed that, contrary to literature reports, almost no decomposition of SeMet takes place. The best recoveries by microwave hydrolysis were obtained at 130°C within 30 min. Higher temperature (150°C) caused decomposition of selenomethionine, resulting in lower yields.

MOLECULAR MODELLING OF HUMAN ADENOSINE A2a RECEPTOR AND ITS INTERACTION WITH LIGANDS

Mahnaz Sofaf, Bahram Habibi-Nezhad and Masoud Mahmodian

Tehran University of Medical Sciences
Pharmacy School, Tehran, Iran

Adenosine is an important neuro-transmitter in various tissues. Its receptors belong to the "G protein-coupled receptors" family which have shown some structural similarity with bacteriorhodopsin. Since these group of receptors are membrane-bound proteins, it is not possible to determine their structures using conventional methods such as X-ray crystallography. Therefore, computer-aided molecular modelling of these receptors could provide valuable information regarding their structure activity relationships. In the present study, various computational techniques were used to determine the possible regions of transmembrane helices of human adenosine A2a receptor. A 3D molecular model of the transmembrane region of this receptor was constructed on the basis of the similarity of these helices with that of bacteriorhodopsin. The atomic coordinates of amino-acid residues of each helix were optimized using molecular mechanics calculation (MMX). Subsequently, the structural models of some ligands such as adenosine, CGS21680, DPCPX and CP-66-713 were docked into the receptor cavity. The results show that important amino-acids are located in the inner cavity and therefore may play an important role in the binding of ligands to the receptor. The constructed model provides the possibility of evaluation of ligand-receptor interaction and may be used for the design of new ligands for A2a receptor.

IDENTIFICATION OF POST TRANSLATIONAL MODIFICATION OF LENS α-CRYSTALLIN MODIFIED BY ASPIRIN USING CONTINUOUS FLOW MASS SPECTROMETRY

Azeemul Hasan,[1] J. B. Smith,[2] Z. H. Zaidi, and D. L. Smith[2]

[1] HEJ Research Institute of Chemistry
University of Karachi, Karachi, Pakistan
[2] Department of Medicinal Chemistry
Purdue University, W. Lafayette, Indiana 47907

Mass spectroscopy is an excellent technique for analyzing the chemical modifications of proteins but certain modifications (acid labile modifications) can not be monitored by conventional mass spectroscopic methods due to prolonged exposure of samples in acid.

We have established a rapid method to solve this problem. Continuous flow FABMS require only 20-30 minutes for analyzing the whole digest so that the chances of cleavage of modifying groups from the sites are minimized and thus they can be detected by mass spectroscopic methods.

ESTIMATION OF GLUCOSE OXIDASE STABILITY AT DIFFERENT TEMPERATURE

Mohammad R. Housaindokht,[1] Ali A. Moosavi- Movahedi,[1] and Jalil Moghadasi[2]

[1] Institute of Biochemistry and Biophysics
Tehran University, Tehran, Iran
[2] Chemistry Department
Shiraz University, Shiraz, Iran

It has been shown that glucose oxidase resist to enzyme denaturants, such as temperature, sodium n-dodecyl sulphate, a potent biological detergent. In order to calculate the stability of glucose oxidase at different temperatures; the reaction with urea has been carried out at 30-80°C. The stability curve shows a maximum at 50°C. This means that at this temperature glucose oxidase has a new conformation which is more stable. Thermal denaturation curve of the enzyme shows, in the presence of urea, a "three state mechanism" the intermediate of which occurs at about 50°C.

COMPUTER AIDED MODELLING OF THE ACTIVE SITE OF DIHYDROOROTATE DEHYDROGENASE

Shahram Khademi and Alireza Khadem

Department of Medicinal Chemistry
Shaheed Beheshti University of Medical Sciences, Tehran, Iran

Dihydroorotate dehydrogenase (DHODase, EC 1.3.3.1) is the fourth enzyme of the de novo pyrimidine biosynthetic pathway. It has been shown that the inhibitors of this enzyme have anti tumor and anti malarial activity. The structure of this enzyme has not been determined. In this study the 3D model of the active site of DHODase has been built and investigated.

The secondary structure of the active site of DHODase was predicted using GARNIER method. Also amino acids which were conserved in the active site were determined by comparing the sequences of DHODase from various sources. The 3D structure of the active site was built and refined using AMBER package. Substrate, intermediate, product and inhibitor was placed at the postulated active site and its energy of interaction was determined by molecular dynamic calculations.

The result of secondary structure prediction showed that the active site of DHODase must be in the form of $\beta\alpha\beta$. The result of molecular modelling of the active site showed that the guanidine group of Arg 289 makes an electrostatic interaction with carboxy group of various ligands. Furthermore the hydrogen bonding between Arg 305, Glu 306 and Ser 313 of the enzyme and the other parts of the ligands contribute to the binding. The His 335 is found to be in the vicinity of the axial hydrogen of C5 atom of the substrate.

AN IBM-PC PROGRAM FOR PREDICTION OF PROTEIN STRUCTURAL CHARACTERISTICS FROM AMINO ACID SEQUENCES

Shahram Khademi and Masoud Mahmoudian

Department of Medicinal Chemistry
Shaheed Beheshti University of Medical Sciences, Tehran, Iran

An IBM-PC program is developed for display of the structural characteristics of a polypepide using its amino acid sequences. This program uses the modified Chou-Fasman parameters to calculate the probability of secondary structure of a polypeptide. The assigned probability of each type of secondary structure as well as hydrophobicity, charge distribution, the presence of cysteine residues and the position of each segment with respect to cell membrane will be plotted against the sequence of the peptide for visual inspection for possible conformation of the peptides. The program will print the final prediction of the conformations of various segments of the polypeptides. The result of application of this program to predict the secondary structure of various peptides such as L, M and H subunits in the photosynthetic reaction center of *Rhodoseudomonas viridis*, cytochrome P450cam and cytochrome P450b, showed a good agreement with the experimental and other theoretical observations.

CHROMATOGRAPHIC SEPARATION OF THE SUBUNITS OF BOVINE-CRYSTALLINS

Riffat Parveen and Zafar-ullah Khan[*]

Department of Chemistry
Bahauddin Zakariya University, Multan, Pakistan

Subunits of bovine-Crystallins i.e., αA_2, αA_1, αB_2 and αB_1, have been separated by DEAE-Cellulose anion exchange chromatography using Tris buffer containing 6M urea. The purity of the subunits was checked by polyacrylamide gel electrophoresis.

ACKNOWLEDGEMENT

Supported by NSRDB Pakistan.

PROTEIN PROFILE OF RHIZOBIA STRAINS INDIGENOUS TO FOOD AND FODDER LEGUME GROWN UNDER VARIOUS AGROCHEMICAL STRESS SITUATIONS

S. A. Hafeez[1] and Y. H. Khan[2]

[1] Applied Microbiology
NARC, Islamabad, Pakistan
[2] Ecotoxicology Research Center
NARC, Islamabad, Pakistan

A variety of rhizobia strains were screened by selective marker techniques. Systemic and nonsystemic pesticides were used as selective markers. Arrested and suppressed growth of pesticide sensitive rhizobia strains and their restricted cell and colony morphology due to high inputs of pesticides is hindering the efficiency of rhizobia responsible for biological nitrogen fixation. Some pesticide resistant rhizobia strains show different response to different amounts and types of pesticides. Some food and fodder legumes e.g. *Lathyrus sativa* cause lathyrism in human beings due to presence of some amino acid. The present investigation was undertaken to identify and group these rhizobia based on their protein profiles. The protein profile done so far is not setting a well defined trend however haphazard protein pattern in several strains indicating a separate group of rhizobia strains. Efforts are also made to make a correlation between nitrogen fixers, protein pattern and associated health problem. Research is in progress and some interesting picture seems to emerge from this study.

PARTIAL PURIFICATION AND SOME PROPERTIES OF *Oxystelma exculantum* ROOT TUBER PHOSPHOLIPASE C

A. N. Memon[*] and A. R. Memon

Department of Biochemistry
Institute of Chemistry
University of Sindh, Jamshoro, Pakistan

Oxystelma esculantum root tuber phospholipase C was purified by acetone precipitation and column chromatography on Sephadex G-100. It showed two fractions (PLC I and II). Fraction II (Phospholipase C II) was found homogeneous on polyacrylamide disc gel electrophoresis. *Oxystelma esculantum* phospholipase C II showed the optimum activity at pH 5.5 and at temperature 30°C. This enzyme was found heat labile and inactivated to the extent of 80% at 80°C in 10 minutes. Enzyme activity was activated in presence of $CoCl_2$, $ZnCl_2$ and $CaCl_2$ whereas completely inhibited with EDTA due to chelation with metal ion.

THE USE OF METAL IONS AS SPECTROSCOPIC PROBES IN THE ELUCIDATION OF pKa's OF SMALL MOLECULES

Bijan Farzami and Farzaneh Mirzajani[*]

Department of Biochemistry
Tehran University of Medical Sciences, Tehran, Iran

The apparent pKs of the functional groups interacted with metal ions in solutions were determined using a technique devised in this laboratory (Farzami, 1992). In these studies several groups of molecules such as coenzyme thiamine diphosphate and its derivatives, dipeptides, amino acids and nucleotides and their bases were studied for the pK values. The results obtained were correlated with that obtained by other techniques such as NMR. In instances where the metal ion could indirectly alter the ionisation properties of a group, a conformational scheme could be drawn.

This method is also applicable to macromolecules (Farzami 1990-1992).

STABILITY ASPECT OF DENATURATION OF ADENOSINE DEAMINASE BY UREA AND GUANIDINE HYDROCHLORIDE

S. H. Moghadamnia and A. A. Moosavi-Movahedi

Institute of Biochemistry and Biophysics
University of Tehran, Tehran, Iran

A useful key towards understanding the stability of proteins in solution is to observe the manner in which these molecules denature. Protein denaturation results in a change in the three-dimensional structure from native to denatured form.

On the molar scale, urea and guanidine hydrochloride are probably the most effective denaturants.

Here, the denaturation of adenosine deaminase as a key enzyme for metabolism of nucleic acid components was carried out by urea and guanidine hydrochloride. This investigation was done spectrophotometrically at two different temperatures of $27°$ and $37°C$ using phosphate buffer (55 mM) pH 7.5.

A simple reversible two state transition, Native $\Leftrightarrow$ Denatured, is used to analyze denaturation process from which the conformational stability can be estimated by using three different methods as indicated in the case of linear extrapolation method (LEM), Tanford's model and denaturant binding method (DBM). A good agreement was observed among these methods.

CHARACTERIZATION OF CAMEL MILK β-CASEIN

Ashiq Mohammad[1] and O. U. Beg[2]

[1] H.E.J. Research Institute of Chemistry
University of Karachi, Pakistani, Karach
[2] Department of Anatomy and Cell Biology
Meharry Medical College, Nashville, Tennessee

β-casein was purified by ion exchange chromatography and its structure was determined by analysis of intact protein and peptides generated by enzymatic cleavage i.e. Lys-C, Arg specific and Asp-N by manual (DABITC) as well as on gas phase sequencer. Camel milk β-casein contains 217 amino acid residues with molecular mass of 28,600 dalton (on the basis of amino acid sequence). The primary structure of camel milk β-casein shows 64%, 63%, 48%, 62% homology to its counterpart from bovine, ovine, buffalo and human respectively. Camel β-casein shows 40% conservation to afore mentioned β-caseins. It shows an extension of 7 residues at its C-terminal when compared to bovine, ovine and buffalo β-casein. It has a number of short tandem repeats. Primary structure of β-casein shows several biologically active peptides i.e. β-casomorphin, angiotensin-1 converting enzyme inhibitor, immunostimulating peptides and a large number of opioid peptides.

STABILITY AND STRUCTURAL STUDY OF BSA IN NON IONIC GRADIENT ELECTROPHORESIS

Hossain Naderimanesh and S. Zahra Moosavinejad

Department of Biochemistry
School of Basic Sciences
Tarbiat Modarres University, Tehran, Iran

The denaturation and renaturation processes of Bovine serum Albumin (BSA) at pH 7.0 by non ionic denaturant gradient-polyacrylamide gel electrophoresis have been studied.

BSA has one free sulphydryl group that is able to form intramolecular thiol-disulphide exchange and intermolecular multimers. In refolding process also monomer molecule can reform polymers.

Free sulphydryl groups of BSA molecule have been blocked and monomers have been separated. Then denaturation and renaturation of blocked and nonblocked monomers at different temperature have been studied by Urea- and Guanidine HCl-gradient PAGE.

PHOSPHOLIPASE A2: A New Simple and Rapid Method of Identification

S. M. Naeem, M. Kamal, and Zafar H. Zaidi

HEJ Research Institute of Chemistry
University of Karachi, Karachi, Pakistan

Phospholipase A2 is used as diagnostic marker in many diseases. Its high serum level is an indication of a number of clinical conditions. High levels of the enzyme are reported in the sera of patients with acute pancreatitis, renal failure, chronic liver diseases and hepatocellular carcinoma. It will therefore be of clinical importance to have an appropriate rapid method for the detection of Phospholipase A2 in serum. In this presentation we describe a new simple and rapid method for qualitative and semiquantitative determination of phospholipase A2. The method is based on the formation of a zone of clearance on egg yolk agar plate, which is due to the action of the enzyme on glycerophospholipids of egg yolk. This method requires less quantity of both the enzyme as well as substrate and gives reproducible results. Results are discussed.

COMPUTER AIDED MODELLING OF PAF RECEPTOR

Shahram Khademi and Nikzad Nikbin

Department of Medical Chemistry
Shaheed Beheshti University of Medical Science, Tehran, Iran

PAF receptor is a G-protein coupled receptor (GPCR). This receptor is a membrane associated protein and its 3D structure is not available. However, the amino acid sequences of this receptor has been determined. In this study the 3D model of PAF receptor has been built from amino acid sequences in order to investigate the mode of receptor/ligand interactions. The amino acids involved in the seven transmembrane domains were identified on the basis of hydropathicity profiles and on the primary sequence homology analysis of PAF and several members of the GPCR family. The seven transmembrane regions were constructed as α-helices and positioned as the trans- membrane domains in the bacteriorhodopsin experimental structure. The relative orientation of the seven helices were adjusted according to Baldwin arrangement. The resulting 3D model were optimized using AMBER force field.

On the basis of structural comparison and conformational analysis of various PAF antagonists, Hodgkin *et al.* recently proposed a binding site which includes two groups with a distance of nearly 10Å that make hydrogen binding with ligand in PAF receptor. There are several pairs of residues in our model that have the above distance, of which the following sites have been verified by the experimental results:

1) HIS (248) in TM6 and SER (104) in TM3$\rightarrow$
2) HIS (248) in TM6 and THR (101) in TM3 $\rightarrow$

Studies are in progress to dock ligand to the above sites and calculate the energy of interactions.

PROTEIN SECONDARY STRUCTURE PREDICTION AND MODIFICATION OF PROBABLISTIC METHODS

Hossain Naderimanesh[1] and Seyyed Hossain Rasta[2]

[1] Faculty of Science
Tarbiat Modarres University, Iran
[2] Institute of Biochemistry and Biophysics
Tehran University, Tehran, Iran

We have constructed an information system in the Fox media for predicting and studying protein secondary structure employing probability methods. This system which is a complement of the Chou and Fasman method has two advantages:

1) The mutual effect of the adjacent residue which was overlooked in the Chou and Fasman method is taken into consideration. 2) This new software has more flexibility to predict protein secondary structures. This enables the user to have information about the distribution of a given amino acid and the probability of its combination with other amino acids (singlets, doublets ND triplet residue effect) in the secondary structure.

The secondary structure of few proteins was predicted using this software (SRNP) and the results were compared with the result obtained by X-ray crystallography and the Chou and Fasman method. In all cases the prediction based on this method proved to be 6-10 percent more accurate.

FREE ENERGY DISTRIBUTION OF THE INTERACTION OF BINDING SITES FOR COOPERATIVE AND ANTICOOPERATIVE SYSTEMS

A. A. Saboury[1] and A. A. Moosavi-Movahedi[2]

[1] Department of Chemistry
University of Tabiat-Modarres, Tehran, Iran
[2] Institute of Biochemistry and Biophysics
University of Tehran, Tehran, Iran

The theory of Weber and Wyman regarding free energy of ligand binding to macromolecule was extended and analysed. Here it is defined a physical model to obtain types of free energies of different order of interactions which is resolved from intrinsic free energies of binding sites and statistical factors.

A serious problem to progress in understanding of cooperativity is the limitation of current methods of formulating complex model either theoretical or experimental for data analysis. Interactions of binding sites for cooperative and anticooperative systems brought better understanding by this model.

THE IMMUNE RESPONSE OF RAINBOW TROUT *Oncorhynchus mykiss* AGAINST A LAMBDA BACTERIOPHAGE

Anwar Iqbal Saifi[1] and John B. Alexander[2]

[1] H.E.J. Research Institute of Chemistry
University of Karachi, Karachi, Pakistan
[2] Biological Sciences
University of Salford, Salford, United Kingdom

The primary and secondary immune response of rainbow trout, *Oncorhynchus mykiss* against injections of a bacteriophage has been investigated.

Enzyme-linked immunosorbent assay was used to detect and measure the antibody titre. For this purpose the optimal antigen concentration was found to be 10 mg ml^{-1}. Trout antibacteriophage serum was used at a ratio of 1:200. It was observed that a high dilution (1:1000) of rabbit anti-trout serum was required to minimize non-specific reactions.

In the primary response, antibodies were first detected 6 days after IP injection and by Day 21 all fish were positive. Maximum mean titres detected on Day 30 which was 3 Days after the fish had been given a second injection. The mean titres had declined by Day 33 (Day 6 after the second injection) and remained constant from Day 39 to Day 45.

Antigen localization in the spleen was studied by an indirect peroxidase method. The distribution varied with time. Initially it occured associated with the ellipsoids but by Day 7 had started to appear associated with melano-macrophages. By Day 28 nearly all the antigen was associated with these centres. Three days after a second injection the antigen was found distributed throughout the spleen but by Day 14 after the second injection it was mainly restricted to the melano-macrophage centres. Therefore after a second injection the movement of antigen to the melano- macrophage centres occured much faster and would indicate the presence of immunological memory.

PHYSICO-CHEMICAL ASPECTS OF D. AMINO ACID OXIDASE BY SODIUM N-DODECYL SULPHATE

B. Shareghi, A. Golestani, and A. A. Moosavi-Movahedi

Institute of Biochemistry and Biophysics
University of Tehran, Tehran, Iran

D- amino acid oxidase (DAO) is a flavoprotein which in the presence of molecular oxygen oxidatively deaminates D- amino acids to corresponding α-keto acids:

$$RCHNH_2COOH + O_2 + H_2O \rightarrow RCOCOOH + NH_4 + H_2O_2$$

Here, the interaction of DAO at monomer state and different concentrations of sodium n - dodecyl sulphate (SDS) was studied by various physical techniques to obtain kinetic and thermodynamic parameters in 0.02M sodium pyrophosphate, pH 8.3 at various temperatures. DAO has a boundary between monomer and dimer at concentration of 0.004 mg/ml and 0.2 mg/ml respectively.

Our studies show that interaction of SDS at low concentration with DAO has no conformational change on enzyme's structure because it is playing as competitive inhibitor but at higher concentration produces conformational changes because of playing mixed action of inhibition and denaturation. These subjects are approved by thermodynamic parameters which corresponds to kinetics data. Our results on specific activity and ΔG (H_2O) also show that the enzyme keeps it's function on native conformation between 37-52°C.

STUDIES ON THE INFLUENCES OF CARCINOGENS ON RIBOSOME-MEMBRANE INTERACTIONS AND EXPORTED GLYCOPROTEINS IN THE CISTERNAL SPACE OF ENDOPLASMIC RETICULUM

Sucheta Sharma[1] and Harinder M. Dani[2]

[1] Department of Biochemistry, Punjab Agricultural University, Ludhiana, India
[2] Department of Biochemistry, Punjab University, Chandigarh, India

In vivo effects of 4-dimethylaminoazobenzene and benzo(a)pyrene followed by administration of phenobarbital for 35 days and 51 days respectively on the components of endoplasmic reticular membranes have been studied. Degranulation of microsomes by these carcinogens occur in their target organs, liver for dimethylaminoazobenzene and lung for benzo(a)pyrene. Histochemical studies indicated depletion of glycoproteins in their specific target organs. Proteins with molecular weight of more than 60,000 disappeared from the cisternal space of endoplasmic reticulum after treatment with dimethylaminoazobenzene. As glycoproteins are components of plasma membranes and gap junctions and control the molecular traffic across the plasma membrane, it is thus concluded that these glycoproteins are acting possibly by their involvement in transmembrane signalling which play important roles in epigenetic carcinogenesis.

MODIFICATION OF AETIOLOGY OF VASCULAR DISEASE THROUGH ALTERATION OF FIBRIN ARCHITECTURE BY HYPOLIPIDAEMIC DRUGS

Abid Azhar, Ambreena Siddiqui, and Fizza Ahmad

Coagulation and Haemostasis Research Unit
Department of Biochemistry
University of Karachi, Karachi, Pakistan

An understanding of basic mechanisms involved in the aetiology of atherosclerosis is crucial to the development of strategies for its intervention and prevention. A new approach in this regard is emerging that deals with the alteration of the structure of fibrin to modify the aetiology of vascular disease. It has been established that the plasma fibrinogen levels correlate with other risk factors for coronary heart disease like ageing, smoking, hypertriglyceridaemia, and elevated LDL-cholesterol levels, Furthermore, the lowering of raised plasma fibrinogen concentration by a number of 'hypolipidaemic drugs' indicates that the lipid and haemostatic systems may act synergestically in pathogenesis of cardiovascular disease. Central to this hypothesis is the suggestion that the fibrinogen synthesis by liver could be modified by the agents affecting lipid levels. A molecular model relating hepatic free fatty acids to fibrinogen is yet to be established.

The current investigation is aimed to study the capacity of lipid lowering drug Bezafibrate, to induce changes in blood coagulation system and the architecture of fibrin network. Blood samples collected before the administration of the drug to the animal models and after the completion of the study period. The samples were analysed for blood parameters like cholesterol, LDL-cholesterol, triglycerides, and fibrin network characteristics. In addition to the changes in lipid profile, significant alterations were observed in the characteristics of the fibrin network by the drugs. The relationship between these changes and the atherogenesis and thrombolysis is explored.

HYDROLYTIC AND SYNTHETIC ACTIVITY OF β-FUCOSIDASE/β-GLUCOSIDEASE FROM THAI ROSEWOOD SEEDS

Chantragan Srisomsap, Sauvarat Khunyoshyeng, Kanokporn Boonpuan, Phannee Sawangareetrakul, Rudee Surarit, and Jisnuson Svasti

Laboratory of Biochemistry
Chulabhorn Research Institute, Bangkok, Thailand
Department of Biochemistry
Mahidol University, Bangkok, Thailand

Hydrolytic activity of purified β-D-fucosidase/β-D-gluco- sidase from Thai Rosewood (*Dalbergia cochinchinensis* Pierre) towards natural substrates has been studied. Within 3 h, the hydrolysis of 10 mM sophorose [β(1,2)] gave 25.7% D-glucose as product, while the hydrolysis of gentiobiose [β(1,6)] and laminaribiose [β(1,3)] gave 14.1% and 15.6% D-glucose. Hydrolysis of cellobiose [β(1,4)] was very much slower, but at prolonged incubations, glucose could be released from cellobiose, cellotriose and cellotetraose. However, no hydrolysis could be observed when the cyanogenic glycoside laminarin or the [β(1,3)] glucose polymer laminarin were used as substrates. When the enzyme was incubated with high concentrations of D-glucose at elevated temperature, disaccharides and trisaccharides could be produced. Relative yields of disaccharides synthesized were 35% for gentiobiose, 11% for sophorose and laminaribiose combined, and 0% for cellobiose. The optimum pH and the optimum temperature for the synthesis with D-glucose were pH 5 and 60°C, respectively, while different concentrations of D-glucose from 20% to 70% (w/w), yields of products were lower, namely 1.1% product with D-fucose and 4.7% product with D-galactose, and no product with L-fucose or D-fructose. Novel peaks were also obtained from synthesis using mixtures of D-glucose and D-fucose.

MULTIPLE ALIGNMENTS OF HDC WITH OTHER PLP-DEPENDENT ENZYMES

Hossein Naderimanesh[1] and Fatemeh-sadat Taha-Nejad[2]

[1] Faculty of Science
Tarbiat Modarres University, Iran
[2] Department of Medicinal Chemistry
Tehran Medical Sciences University, Tehran, Iran

Multiple sequence alignments of morganella Am-15 Histidine decarboxylase (HDC) and Aspartate amino transferases have been studied. Previous studies have defined the role of certain residues in enzymes- substrate interactions.

This work suggests that His-119 and Ser-115 in HDC structure could be involved in hydrogen binding of enzyme-substrate interaction, which could lead us to understand the nature of HDC active site and ultimately development of a specific inhibitor of the enzyme in future.

DETERMINATION OF CHARACTERISTICS OF PYRIDOXAL 5'-PHOSPHATE DEPENDENT HISTIDINE DECARBOXYLASE ACTIVE SITE

Hossain Naderimanesh,[1] Fatemeh-Sadat Taha-Nejad,[2] and Shahram Khademi[3]

[1] School of Science
Tarbiat Modarres University, Tehran, Iran
[2] Department of Medical Chemistry
Tehran University of Medical Sciences, Tehran, Iran
[3] Department of Medical Chemistry
Shaheed Beheshti University of Medical Sciences, Tehran, Iran

Histidine decarboxylase (HDC) catalyzes the decarboxylation of l-histidine to histamine, which involves in inflamation, allergic reactions, and gastric acid secretion. The 3D-structure of HDC has not been determined yet.

In this study multiple sequence alignments have been performed to HDC from *Morganella Morganii* and Aspartate Amino-transferases as a whole, for identifying the conserved residues, which could lead to an understanding of the nature of HDC active site and ultimately, development of a specific inhibitor of the enzyme in the future.

The results show that His-119 and Ser-115 in HDC, which are conserved in the active site of Aspartate Amino-transferases, could be involved in enzyme-substrate interactions.

IDENTIFICATION OF AN ALDEHYDE DEHYROGENASE WITH LOW K_M FOR ACETALDEHYDE IN RAT LIVER CYTOSOL

Muhammad Kalim Tahir

Department of Chemistry
University of Azad Jammu and Kashmir
Muzaffarabad, Azad Jammu and Kashmir

A survey of aldehyde dehydrogenases in rat liver has uncovered the presence of a new cytosolic enzyme with a low K_M for acetaldehyde. The enzyme has been purified from soluble fractions by using different chromatographic steps based on gel permeation, ion exchange chromatography, and affinity chromatography on α-cyano-cinnamate Sepharose and 5' AMP-Sepharose. Analysis by SDS-PAGE demonstrated that the enzyme was about 95% pure. The enzyme has a subunit mass of 57 KDa and is composed of two subunits. Some of the aldehydes that have been used as substrates include acetaldehyde, benzaldehyde, p-nitro- benzaldehyde and 6-dimethylamino-2-Naphthaldehyde. The enzyme also displays esterase activity, when p-nitrophenyl acetate is used as a substrate.

INTERACTION OF CORE WITH FOLDED AND INTACT HISTONE H1 IN THE PRESENCE OF SODIUM N-DODECYL SULPHATE

F. Tasht-Zarin and A. A. Moosavi-Movahedi

Institute of Biochemistry and Biophysics
University of Tehran, Tehran, Iran

Interaction between core (complexes of DNA and histones octamers) and histone H_1 was studied in the presence and absence of sodium n-dodecyl sulphate (SDS) in 2.5 mM phosphate buffer, pH 6.4 at temperature of 27 and 37oC by uv spectrophotometer, equibibrium dialysis and titration.

The presence of 1.33 mM SDS caused the folding of H_1 and more interaction with core which has marked difference with complexes of intact H_1-DNA. This indicates the role of H_1 alone (without core) in compacted DNA which has higher interaction than histones octamers.

Using binding data free energy and enthalpy of interaction was calculated in terms of Wyman theoretical model. The enthalpy of interacton obtained by Pace method to Wyman model, which was consistent together, is also compared.

The thermodynamic parameters and precipitation percent of core-H_1 complexes indicate the folded H_1-core having more interaction than native H1-core complexes.

ISOLATION AND PURIFICATION OF TOXIC PEPTIDES FROM THE IRANIAN SCORPION VENOM *Odontobuthus doriae* AND THEIR EFFECTS ON GUINEA PIG'S HEART ATRIUM

Hassan Tehrani,[1] Mohammad Ghaemmaghami,[1] Masoud Mahmodian,[2] Mohammad A. Daneshmehr,[1] Abbas Zareh,[3] Abolfazl Akbari,[3] and A. Mirkhani[2]

[1] Pharmacy Department
Shaheed Beheshti University of Medical Sciences, Iran
[2] Pharmacology Department
Iran Medical Sciences University, Tehran, Iran
[3] Razi Institute
Karaj, Iran

Toxic peptides were isolated from the venom of the Iranian scorpion, *Odontobuthus doriae*, by gel filtration on sephadex G-50 and purified by reverse phase high pressure liquid chromatography using C-18 analytical column. Of the four toxic fractions determined by LD_{50} measurements in mice, two showed an increase of guinea pig's cardiac contraction force with no effect on heart rate.

Further studies will be the determination of amino acid sequences of these toxins and compare them with those of other scorpion toxins in order to achieve structure-activity relationships data.

MODULATION OF ACETYLCHOLINE RELEASE BY INTERACTION OF THREE HOMOLOGOUS SCORPION TOXINS WITH K+ CHANNELS

H. Vatanpour,[1] E. G. Rowan,[2] and A. L. Harvey[2]

[1] Department of Toxicology and Pharmacology, Medical Sciences
University of Shaheed Beheshti, Tehran, Iran
[2] Department of Physiology and Pharmacology
University of Strathclyde, Glasgow, Scotland

The effects of three scorpion toxins charybdotoxin (CTX), iberiotoxin (IbTX), and noxiustoxin (NTX) have been studied on acetylcholine (Ach) release and on K^+ channels by means of twitch tension and electrophysiological recording technique using isolated skeletal muscle preparations and by a radioligand binding assay using ^{125}I-labelled dendrotoxin I<D>(DpI) and rat brain synaptosomal membranes (RBSM). On chick biventer cervics preparation CTX and IbTX (100nM) augmented the twitch responses to indirect and direct muscle stimulation. However, the increase in response to indirect stimulation (about 80%) was larger. Further, the interest was fast in onset reaching a maximum within 25-30 min. NTX at 100nM produced a slow augmentation of the twitch response to indirect muscle stimulation, with the maximum response being seen after 40-50 min. On mouse triangularis stemi preparations, CTX (300nM after 35-40 min.) and IbTX (100nM after 15 min.) increased quantal content of the evoked endplate potentials by about two fold. However, NTX caused only a small increase in e.p.p. amplitude followed by repetitive e.p.p.s in response to single shock nerve stimulation after 40-50 min. Extracellular recording of nerve terminal current waveforms in triangularis stemi preparations revealed that CTX and IbTX, but not NTX (even at 10 times higher concentration), block the Ca^{2+} activated K^+ current, I_{K-Ca}. However, there was no major change in the portion of the nerve terminal waveform associated with voltage-dependent K^+ currents, I_{KV}. In the radioligand binding assay NTX displaced labelled^{125}I-DPI very potent, whereas CTX produced only partial displacement. However, IbTX did not displace^{125}I-DPI from its binding sites on RBSM. We conclude that these three structurally homologous scorpion toxins act on different K^+ channels and that this leads to different patterns of facilitation of Ach release. IbTX acts selectively on high conductance I_{K-Ca} channels, and that this increase the amplitude of e.p.p.s without any other changes. NTX acts on I_{KV} channels that are sensitive to dendrotoxin and causes repetitive e.p.p.s. CTX shares amino acid residues that exist in the structures of IbTX and NTX; CTX acts on both I_{K-Ca} and I_{KV} channels.

CHARACTERIZATION OF gp63, A MAJOR SURFACE GLYCOPROTEIN, ON *Leishmania* ISOLATES OF PAKISTAN

M. Masoom Yasinzai

Institute of Biochemistry
University of Baluchistan, Quetta, Pakistan

Despite the diversity of clinical manifestations, the life cycle of all species of leishmania is remarkably similar. Glycoconjugates on the parasite surface form the interface between the parasites and the microenvironments provided by their hosts and vectors. These glycoconjugates are thought to play a vital role in parasite invasion of macrophages and their survival in the hostile environment of the phagolysosomal compartment. gp63, a glycoprotein and a major surface antigen, is conserved in almost all pathogenic species of *Leishmania*.

We took three representative isolates of Pakistan Leishmania, speciated as *L. major, L. tropica* and *L. infantum*. Using polyclonal antibodies raised against new world *Leishmania*, we were able to show clearly by western blotting the presence of gp63 on all representative isolates. The ectoprotein nature of this metalloproteinase is shown by its proteolytic activity in gelatin and fibrinogen gels. This is also clearly demonstrated that the proteolytic activity of the parasite is directly related to the virulence of the parasite.

SEPARATION AND CHARACTERIZATION OF HUMAN LENS CRYSTALLINS

Sabira Naqvi, Mustafa Kamal, Atiya Abbasi, and Zafar H. Zaidi

HEJ Research Institute of Chemistry
University of Karachi, Karachi, Pakistan

Cataract, the loss of lens transparency is a major cause of visual disability and blindness throughout the world. Crystallin form the major portion of lens contributing to 90% of the soluble proteins. These are further divided into α, β and γ crystallins on the basis of their molecular mass. α and β crystallins are multimeric proteins having different subunits while γ crystallins exist in monomeric form. Owing to the high concentration it has been suggested that crystallins play major role in maintaining transparency in the lens.

We have undertaken a study in order to investigate water soluble and insoluble proteins from human cataractous lens. Present communication describes our initial studies on insoluble fractionation of lens proteins which has not been well studied using different chromatographic techniques.

STRUCTURAL STUDIES ON γs CRYSTALLIN

S. Zarina,[1] C. Slingsby,[2] R. Jaenicke,[3] Zafar H. Zaidi,[1] H. Driessen,[2] and
N. Srinivasan[2]

[1] HEJ Research Institute of Chemistry
University of Karachi, Karachi, Pakistan
[2] Laboratory of Molecular Biology and Imperial Cancer Research Fund Unit
Department of Crystallography
Birkbeck College, London, United Kingdom
[3] Institut für Biophysik und Physikalische Biochemie
Universitat Regensburg, Regensburg, Germany

Cataract, the lens opacity is a common disorder of older age being most prevalent in third world countries. Loss of lens transparency is believed to result from spatial fluctuations in the refractive index which in turn is provided by unique arrangement of a diverse group of proteins, the crystallins. Hence it is essential to study lens proteins at primary, secondary and tertiary level. Crystallins are classified into two major families, the α and $\beta\gamma$ crystallins. β crystallins are oligomeric in nature having acidic and basic subunits which are 45 % related. The γ crystallins are monomers with 80 % sequence identity. Three dimensional studies have revealed that members of $\beta\gamma$ superfamily are built up of repeating unit of "Greek Key" motifs arranged in two similar domains. However, the β and γ crystallins differ in the conformation of their connecting peptide resulting in dimeric or monomeric association of the domain. Present study describes a model of γs crystallin, a member of $\beta\gamma$ superfamily, constructed using comparative molecular modeling approach. The model shows a bias towards monomeric γB crystallin which is structurally more similar.

INDEX